Modelling, Dimensioning and Optimization of 5G Communication Networks, Resources and Services

Modelling, Dimensioning and Optimization of 5G Communication Networks, Resources and Services

Editors

Ioannis D. Moscholios
Mariusz Głąbowski
Panagiotis Sarigiannidis
Michael D. Logothetis

MDPI • Basel • Beijing • Wuhan • Barcelona • Belgrade • Manchester • Tokyo • Cluj • Tianjin

Editors
Ioannis D. Moscholios
University of Peloponnese (UoP)
Greece

Mariusz Głąbowski
Poznan University of Technology
Poland

Panagiotis Sarigiannidis
University of Western Macedonia
Greece

Michael D. Logothetis
University of Patras
Greece

Editorial Office
MDPI
St. Alban-Anlage 66
4052 Basel, Switzerland

This is a reprint of articles from the Special Issue published online in the open access journal *Applied Sciences* (ISSN 2076-3417) (available at: https://www.mdpi.com/journal/applsci/special_issues/5G_Communication_Networks).

For citation purposes, cite each article independently as indicated on the article page online and as indicated below:

LastName, A.A.; LastName, B.B.; LastName, C.C. Article Title. *Journal Name* **Year**, *Volume Number*, Page Range.

ISBN 978-3-0365-5575-1 (Hbk)
ISBN 978-3-0365-5576-8 (PDF)

Contents

About the Editors vii

Ioannis D. Moscholios, Mariusz Głąbowski, Panagiotis G. Sarigiannidis and Michael D. Logothetis
A Special Issue on Modeling, Dimensioning, and Optimization of 5G Communication Networks, Resources, and Services
Reprinted from: *Appl. Sci.* **2022**, *12*, 1859, doi:10.3390/app12041859 1

Ivo Sousa, Nuno Sousa, Maria Paula Queluz and António Rodrigues
Fronthaul Design for Wireless Networks
Reprinted from: *Appl. Sci.* **2020**, *10*, 4754, doi:10.3390/app10144754 5

Woosik Lee, Eun Suk Suh, Woo Young Kwak and Hoon Han
Comparative Analysis of 5G Mobile Communication Network Architectures
Reprinted from: *Appl. Sci.* **2020**, *10*, 2478, doi:10.3390/app10072478 31

Kaleem Arshid, Iftikhar Hussain, Muhammad Khawar Bashir, Shahid Naseem, Allah Ditta, Natash Ali Mian, Misha Zahid and Israr Ali Khan
Primary User Traffic Pattern Based Opportunistic Spectrum Handoff in Cognitive Radio Networks
Reprinted from: *Appl. Sci.* **2020**, *10*, 1674, doi:10.3390/app10051674 49

Abdulraqeb Alhammadi, Mardeni Roslee, Mohamad Yusoff Alias, Ibraheem Shayea and Abdullah Alquhali
Velocity-Aware Handover Self-Optimization Management for Next Generation Networks
Reprinted from: *Appl. Sci.* **2020**, *10*, 1354, doi:10.3390/app10041354 69

Zbigniew Zakrzewski
D-RoF and A-RoF Interfaces in an All-Optical Fronthaul of 5G Mobile Systems
Reprinted from: *Appl. Sci.* **2020**, *10*, 1212, doi:10.3390/app10041212 83

Xin Li, Jun Xu, Lav Gupta and Raj Jain
Towards Efficiently Provisioning 5G Core Network Slice Based on Resource and Topology Attributes
Reprinted from: *Appl. Sci.* **2019**, *9*, 4361, doi:10.3390/app9204361 111

Muhammad Waqar and Ajung Kim
Performance Improvement of Ethernet-Based Fronthaul Bridged Networks in 5G Cloud Radio Access Networks
Reprinted from: *Appl. Sci.* **2019**, *9*, 2823, doi:10.3390/app9142823 133

Mariusz Głąbowski, Adam Kaliszan and Maciej Stasiak
A Palm-Jacobaeus Loss Formula for Multi-Service Systems with Separated Resources
Reprinted from: *Appl. Sci.* **2020**, *10*, 4019, doi:10.3390/app10114019 151

Panagiotis I. Panagoulias, Ioannis D. Moscholios, Panagiotis G. Sarigiannidis, Mariusz Głąbowski and Michael D. Logothetis
An Analytical Framework in OFDM Wireless Networks Servicing Random or Quasi-Random Traffic
Reprinted from: *Appl. Sci.* **2019**, *9*, 5376, doi:10.3390/app9245376 165

Qiuming Zhu, Wei Huang, Kai Mao, Weizhi Zhong, Boyu Hua, Xiaomin Chen and Zikun Zhao
A Flexible FPGA-Based Channel Emulator for Non-Stationary MIMO Fading Channels
Reprinted from: *Appl. Sci.* **2020**, *10*, 4161, doi:10.3390/app10124161 . . . . **181**

Byoungjin Seok, Jose Costa Sapalo Sicato, Tcydenova Erzhena, Canshou Xuan, Yi Pan and Jong Hyuk Park
Secure D2D Communication for 5G IoT NetworkBased on Lightweight Cryptography
Reprinted from: *Appl. Sci.* **2020**, *10*, 217, doi:10.3390/app10010217 . . . . **195**

Khalid Sheikhidris Mohamed, Mohamad Yusoff Alias and Mardeni Roslee
Interference Avoidance Using TDMA-Beamforming in Location Aware Small Cell Systems
Reprinted from: *Appl. Sci.* **2019**, *9*, 4979, doi:10.3390/app9234979 . . . . **209**

Zeng-You Sun and Yu-Jie Zhao
A Novel Self-Interference Cancellation Method Using an Optimized LMS Algorithm in CCFD Systems for a 5G Communication Network
Reprinted from: *Appl. Sci.* **2019**, *9*, 3308, doi:10.3390/app9163308 . . . . **221**

Rana M. Aly, Amira Zaki, Waleed K. Badawi and Moustafa H. Aly
Time Coding OTDM MIMO System Based on Singular Value Decomposition for 5G Applications
Reprinted from: *Appl. Sci.* **2019**, *9*, 2691, doi:10.3390/app9132691 . . . . **235**

Lei Wen, Razieh Razavi and Jing Lei
Intrinsic Interference Use for FBMC-IOTA Systems
Reprinted from: *Appl. Sci.* **2019**, *9*, 3210, doi:10.3390/app9163210 . . . . **247**

Thi-Minh-Dung Tran, Luu Ngoc An and Ngoc Chi Nam Doan
Concensus-Based ALADIN Method to Faster the Decentralized Estimation of Laplacian Spectrum
Reprinted from: *Appl. Sci.* **2020**, *10*, 5625, doi:10.3390/app10165625 . . . . **257**

About the Editors

Ioannis D. Moscholios

Ioannis D. Moscholios received his Dipl. Eng. degree in Electrical Computer Engineering from the University of Patras, Patras, Greece in 1999; his M.Sc. degree in Spacecraft Technology Satellite Communications from the University College London, UK in 2000; and his Ph.D. in Electrical Computer Engineering from the University of Patras in 2005. He is an Associate Professor in the Dept. of Informatics Telecommunications, University of Peloponnese, Tripolis, Greece. His research interests include teletraffic engineering and simulation and performance analysis of communication networks. He has published over 190 papers in international journals/conferences and is co-author of the book: Efficient M ultirate T eletraffic Lo ss Mo dels Be yond Er lang (I EEE Pr ess, Wi ley, April 2019). He has served as a Guest Editor in: (a) IET Communications, (b) IET Networks, (c) Applied Sciences, and d) Mobile Information Systems. He is an IARIA Fellow and a member of the Technical Chamber of Greece (TEE).

Mariusz Głąbowski

Mariusz Głąbowski received his M. Sc., Ph. D. and D. Sc. (Habilitation) degrees in telecommunication from Poznan University of Technology, Poland, in 1997, 2001, and 2010, respectively. In 2020, he was nominated as a full professor. Since 2020, he has worked in the Faculty of Computing and Telecommunications, Poznan University of Technology. He is engaged in research and teaching of performance analysis and modelling of multiservice networks and switching systems. Prof. Mariusz Głąbowski is the author/co-author of 4 books, 12 book chapters, and over 150 papers, which have been published in communication journals and presented at national and international conferences. Mariusz Głąbowski is an Associate Editor of the IEICE Communications Express and an Editor-in-Chief of the *International Journal on Advances in Internet Technology*. He has served as a Guest Editor in *Optical Switching and Networking, Mobile Information Systems, IET Networks, IEICE Transactions on Communications, Journal of Telecommunications* and *Information Technology*. He is a senior member of IEEE, an IARIA Fellow, a senior member of IEICE, and a member of FITCE.

Panagiotis Sarigiannidis

Panagiotis Sarigiannidis has been an Associate Professor in the Department of Electrical and Computer Engineering in the University of Western Macedonia, Kozani, Greece since 2016. He received his B.Sc. and Ph.D. degrees in computer science from Aristotle University of Thessaloniki, Thessaloniki, Greece, in 2001 and 2007, respectively. He has published over 180 papers in international journals and conferences, and numerous book chapters. He has been involved in several national, European and international projects. He is currently the project coordinator of three H2020 projects (SPEAR, EVIDENT, TERMINET) while he coordinates two operational project (MARS, SmartROOT). He also serves as the principal investigator in one H2020 project (SDN-microSENSE) and three more Erasmus+ KA2 projects (Arrange-ICT, JAUNTY, STRONG). His research interests include telecommunication networks, Internet of Things and network security. He is an IEEE member and participates in the Editorial Boards of various journals, including *International Journal of Communication Systems* and *EURASIP Journal on Wireless Communications and Networking*.

Michael D. Logothetis

Michael D. Logothetis (SM 08) received the Dipl.Eng. and Ph.D. degrees in electrical engineering from the University of Patras, Patras, Greece, in 1981 and 1990, respectively. Since 2009, he has been a (Full) Professor in the Electrical and Computer Engineering Department, University of Patras, Patras, Greece. He has authored over 230 conference/journal papers and is a co-author of the book: *Efficient Multirate Teletraffic Loss Models Beyond Erlang* (IEEE Press, Wiley, April 2019). His current research interests include teletraffic theory and engineering, traffic/network control, simulation, and performance optimization of telecommunications networks. He was General Chair of the IEEE/IET CSNDSP 2006 and IEICE ICTF 2016. He has been the Guest Editor of five journals: *Mediterranean Journal of Electronics and Communications, Mediterranean Journal of Computers and Networks, IET Circuits, Devices & Systems, IET Networks*, and *Ubiquitous Computing and Communication Journal*. He is a senior member of IEEE, a member of the IARIA (Fellow), the IEICE, the FITCE, and the Technical Chamber of Greece.

applied sciences

MDPI

Editorial

A Special Issue on Modeling, Dimensioning, and Optimization of 5G Communication Networks, Resources, and Services

Ioannis D. Moscholios [1,*], Mariusz Głąbowski [2], Panagiotis G. Sarigiannidis [3] and Michael D. Logothetis [4]

1 Department Informatics & Telecommunications, University of Peloponnese, 22131 Tripolis, Greece
2 Faculty of Computing and Telecommunications, Poznan University of Technology, 60965 Poznan, Poland; mariusz.glabowski@put.poznan.pl
3 Department Electrical and Computer Engineering, University of Western Macedonia, 50100 Kozani, Greece; psarigiannidis@uowm.gr
4 Department Electrical and Computer Engineering, University of Patras, 26504 Patras, Greece; mlogo@upatras.gr
* Correspondence: idm@uop.gr

1. Introduction

Designing, dimensioning, and optimization of communication networks resources and services have been inseparable parts of the development of telecommunications since the very beginning of their existence. These networking problems have changed dramatically in recent years as a result of the changes in users' requirements for converging wired and wireless multiservice communications networks.

The global network of 5G remains a network of many heterogeneous systems interacting with each other, while widely used broadband mobile devices and cloud computing have given rise to tremendous growth in network traffic. Contemporary telecom networks must convey a large volume of traffic and provide service to traffic streams with highly differentiated requirements, in terms of bit rate and service time as well as required quality of both service and experience parameters. In such a communication infrastructure, there are important challenges, such as the study of necessary multilayer cooperation, new protocols, performance evaluation of different network parts, low layer network design, network management and security issues, and new technologies in general.

Each new type of network technology that is introduced in 5G is followed by a substantial increase in both the number and complexity of problems that need to be resolved by theoreticians and engineers. No matter what these developing changes may bring, the essential tasks for modern communication networks remain the same over the years—namely, (1) to develop new technologies offering increasing network capacity, (2) to determine and evaluate the relationship between the quality of service (quality of experience) parameters and the parameters characterizing traffic sources (services), (3) to control and optimize the usage of network resources, and (4) to enhance the capabilities of data transport, transmission, and reception between end users and the core network. These tasks provide a basis for developing engineering algorithms and tools used for designing, analysis, dimensioning, and optimization of wired and wireless transmission systems and networks.

2. 5G Communication Networks Modeling, Dimensioning, and Optimization

Based on the above, this Special Issue aimed to amass state-of-the-art research contributions that address challenges in the design, dimensioning, and optimization of emerging 5G networks. After a rigorous review process, 16 papers were ultimately accepted for publication in this Special Issue, covering a range of important and well-timed subject areas.

The first paper, authored by I. Sousa, N. Sousa, M. Queluz, and A. Rodrigues, proposes a methodology for the evaluation and comparison of the performance of microwave radio transmission, free-space optics, and fiber optic technologies in the Cloud Radio Access

Citation: Moscholios, I.D.; Głąbowski, M.; Sarigiannidis, P.G.; Logothetis, M.D. A Special Issue on Modeling, Dimensioning, and Optimization of 5G Communication Networks, Resources, and Services. *Appl. Sci.* **2022**, *12*, 1859. https://doi.org/10.3390/app12041859

Received: 14 December 2021
Accepted: 31 January 2022
Published: 11 February 2022

Network (C-RAN) fronthaul segment, assuming different weather conditions, link lengths, and bit rate requirements [1]. Such a methodology enables the determination of the most cost-effective solution for each remote radio head (RRH)–baseband unit (BBU) link and the computation of the required number of BBUs. The second paper, authored by W. Lee, E. Suk Suh, W. Kwak, and H. Han, analyzes via simulation three different 5G communication network architectures—namely, the centralized cloud computing architecture, the multiaccess edge computing architecture, and the proposed Hybrid Cloud Computing (HCC) architecture, in terms of their response capability under normal and disruptive data traffic conditions [2]. The behavior of the abovementioned architectures was assessed using actual data traffic patterns obtained from the field, while the results obtained were used to quantify the advantages and disadvantages of each architecture and demonstrated the benefit of the new HCC architecture. The third paper, authored by K. Arshid, I. Hussain, M. Bashir, S. Naseem, A. Ditta, N. Mian, M. Zahid, and I. Khan, proposes a primary user traffic pattern-based opportunistic spectrum handoff (PUTPOSH) scheme for cognitive radio networks (CRNs) [3]. The proposed scheme permits a cognitive radio user to sense the primary user's traffic through energy-efficient sensing and then adopt a proactive or reactive handoff strategy according to the interarrival rate. Simulation results show that PUTPOSH maximizes the channel utilization and the throughput and, at the same time, minimizes the overall service time and the number of handoffs in CRNs. The fourth paper, authored by A. Alhammadi, M. Roslee, M. Alias, I. Shayea, and A. Alquhali, proposes a velocity-based self-optimization algorithm to adjust the handover control parameters in 4G/5G networks [4]. The proposed algorithm utilizes the user's received power and speed to adjust the handover margin and the time to trigger during the user's mobility in the network. Based on simulation results, the authors show that the proposed self-optimization algorithm reduces the rate of ping-pong handovers and radio link failure, compared with other existing algorithms. The fifth paper, authored by Z. Zakrzewski, proposes a solution for enabling the coexistence of digitized radio-over-fiber and analog radio-over-fiber interfaces operating in the optical fronthaul of 5G mobile systems [5]. The sixth paper, authored by X. Li, C. Guo, J. Xu, L. Gupta, and R. Jain, studies the interesting subject of 5G network slices provisioning by taking into account both the slice node provisioning and the slice link provisioning aspects [6]. More specifically, the authors propose a two-stage heuristic slice provisioning algorithm for the 5G core network by jointly considering network resource attributes and topology attributes. Extensive simulation results show that the proposed algorithm increases the slice request acceptance ratio and, consequently, the revenue of the network infrastructure provider. The seventh paper, authored by M. Waqar and A. Kim, proposes and studies via simulation the end-to-end latency-aware path computation scheme, to improve the link utilization of capacity constraint Ethernet-based fronthaul bridged networks (EFBNs) in 5G C-RANs [7]. More specifically, the authors consider the queuing delays and end-to-end latencies for the selection of the optimal paths for individual enhanced Common Public Radio Interface (eCPRI) traffic streams in the EFBNs. In addition, they propose a packet forwarding mechanism that maximizes the transmissions of multiple eCPRI streams at tolerable latencies and improves the link distances between the RRHs and the BBU pool. The eighth paper, authored by M. Głąbowski, A. Kaliszan, and M. Stasiak, proposes a method for the determination of the probability of strictly determined multiservice resources in a group of resources [8]. The proposed method allows the authors to elaborate a new formula for calculating the blocking probability of strictly determined resources in systems servicing multiservice traffic, e.g., in a group of cells in 4G and 5G systems. The proposed formula, whose accuracy is verified via simulation, is an extension of the well-known Palm–Jacobaeus loss formula elaborated for systems with single-rate traffic. The ninth paper, authored by P. Panagoulias, I. Moscholios, P. Sarigiannidis, M. Głąbowski, and M. Logothetis, studies both analytically and through simulation the downlink of an orthogonal frequency division multiplexing based cell that services multirate traffic [9]. The call arrival process is random (Poisson) or quasi-random. To determine congestion probabilities and resource utilization, the cell is modeled as a multirate loss model, while the call admission is based on three different policies—restricted accessibility, bandwidth reservation, and complete sharing policies. In all three policies, recursive formulas are proposed for the determination of the various performance measures. The 10th paper, authored by Q. Zhu, W. Huang, K. Mao, W. Zhong, B. Hua, X. Chen, and Z. Zhao, proposes a discrete, non-stationary, multiple-input–multiple-output (MIMO) channel model suitable for the fixed-point realization on the field-programmable gate array (FPGA) hardware platform [10]. To this end, the authors develop a flexible hardware architecture with configurable channel parameters and implement it on a non-stationary MIMO channel emulator in a single FPGA chip. In addition, an improved non-stationary channel emulation method is employed to guarantee accurate channel fading and phase, and the schemes of other key modules are also illustrated and implemented in a single FPGA chip. Hardware tests presented by the authors show that the output statistical properties of the proposed MIMO channel emulator agree well with the corresponding theoretical ones. The 11th paper, authored by B. Seok, J. Sicato, T. Erzhena, C. Xuan, Y. Pan, and J. Park, studies and analyzes existing results about secure device-to-device (D2D) communication systems in terms of their security considerations and limitations [11]. In addition, a secure D2D communication system is proposed with the aim to address the security challenges and limitations of the existing results. The proposed secure D2D communication system was designed based on elliptic curve cryptography and lightweight authenticated encryption with associated data ciphers to cover resource-constrained IoT devices. The 12th paper, authored by K. Mohamed, M. Alias, and M. Roslee, studies the applicability of time division multiple access beamforming for future communication systems [12]. The authors propose the Sl-beamforming scheme for line-of-sight small cell systems in the LTE architecture. Simulation results show that the proposed scheme enhances the network signal to interference plus noise ratio and user average throughput, compared with non-dominantly interfered region schemes. The 13th paper, authored by Z. Sun and Y. Zhao, proposes a novel time-varying least mean square adaptive filtering algorithm, with the aim to cancel the self-interference of near-end transceivers in co-frequency co-time full-duplex (CCFD) systems of the radio frequency domain [13]. The proposed algorithm solves the problem of strong self-interference of the radio frequency domain in CCFD systems by cooperatively controlling the autocorrelation values between the time factor and the error signal, to update the step size of the algorithm and provide a better quality of service for 5G communication networks. The 14th paper, authored by R. Aly, A. Zaki, W. Badawi, and M. Aly, proposes a space-time, coded, orthogonal transform division multiplexing (STC OTDM) technique for 5G applications [14]. The proposed system enhances the data rate and performance of the orthogonal transform division multiplexing (OTDM) technique and is based on using space-time coding (STC) with OTDM, to increase the system diversity and consequently the system performance. Simulation results provided by the authors show that the proposed technique achieves better performance when compared with other multicarrier techniques. The 15th paper, authored by L. Wen, R. Razavi, and J. Lei, analyzes the intrinsic interference of filter bank-based multicarrier systems with isotropic orthogonal transfer algorithm pulse-shaping [15]. Such intrinsic interference is treated as a parity symbol and an iterative soft-in-soft-out detector, which is based on a message-passing algorithm, is proposed to exploit the useful information of the intrinsic interference. Finally, the last paper, authored by T. Tran, A. Ngoc, and N. Doan, presents an augmented Lagrangian-based alternating-direction inexact Newton (ALADIN) method in order to estimate the Laplacian spectrum in a decentralized scheme for dynamic controlled networks [16]. The key feature of this paper is a direct solution to non-convex optimization for Laplacian spectrum estimation using the proposed ALADIN method.

Funding: This research received no external funding.

Acknowledgments: We would like to thank and congratulate all the authors for their valuable contributions to this Special Issue. We also thank the peer reviewers for their comments and suggestions, which helped the authors improve their papers. Finally, we would like to express our gratitude to the editorial team of MDPI *Applied Sciences* for their continuous support.

Conflicts of Interest: The authors declare no conflict of interest.

References

1. Sousa, I.; Sousa, N.; Queluz, M.P.; Rodrigues, A. Fronthaul Design for Wireless Networks. *Appl. Sci.* **2020**, *10*, 4754. [CrossRef]
2. Lee, W.; Suh, E.S.; Kwak, W.Y.; Han, H. Comparative Analysis of 5G Mobile Communication Network Architectures. *Appl. Sci.* **2020**, *10*, 2478. [CrossRef]
3. Arshid, K.; Hussain, I.; Bashir, M.K.; Naseem, S.; Ditta, A.; Mian, N.A.; Zahid, M.; Khan, I.A. Primary User Traffic Pattern Based Opportunistic Spectrum Handoff in Cognitive Radio Networks. *Appl. Sci.* **2020**, *10*, 1674. [CrossRef]
4. Alhammadi, A.; Roslee, M.; Alias, M.Y.; Shayea, I.; Alquhali, A. Velocity-Aware Handover Self-Optimization Management for Next Generation Networks. *Appl. Sci.* **2020**, *10*, 1354. [CrossRef]
5. Zakrzewski, Z. D-RoF and A-RoF Interfaces in an All-Optical Fronthaul of 5G Mobile Systems. *Appl. Sci.* **2020**, *10*, 1212. [CrossRef]
6. Li, X.; Guo, C.; Xu, J.; Gupta, L.; Jain, R. Towards Efficiently Provisioning 5G Core Network Slice Based on Resource and Topology Attributes. *Appl. Sci.* **2019**, *9*, 4361. [CrossRef]
7. Waqar, M.; Kim, A. Performance Improvement of Ethernet-Based Fronthaul Bridged Networks in 5G Cloud Radio Access Networks. *Appl. Sci.* **2019**, *9*, 2823. [CrossRef]
8. Głąbowski, M.; Kaliszan, A.; Stasiak, M. A Palm-Jacobaeus Loss Formula for Multi-Service Systems with Separated Resources. *Appl. Sci.* **2020**, *10*, 4019. [CrossRef]
9. Panagoulias, P.I.; Moscholios, I.D.; Sarigiannidis, P.G.; Głąbowski, M.; Logothetis, M.D. An Analytical Framework in OFDM Wireless Networks Servicing Random or Quasi-Random Traffic. *Appl. Sci.* **2019**, *9*, 5376. [CrossRef]
10. Zhu, Q.; Huang, W.; Mao, K.; Zhong, W.; Hua, B.; Chen, X.; Zhao, Z. A Flexible FPGA-Based Channel Emulator for Non-Stationary MIMO Fading Channels. *Appl. Sci.* **2020**, *10*, 4161. [CrossRef]
11. Seok, B.; Sicato, J.C.S.; Erzhena, T.; Xuan, C.; Pan, Y.; Park, J.H. Secure D2D Communication for 5G IoT Network Based on Lightweight Cryptography. *Appl. Sci.* **2020**, *10*, 217. [CrossRef]
12. Mohamed, K.S.; Alias, M.Y.; Roslee, M. Interference Avoidance Using TDMA-Beamforming in Location Aware Small Cell Systems. *Appl. Sci.* **2019**, *9*, 4979. [CrossRef]
13. Sun, Z.-Y.; Zhao, Y.-J. A Novel Self-Interference Cancellation Method Using an Optimized LMS Algorithm in CCFD Systems for a 5G Communication Network. *Appl. Sci.* **2019**, *9*, 3308. [CrossRef]
14. Aly, R.M.; Zaki, A.; Badawi, W.K.; Aly, M.H. Time Coding OTDM MIMO System Based on Singular Value Decomposition for 5G Applications. *Appl. Sci.* **2019**, *9*, 2691. [CrossRef]
15. Wen, L.; Razavi, R.; Lei, J. Intrinsic Interference Use for FBMC-IOTA Systems. *Appl. Sci.* **2019**, *9*, 3210. [CrossRef]
16. Tran, T.-M.-D.; Ngoc An, L.; Doan, N.C.N. Concensus-Based ALADIN Method to Faster the Decentralized Estimation of Laplacian Spectrum. *Appl. Sci.* **2020**, *10*, 5625. [CrossRef]

applied sciences

MDPI

Article

Fronthaul Design for Wireless Networks

Ivo Sousa *, Nuno Sousa, Maria Paula Queluz and António Rodrigues

Instituto de Telecomunicações, IST, University of Lisbon, 1049-001 Lisbon, Portugal;
nunoguerreirodesousa@gmail.com (N.S.); paula.queluz@lx.it.pt (M.P.Q.); antonio.rodrigues@lx.it.pt (A.R.)
* Correspondence: ivo.sousa@lx.it.pt; Tel.: +351-218-418-454

Received: 12 June 2020; Accepted: 8 July 2020; Published: 10 July 2020

Abstract: Cloud Radio Access Network (C-RAN) architectures have arisen as an alternative to traditional wireless network architectures, notably by taking advantage of the functional split between the multiple distributed Remote Radio Heads (RRHs) and the centralized Baseband Units (BBUs), through the creation of a new connectivity segment—the fronthaul. In order to maximize the investment return, it is important to find out, for this C-RAN segment, which technologies provide cost-effective solutions. This paper addresses this issue by evaluating and comparing the performance of Microwave Radio Transmission (MRT), Free Space Optics (FSO), and Fiber Optics (FO) technologies when applied to the fronthaul. First, a methodology is provided to determine the most cost-effective solution for each RRH–BBU link, as well as to compute the required number of BBUs and where they should be positioned in order to minimize the overall network costs. Next, a cost-effectiveness comparison of the aforementioned communication technologies is presented for individual fronthaul segments under different weather conditions, link lengths, and bit rate requirements. Moreover, an assessment is performed regarding the impact of the RRH density on the selection of cost-effective communication technologies for C-RANs. The obtained results allow concluding that fronthaul expenses are significantly affected by the performance of FSO systems, which in turn is affected by weather conditions; this highlights the relevance of having accurate climate statistics and forecasts in order to get the most out of the FSO technology and, consequently, lowering the overall network costs.

Keywords: Cloud Radio Access Network (C-RAN); fronthaul; Microwave Radio Transmission (MRT); Free Space Optics (FSO); Fiber Optics (FO)

---

1. Introduction

Mobile phones are currently the most-often used means of accessing the Internet, and the number of mobile communication subscriptions continues to grow [1]. This results from the fact that consumers now expect to access data when and where they need them, as well as to have higher throughput while paying the same (or even less) in the future. On the other hand, the high consumption of mobile data represents a critical revenue driver for operators [2]; in fact, as data traffic becomes more and more dominant over time, the revenues become flat [3], thus putting an immense amount of pressure on the mobile network providers, which need to optimize costs while providing high-capacity and high-quality services to their subscribers.

It is generally recognized that an increase of the mobile cell density is an effective way to attain vastly high data rate coverage. However, the respective deployment costs may hinder the Radio Access Network (RAN) improvement; hence, cost-effective solutions are needed for this part of the mobile telecommunication system, so that the investments can be supported by the revenues. Mobile network virtualization, notably the concept of Cloud-RAN (C-RAN) [4,5], arose as a path to lower Capital Expenditures (CAPEX) and Operational Expenditures (OPEX), as it enables the sharing of infrastructures, which can help to reduce the overall expenses of deployment and operations. Nevertheless, the costs of fronthauling the mobile cells—i.e., the expenses to establish connections between the multiple distributed Remote Radio Heads (RRHs) and the centralized Baseband Units (BBUs)—still pose an important challenge.

This paper considers the C-RAN concept and aims at finding out, for a wide range of scenarios, which underlying fronthaul technologies are cost-effective; the key technologies examined are Microwave Radio Transmission (MRT), Free Space Optics (FSO), and Fiber Optics (FO). The following contributions are made by this paper: (1) a cost-effectiveness comparison of the aforementioned communication technologies is performed for individual fronthaul segments under different weather conditions, lengths of the links, and bit rate requirements; (2) an assessment is also performed regarding the impact of the RRH density on the selection of cost-effective communication technologies for C-RANs. These studies were made possible by the development of a methodology that enables determining the most cost-effective solution for each RRH–BBU link, as well as to compute the required number of BBUs and where they should be positioned in order to minimize the overall network costs. In a nutshell, the provided methodology along with the findings of this work aim at serving as useful guidelines when designing fronthaul networks.

This paper is organized as follows. After the introduction, Section 2 provides a background on the topics of interest for this work, namely C-RAN and fronthaul communication technologies, as well as a review of related work. Section 3 describes the developed methodology to evaluate and compare the performance of MRT, FSO, and FO technologies when applied to the fronthaul. Section 4 provides and discusses the results of the studies performed herein, namely an evaluation of which technology (MRT, FSO, or FO) is the most cost-effective choice for an individual RRH–BBU link regarding the conditions of its implementation, as well as a cost-effectiveness analysis of communication technologies for C-RANs with different RRH densities. Finally, Section 5 concludes the paper.

2. Background on C-RAN and Fronthaul Communication Technologies and Related Work

In order to contextualize the topics address in this paper, this section presents a description of the C-RAN architecture, which enables pointing out the role of the fronthaul segment within the mobile network. Moreover, typical fronthaul communication technologies are also overviewed and compared, namely MRT, FSO, and FO. The section ends with a review of related work.

2.1. Cloud Radio Access Network and Fronthaul

In its traditional architecture, the RAN comprises costly stand-alone Base Stations (BSs), each one serving a certain area and thereby only handling transmitted and received signals of the User Equipment (UE) within that same area. In order to attain high data rate coverage, a large number of BSs is required, which also poses some challenges, such as the costly initial investment, site rental, and site support; thus, an increase in the number of BSs gives rise to a significant increase in CAPEX and OPEX. On the other hand, traditional BSs are configured to fulfill peak demands in order to reduce outages; however, since the average network load is usually far lower than that in peak load, the BS utilization rate ends up being low. Noticing that the processing capacity of each BS cannot be shared with others under the traditional RAN architecture, reducing the number of BSs or their processing resources is not an option to optimize their utilization efficiency, as it would lead to many and unacceptable congestion events during peak hours, or even a lack of coverage issues.

In order to reduce the overall expenses of deployment and operations, the C-RAN architecture was proposed: a radio access network model in which baseband resources are pooled so that BSs can share them. This is achieved by separating the traditional BS into an RRH, which is responsible for power amplification and analog processing, and a BBU, which manages the common digital baseband functions, such as modulation and coding. The optimization of the BBU usage is reached through the virtualization of a BBU pool, which can then be shared among cell sites, i.e., RRHs. Figure 1 depicts a typical C-RAN architecture, which highlights the backhaul segment, i.e., the connection between the core network and the BBU, and also the new connectivity segment—the fronthaul—introduced by the C-RAN architecture between the multiple distributed RRHs and the BBU. According to the related literature [4,5], a C-RAN presents the following benefits relative to a traditional RAN: energy efficiency, decrease in CAPEX and OPEX, throughput improvement, adaptability to non-uniform traffic, and smart Internet traffic offload.

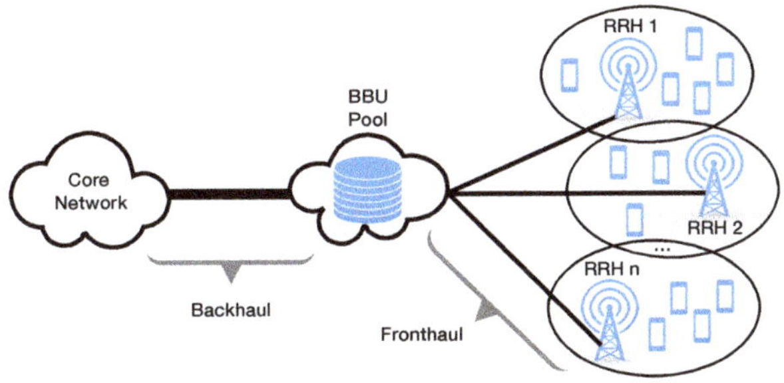

Figure 1. C-RAN architecture. BBU, Baseband Unit; RRH, Remote Radio Head.

One of the main technical challenges regarding the C-RAN architecture is the large amounts of baseband sampling data that have to be carried out in real time between RRHs and the BBU pool. Besides strict latency requirements, the wideband required by some radio access systems, like Long-Term Evolution (LTE) or 5G, implies that each fronthaul link should support substantial bit rates, e.g., 1 Gbps [6].

2.2. Typical Fronthaul Communication Technologies

The adopted technologies for the fronthaul links and associated transmission medium, which can be wired or wireless, must be able to fulfill the requirements highlighted previously. Typically, the fronthaul segment uses terrestrial communication technologies, namely MRT, FSO, and FO—for instance, the MRT technology was considered as an option for the fronthaul in [4–7], whereas FSO systems were considered in [8,9], and the FO technology was contemplated in [4–7,10,11].

MRT is a line-of-sight wireless communication technology that uses microwave radio waves in order to provide high speed wireless connections, which generally can go up to several Gbps [12]. MRT is widely used for point-to-point communications, as the small wavelength allows the use of conveniently-sized antennas to create highly directional beams, making their installation rather simple and suitable for any terrain. Moreover, since the associated equipment can be easily disassembled, relocated, and reused, MRT systems are also a good easy-to-deploy option in the case of a catastrophe or emergency, especially for remote locations. Multi-path fading is a significant factor degrading the performance of MRT links operating below 10 GHz; however, when considering frequencies above 10 GHz, weather conditions represent the dominant cause of communication disruption, notably rain (which becomes a crucial attenuation factor for this technology) and atmospheric absorption [13]. It is also worth mentioning that relay stations may be used to overcome the line-of-sight limitation of MRT systems (notably for link lengths longer than the visual horizon). The installation costs of MRT systems are independent of each hop length (i.e., the distance between two MRT stations) if the same equipment is used; on the other hand, the total costs of an MRT link comprising multiple hops is greatly dependent on the number of hops used [14]. Furthermore, and in general, MRT systems operate in licensed frequency bands; hence, individual MRT systems are subject to license (on a case-by-case basis to avoid interference between adjacent systems) by the respective national regulatory authority for communications, which normally entails the payment of an annual fee (an operational cost that is usually square-root dependent on the link length) [15,16].

The FSO technology uses infrared radiation in free space (i.e., air, outer space, or vacuum) to wirelessly transmit data; for that reason, there has been growing recognition in recent years of the importance of FSO systems in the ongoing development of fixed telecommunication systems, mainly because not only FSO links can be used in conjunction with MRT links without causing interference, but also the FSO technology is well suited for environments where the radio spectrum is already crowded (e.g., dense urban areas). In particular, FSO is a point-to-point line-of-sight communication technology like MRT, but it allows higher capacities, namely connections of up to several dozen Gbps [17]. An FSO system consists of optical transceivers at the stations; it has a low initial investment, and it is easy to deploy (similar to MRT systems). One characteristic of the FSO technology is the low diffraction of the laser generated beam; on the one hand, this provides high security of the transmitted data, as an interceptor would have to be perfectly aligned with the transmitter in order to obtain the data; on the other hand, the very narrow optical beam makes FSO a sensitive system, as physical obstructions (like trees or even flying birds) can block the signal and cause interruptions in a link. A major limitation associated with FSO links is the attenuation caused by fog, which can reach values as high as 480 dB/km in extreme situations [18]. Other attenuation factors are atmospheric turbulence, i.e., changes in the atmosphere refractive index (e.g., due to high temperatures), which can cause deviations in the beam, and interference from background light sources (including the Sun). The deployment costs of FSO systems are independent of each hop length (if the same equipment is used). Moreover, these systems have the advantage of being license free, and no planning permission is required as long as they are eye safe. Nevertheless, the maximum achievable range per hop is lower than in MRT [17].

FO communication systems are light wave systems that employ optical fibers for information transmission, which means that these are not wireless systems (as opposed to MRT and FSO technologies) and, therefore, do not require line-of-sight. The capacity of current optical fibers can reach values up to several hundred Gbps (per wavelength channel) [19], being mainly conditioned by optical pulse widening due to fiber dispersion, which limits the maximum bit rate $\times$ distance product (a metric that describes the fiber performance in terms of transmission capability). Since the optical fiber cable is usually installed underground, FO systems may be damaged by floods. Moreover, accessibility issues regarding the path of installation of optical fibers, such as those in mountainous regions, might lead to FO being regarded as an unsuitable option. The deployment costs of an FO system vary linearly (roughly) with the considered connection length [14]. Accordingly, the installation of FO systems is usually more expensive than MRT and FSO systems' installation.

A summary of the main differences between MRT, FSO, and FO is presented in Table 1. As can be inferred from this table, the best choice regarding the fronthaul communication technology is not straightforward, as it depends on the required bit rate, terrain characteristics, and typical weather conditions of the installation sites, among others.

Table 1. Comparison of typical fronthaul communication technologies.

	MRT	FSO	FO
Type	Wireless (requires line-of-sight)		Wired
Capacity	Up to several Gbps	Up to several dozen Gbps	Up to several hundred Gbps
Installation	Easy to deploy and suitable for any terrain		Unsuitable for difficult terrain (e.g., mountains)
Climate influence	Notably rain	Notably fog	Not influenced by climate (except floods)
Other	MRT maximum range per hop is higher than FSO		Installation is usually more expensive

2.3. Related Work

Since the fronthaul optimization can lead to a large decrease of CAPEX and OPEX of the C-RAN, it is of utmost importance to study and evaluate different approaches regarding the fronthaul architecture. The main objective of this paper is to evaluate and compare the use of MRT, FSO, and FO systems in the fronthaul, namely to study their influence in terms of cost-effective solutions for RRH–BBU links.

There are few works, available in the related literature, that tackle topics similar to the one of this paper. The authors of [20] analyzed operational and deployment costs of C-RANs and proposed a mathematical formulation for fronthaul planning, taking into account the aforementioned costs. Solutions were also presented regarding their optimization problem, which aim at minimizing the C-RAN costs subject to traffic demand constraints. Unfortunately, this work was limited in the sense that it considers only a single technology (namely FO) for the fronthaul.

In [21], different fronthaul technologies were discussed, including MRT, FSO, and FO, and potential solutions were suggested in order to achieve an efficient C-RAN. Nonetheless, the authors only provided a qualitative costs comparison of the different fronthaul technologies, without considering any specific scenario, as they focused their study on optical technologies (FSO and FO) in order to present feasible means of reducing the system complexity, costs, bandwidth requirement, and latency in the fronthaul.

A comparative costs study regarding the technologies addressed herein (MRT, FSO, and FO) along with networked flying platforms providing FSO links ("vertical FSO") was performed in [22] for a fronthaul/backhaul scenario. However, only a very high dense deployment of mobile cells was considered, and the technologies were compared in terms of the whole network costs when a single technology was adopted. Nevertheless, it can be concluded from their work that although the "vertical FSO" was a valid approach, it was a very expensive option—in fact, not only "vertical FSO" was shown to be the most expensive option, but also the associated costs were estimated to be more than twice the second most expensive solution; furthermore, its implementation posed additional problems, including safety and regulatory ones. For these reasons, the "vertical FSO" approach was not considered in our study.

Recently, other authors addressed rural connectivity and investigated the costs of solutions for the backhaul segment [23], namely by considering MRT, FSO (including the "vertical FSO" approach), and FO, in addition to satellite. In this case, the studied scenario is very specific—remote rural areas where the backhaul link needs to traverse long distances (e.g., 100 km); hence, its results cannot be generalized to the more common fronthaul scenario of RRH–BBU links separated by a few kilometers. Nevertheless, it could be concluded from their work that the "vertical FSO" solution was, again, more costly than the other options; on the other hand, their work showed that the cost-effectiveness of the MRT and FSO solutions depended, among other things, on the tower separation (namely, they compared hops of 3 km vs. 5 km ones), thus suggesting that a thorough study of this costs dependence on the hop distance, notably considering different equipment and communication technologies, would be of great value.

In light of the above, this paper advances the state-of-the-art by providing a quantitative cost-effectiveness comparison of three communication technologies, namely MRT, FSO, and FO, for individual fronthaul segments under different weather conditions, link lengths, and bit rate requirements. Furthermore, an assessment is also provided regarding the impact of low and high RRH densities on the selection of cost-effective communication technologies for C-RANs. In addition to the studies presented and discussed herein, this paper provides a generic and versatile methodology to determine the most cost-effective solution for each RRH–BBU link, as well as to compute the required number of BBUs and where they should be positioned in order to minimize the overall network costs.

3. Fronthaul Design Methodology

This section details the developed methodology to evaluate and compare the performance of MRT, FSO, and FO technologies when applied to the fronthaul. This methodology comprised a software tool built from scratch (based on the MATLAB programming language), which had the following purposes: (i) to determine the most cost-effective solution to connect two points regarding an RRH–BBU link, under user-specified equipment characteristics and link conditions; (ii) to find a fronthaul topology for wireless networks optimally, given the RRHs location—namely, the required number of BBUs and where they should be positioned in order to minimize the overall network costs. The algorithms associated with these goals, designated as the link design algorithm and network planning algorithm, respectively, are presented next. The software tool, including its source code, is freely available online [24]; hence, besides being a generic tool, since it can be used for a wide range of fronthaul projects, it is also a versatile tool, in the sense that it can be modified by anyone to incorporate other features that are specific for a certain project.

3.1. Link Design Algorithm

The link design algorithm determines, out of all the user-specified equipment characteristics associated with one or more of the communication technologies (MRT, FSO, and FO), which is the most cost-effective solution to connect two points under user-specified link conditions; for a set of specifications, namely the equipment features and other installation aspects (such as link distance, required bit rate, and surrounding environment characteristics), this algorithm computes the cheapest solution that is able to deliver the necessary bit rate, while satisfying a certain link margin and error criteria.

With respect to the wireless technologies addressed herein (MRT and FSO), it is important to mention that only single-hop links are considered by the algorithm, because the inclusion of relay stations is not straightforward (e.g., sites for relay deployment may not be available) and leads to more complex business models (e.g., the addition of rental expenses for the extra sites). Accordingly, link distance is henceforth regarded as hop distance when MRT and FSO technologies are under consideration.

The following subsections present the details of the link design algorithm, namely the adopted models for the communication technologies, the economic analysis methodology, and finally, the workflow of the algorithm.

3.1.1. Communication Technologies Models

For wireless communication technologies such as MRT and FSO, the received signal power, P_{Rx}, is given by (in logarithmic units):

$$P_{Rx} = P_{Tx} + G_{Tx} + G_{Rx} - A_0 - A_{equi} - A_{sys}, \tag{1}$$

where P_{Tx} corresponds to the transmitted power (in dBW or dBm), G_{Tx} and G_{Rx} stand for the transmitter and receiver antenna gains (in dBi), respectively, and A_0 represents the free-space path loss (in dB), i.e.,

$$A_0 = 92.4 + 20\log_{10}(d_{[\mathrm{km}]}) + 20\log_{10}(f_{[\mathrm{GHz}]}), \tag{2}$$

where d denotes the link distance and f refers to the carrier frequency, A_{equi} denotes the losses (in dB) related to equipment like cables, modulators, etc. (which are typically lower than 3 dB), and A_{sys} corresponds to other losses (in dB) related to specific attenuation factors regarding the considered communication technology.

With respect to MRT systems, the term A_{sys} incorporates the attenuation caused by obstacles in the line-of-sight path (A_{obs}), the attenuation induced by atmospheric gases (A_{gas}), and the attenuation due to rain (A_{rain}). All these attenuation factors can be computed as described in the related literature [13]; a summary is given as follows. The attenuation due to obstacles can be computed as (in dB):

$$A_{obs} = \max\left\{0\,;\, 6.9 + 20\log_{10}\left(\sqrt{(\nu-0.1)^2+1} + \nu - 0.1\right)\right\}, \tag{3}$$

where ν is proportional to the amount of the Fresnel ellipsoid that is obstructed by the obstacle, i.e.,

$$\nu = \frac{h_{obs}}{17.32}\sqrt{8\frac{f_{[\mathrm{GHz}]}}{d_{[\mathrm{km}]}}}, \tag{4}$$

where h_{obs} denotes the height of the top of the obstacle above the straight line joining the two ends of the link (if the height is below this line, then h_{obs} is negative). The attenuation due to atmospheric gases (namely uncondensed water vapor and oxygen) can be computed as (in dB):

$$A_{gas} = (\gamma_o + \gamma_w) \times d, \tag{5}$$

where γ_o and γ_w represent, respectively, the attenuation caused by oxygen and water by units of length; values for these parameters can be extracted from nonlinear curves as a function of the carrier frequency. With respect to A_{rain}, it should be pointed out that since rain is highly variable over time and differs from place to place, the respective attenuation factor depends on the desired time availability for the MRT link. More specifically, the attenuation due to rain has to be computed taking into account the value of rain intensity that is not exceeded in a certain percentage of the time in the location of interest. Accordingly, and having in mind the Service Level Requirements (SLR)—which address service times, maintenance, availability, performance, etc.—an MRT system designer must first stipulate the minimum time availability of the link (e.g., 99.9% of the time, as adopted in this work); afterwards, by using the method suggested by an ITU-R (International Telecommunication Union – Radiocommunication Sector) recommendation [25], the rain attenuation is computed, thus ensuring that the planned MRT link may still suffer from rain outage, but no longer than the maximum percentage of time unavailability of the link (U_{max}), with $U_{max} = 100\% -$ minimum link availability percent. Formally and given the rain intensity not exceeded in 0.01% of the time ($R_i^{(0.01\%)}$), the rain attenuation for that percentage of the time is given by (in dB):

$$A_{rain}^{0.01\%} = \beta \times \left(R_i^{(0.01\%)}\right)^{\alpha} \times d_{ef}, \tag{6}$$

where β and α denote coefficients that depend on the carrier frequency and on the considered temperature and d_{ef} corresponds to the effective distance through a rainy path, which is computed as:

$$d_{ef} = \max\left\{2.5\,d\,;\ \frac{d}{0.477\,d_{[\mathrm{km}]}^{0.633} \times \left(R_i^{(0.01\%)}\right)^{0.073\,\alpha} \times f_{[\mathrm{GHz}]}^{0.123} - 10.579\left(1 - e^{-0.024\,d_{[\mathrm{km}]}}\right)}\right\}; \tag{7}$$

finally, the rain attenuation exceeded for a percentage of time U_{max} other than 0.01% is obtained as:

$$A_{rain} = A_{rain}^{0.01\%} \times 0.12\,U_{max}^{-(0.546+0.043\log_{10} U_{max})}. \tag{8}$$

Turning the attention now to FSO systems, the term A_{sys} encompasses the attenuation induced by atmospheric absorption (A_{abs}), the attenuation due to atmospheric turbulence (A_{turb}), and the attenuation caused by scattering (A_{sca}). All these attenuation factors can be computed as described in the related literature [26,27]; a summary is given as follows. Considering the atmospheric absorption (which is mainly caused by the presence of gaseous molecules), it can be given by (in dB):

$$A_{abs} = \gamma_{abs} \times d, \tag{9}$$

where γ_{abs} stands for the absorption attenuation coefficient, which depends on the considered temperature and on the relative humidity. The attenuation due to atmospheric turbulence (i.e., small and random variations of the refractive index of the Earth's atmosphere, which are responsible for wave front distortion) can be described by (in dB):

$$A_{turb} = 2\,\sigma_{scin}, \tag{10}$$

where σ_{scin} refers to the scintillation index, which can be expressed as:

$$\sigma_{scin} = \sqrt{1.23\, C_n^2 \times \left(\frac{2\pi}{\lambda_{[m]}}\right)^{\frac{7}{6}} \times d_{[m]}^{\frac{11}{6}}}, \tag{11}$$

where λ corresponds to the operating wavelength and C_n^2 represents the index of refraction structure parameter, which is computed as (in $m^{-2/3}$):

$$C_n^2 = 9.8583 \times 10^{-18} + 4.9877 \times 10^{-16} \times e^{-\frac{h_{a[m]}}{300}} + 2.9228 \times 10^{-16} \times e^{-\frac{h_{a[m]}}{1200}}, \tag{12}$$

where h_a denotes the transmitter altitude. With respect to A_{sca}, which is caused by the occasional presence of fog (including mist and haze) and rain, it is important to stress that fog is the major contributor regarding the attenuation due to scattering; hence, and noticing that the respective attenuation coefficient is computed as a function of the visibility, this type of attenuation also depends on the desired time availability for the FSO link. More specifically, one has to take into account the value of the visibility that is not exceeded for a given percentage of the time in the location of interest. Since the ITU-R recommendations for FSO systems' design do not provide a metric to compute the visibility distribution, one alternative is to use one of the visibility distribution models presented in a related work [28]—e.g., the simplified model introduced therein and adopted in this methodology—which rely on the average number of foggy days (per year) and the average duration of fog events (in hours). Accordingly, the visibility value can be obtained and used to compute the respective attenuation that ensures that the planned FSO link may still suffer from fog outage, but no longer than the maximum percentage of the time regarding tolerated link unavailability, thus enabling fulfilling the SLR; it is important to mention that, in order to ensure the fulfillment of the same SLR regardless of the specific adopted wireless technology, the software tool considers the same percentage of the time for link unavailability regarding the attenuations related to both rain and fog. Formally, the attenuation caused by scattering is given by (in dB):

$$A_{sca} = (\gamma_{fog} + \gamma_{rain}) \times d, \tag{13}$$

where γ_{fog} and γ_{rain} represent, respectively, the attenuation caused by fog and water by units of length. The former parameter can be expressed as (in dB/km):

$$\gamma_{fog} = \frac{3.91}{V_{[km]}} \times \left(\frac{\lambda_{[nm]}}{550}\right)^{-q}, \tag{14}$$

where V stands for the visibility and q refers to a coefficient that is dependent on the size distribution of the scattering particles, which is given by:

$$q = \begin{cases} 1.6 & V > 50 \text{ km} \\ 1.3 & 6 \text{ km} < V < 50 \text{ km} \\ 0.16\, V_{[km]} + 0.34 & 1 \text{ km} < V < 6 \text{ km} \\ V_{[km]} - 0.5 & 0.5 \text{ km} < V < 1 \text{ km} \\ 0 & V < 0.5 \text{ km} \end{cases}. \tag{15}$$

With respect to the visibility, it can be computed as (in km):

$$V = \frac{U_{max}}{100} \times \frac{365.25}{\overline{N}_{fog[days/year]}} \times \frac{24}{\overline{D}_{[h]}}, \tag{16}$$

where $\overline{N_{fog}}$ and $\overline{D}$ refer to the average number of foggy days and the average duration of fog events, respectively. Finally, the parameter related to the attenuation caused by water can be expressed as (in dB/km):

$$\gamma_{rain} = 1.076 \left(R_i^{(0.01\%)}\right)^{0.67} \times 0.12\, U_{max}^{-(0.546+0.043\log_{10} U_{max})}. \tag{17}$$

After computing the received signal power, the wireless link margin, W_{link}, is obtained as (in dB):

$$W_{link} = P_{Rx} - S_{Rx}, \tag{18}$$

where S_{Rx} refers to the sensitivity of the receiver. In order to consider a wireless connection as viable and since the higher the link margin, the more robust the wireless link will be, as it will be prepared for potential extra attenuations, W_{link} must be greater than a user-specified minimum accepted link margin (e.g., 3 dB, as adopted in this work for both MRT and FSO links).

Once the link margin requirement is satisfied, another requirement, namely the Bit Error Rate (BER), must be fulfilled in order to ensure that the wireless link is feasible (e.g., in this work, the BER must be lower than 10^{-6}). The BER can be extracted from mapping curves as a function of the Signal-to-Noise Ratio (SNR) of the MRT link [13] or the FSO link [29]. With respect to the SNR of an MRT link, it can be given by (in dB):

$$\text{SNR}^{(\text{MRT})} = P_{Rx} - N_f - N_0, \tag{19}$$

where N_f corresponds to the noise figure of the receiver and N_0 denotes the thermal noise, which can be computed as (in dBW):

$$N_0 = -204 + 10\log_{10}(b_{w[\text{Hz}]}), \tag{20}$$

where b_w represents the noise equivalent bandwidth of the receiver; it can be expressed as:

$$b_w = \frac{B_{link}}{\log_2(M)}, \tag{21}$$

where B_{link} and M stand for the link bit rate and the QAM (Quadrature Amplitude Modulation) signal constellation size, respectively. Considering an FSO link and making the typical assumption of a shot-noise-limited operation with On-Off Key (OOK) modulation, the SNR can be obtained as (in dB):

$$\text{SNR}^{(\text{FSO})} = P_{Rx} - \frac{P_{Rx} + A_{turb}}{2} - 5\log_{10}\left(2\,h \times \frac{c}{\lambda} \times B_{link}\right), \tag{22}$$

where h and c refer to the Planck constant and the speed of light, respectively.

Considering now FO systems, the associated technology differs from the two wireless communication systems previously discussed (MRT and FSO) as it does not use the atmosphere as the propagation medium. More specifically, since the beam is confined to the fiber, there are no outside weather conditions that need to be taken into consideration when planning point-to-point transmission using the FO technology. Accordingly, evaluating a link budget for FO is equivalent to computing the total loss, suffered by a transmitted signal across various components and along the optical fiber, with reference to the minimum receiver power required to maintain normal operation.

In mathematical terms [30], the FO link budget, L_B, is given by (in dB):

$$L_B = Tx_{min} - Rx_{min}, \tag{23}$$

where Tx_{min} and Rx_{min} correspond to the minimum transmit power (at the transmitter) and minimum received power required (at the receiver), respectively (both in dBW or dBm). The total loss suffered by the transmitted signal along the link, T_L, is given by (in dB):

$$T_L = L + (d \times F_L), \tag{24}$$

where L stands for the losses in optical connectors (in dB) and F_L corresponds to the normalized fiber loss (in dB per units of distance); typical values for these parameters can be found in the related literature [30]. Finally, an FO link is assumed to be feasible if the FO link margin, i.e., $L_B - T_L$, is greater than a user-specified minimum accepted link margin (e.g., 3 dB, as adopted in this work), and if the link bit rate × distance product (i.e., the required bit rate times the link length) does not exceed the maximum bit rate × distance product of the fiber [10].

3.1.2. Link Costs Analysis

When planning a link, the costs associated with the project will always go beyond the costs of the equipment itself; it is important to mention that when referring to the equipment of a certain technology (MRT, FSO, or FO), it includes all the necessary items for installing and operating the respective link. In particular, two different types of costs have to be considered in the scope of these projects:

- CAPEX: These include the fixed costs related to the network infrastructure, such as equipment and respective deployment, spare parts, and project studies. With respect to wireless technologies, one has to consider emitter and receiver costs, as well as the costs of cables, stands, and in the case of FSO systems, an auto-tracker that allows the receptor to align with the received signal (in order to reduce the impact of atmospheric turbulence). Moreover, as stated in Section 2, the costs associated with CAPEX for MRT and FSO technologies do not depend on the link distance (if no repeaters are considered), unlike FO links. Accordingly, CAPEX related to FO systems can be divided into a fixed term, which accounts for Optical Line Terminations (OLTs), Optical Network Units (ONUs), and other miscellaneous electronics, plus a variable term that corresponds to the costs of the fiber itself and the costs of deploying it, which varies with the length of the link, d.
- OPEX: These do not contribute to the infrastructure itself, since they include operational expenses, such as maintenance costs, energy consumption, government taxes, and repayments. Accordingly, any economical analysis regarding OPEX is usually performed taking into account the lifetime of the communication link—e.g., a period of 10 years, as adopted in this work.

After gathering the values for both CAPEX and $\text{OPEX}_{\text{lifetime}}$ (for a given lifetime) regarding the use of a certain equipment, the total costs of the link project are given by the summation of these values, i.e.,

$$\text{Total Costs} = \text{CAPEX} + \text{OPEX}_{\text{lifetime}}. \tag{25}$$

Based on what was previously mentioned and recalling the information provided in Section 2, the total costs depend on the link length in the case of MRT and FO systems, whereas the total costs can be regarded as a fixed value (i.e., independent of the link length) for FSO systems; thus, Expression (25) can be rewritten with respect to the different technologies as:

$$\text{Total Costs}^{(\text{MRT})} = \text{F.Costs}^{(\text{MRT})} + \text{V.Costs}^{(\text{MRT})} \times \sqrt{d}, \tag{26}$$

$$\text{Total Costs}^{(\text{FSO})} = \text{F.Costs}^{(\text{FSO})}, \tag{27}$$

$$\text{Total Costs}^{(\text{FO})} = \text{F.Costs}^{(\text{FO})} + \text{V.Costs}^{(\text{FO})} \times d, \tag{28}$$

where F.Costs and V.Costs stand for fixed costs and variable costs, respectively, with respect to the associated technology (MRT, FSO, or FO); please note that these costs (i.e., $\text{F.Costs}^{(\text{MRT})}$, $\text{F.Costs}^{(\text{FSO})}$, $\text{F.Costs}^{(\text{FO})}$, $\text{V.Costs}^{(\text{MRT})}$, and $\text{V.Costs}^{(\text{FO})}$) also represent user-specified inputs of the link design algorithm.

3.1.3. Link Design Algorithm Workflow

Figure 2 depicts the flowchart of the link design algorithm. The first step is to read and store the information contained in the "MRT.dat", "FSO.dat", and "FO.dat" files. These ".dat" files are text files that contain data about the user-specified equipment being tested, as well as about the respective associated costs; in general, these necessary inputs are provided by the equipment manufacturers and, concerning costs, by taking into account the CAPEX and OPEX items listed in the previous section. The user can test and compare, at the same time, as many different equipment as desired, as the algorithm is able to process a variable amount of equipment; in this manner, the user can, for example, test and compare different solutions from different providers in a single run of the algorithm. More specifically, each line of the ".dat" file associated with the respective communication technology (MRT, FSO, and FO) corresponds to a different equipment of that technology. The structure of each line of these ".dat" files, namely the required inputs for each communication technology equipment (which should be separated by commas), is given in Table 2, where ID refers to the identifier (number or word) of an equipment, B stands for the maximum bit rate that an equipment can offer for the specified inputs, whereas BxD corresponds to the maximum bit rate $\times$ distance product of an optical fiber.

Table 2. Required inputs for the "MRT.dat", "FSO.dat", and "FO.dat" files.

MRT .dat	ID	B (Mbps)	f (GHz)	P_{Tx} (dBW)	G_{Tx} (dBi)	G_{Rx} (dBi)	A_{equi} (dB)	S_{Rx} (dBW)	N_f (dB)	M (M-QAM)	F.Costs$^{(MRT)}$ (¤)	V.Costs$^{(MRT)}$ (¤/$\sqrt{\text{km}}$)
FSO .dat	ID	B (Mbps)	λ (nm)	P_{Tx} (dBW)	G_{Tx} (dBi)	G_{Rx} (dBi)	A_{equi} (dB)	S_{Rx} (dBW)	F.Costs$^{(FSO)}$ (¤)			
FO .dat	ID	B (Mbps)	BxD (Mbps·km)		Tx_{min} (dBW)	Rx_{min} (dBW)	L (dB)	F_L (dB/km)			F.Costs$^{(FO)}$ (¤)	V.Costs$^{(FO)}$ (¤/km)

In addition, the user must provide data (such as link length, required bit rate, and climate information) regarding the link deployment scenario. This information is given in the "Scenario.dat" text file and follows a line structure (separated by commas), where the required inputs are indicated in Table 3; these correspond respectively to link length (d), required bit rate (B_{min}), maximum percentage of the time regarding tolerated link unavailability (U_{max}), temperature (T), rain intensity not exceeded in 0.01% of the time ($R_i^{(0.01\%)}$), relative humidity (H), transmitter altitude (h_a), height difference between the top of an obstacle and the line-of-sight (h_{obs}), average number of foggy days ($\overline{N_{fog}}$), average duration of fog events ($\overline{D}$), and minimum accepted link margins for MRT, FSO, and FO links ($Ml_{min}^{(MRT)}$, $Ml_{min}^{(FSO)}$, and $Ml_{min}^{(FO)}$, respectively).

From here, the algorithm becomes independent of the user. It will go through all the N different user-specified equipment, and for each one, the algorithm first checks if the required bit rate of the link is met by the equipment; if that condition is satisfied, then it is evaluated if the link is feasible with that equipment (taking into account the communication technologies models); in other words, if the required link distance is shown to be too long to be accommodated in a single-hop by the equipment that is under consideration in each iteration, then that equipment is ignored in the remaining analysis. Finally, the algorithm returns the cheapest working solution (taking into account the link costs analysis), namely the respective equipment ID and total costs. Please note that if none of the N different user-specified pieces of equipment meet the link requirements, then the algorithm returns (positive) infinity for the total costs and no ID.

Table 3. Required inputs for the "Scenario.dat" file.

Scenario.dat	d (km)	B_{min} (Mbps)	U_{max} (%)	T (°C)	$R_i^{(0.01\%)}$ (mm/h)	H (%)	h_a (m)	h_{obs} (m)	$\overline{N_{fog}}$ (days/year)	$\overline{D}$ (h)	$Ml_{min}^{(MRT)}$ (dB)	$Ml_{min}^{(FSO)}$ (dB)	$Ml_{min}^{(FO)}$ (dB)

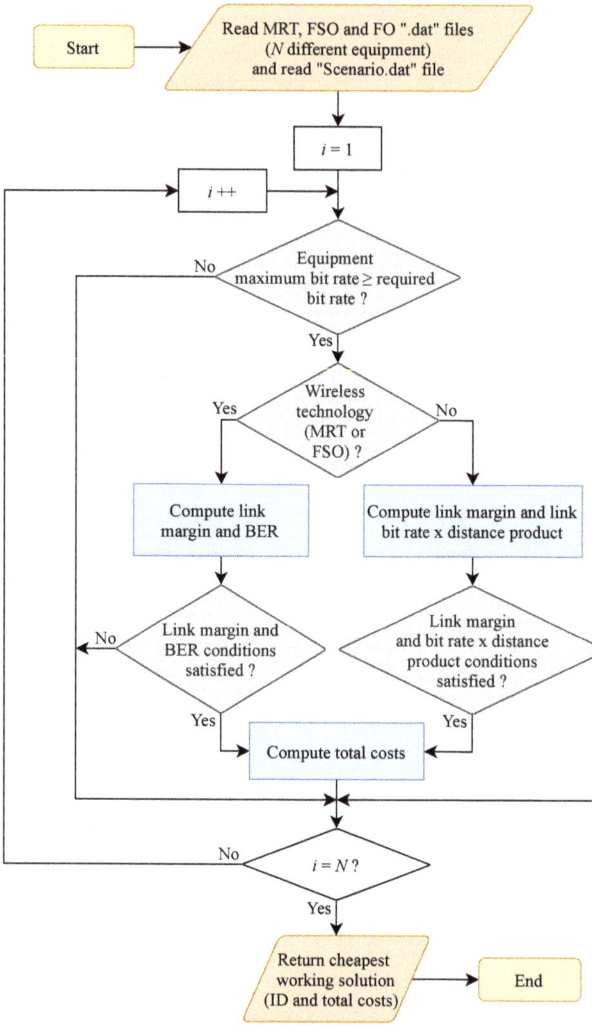

Figure 2. Flowchart of the link design algorithm.

3.2. Network Planning Algorithm

The network planning algorithm has the goal of finding the optimal number of BBUs to be deployed for a certain environment and where to place them, given the positions of the RRHs and their bit rate needs. Using the previously described link design algorithm and considering the costs associated with a BBU, this algorithm is able to compute fronthaul topologies that minimize the total costs of the network.

The following subsection details the network planning algorithm, namely the considered economic aspects and the algorithm workflow.

3.2.1. Network Costs Analysis

With respect to the total costs of the network, there are two main aspects to take into consideration: the costs of each RRH–BBU link and the costs of a BBU. Accordingly, the network total costs are given by the sum of the global costs of the RRH–BBU links plus the costs of the total number of BBUs used in the network.

The costs of each RRH–BBU link can be obtained with the link design algorithm. It is important to notice that when dealing with the acquisition of equipment, as is the case when designing the fronthaul network, some vendors might be able to provide substantial discounts on their prices for purchases involving larger amounts of equipment. Accordingly, these discounts could make one technology preferable to another, cost-wise. Nevertheless, even though this is an important consideration, it is extremely difficult to quantify these discounts as constraints within the algorithm in a universal manner. Therefore, it was assumed that the unitary price of the equipment remained unaltered for large quantities, which means that the global costs of the RRH–BBU links considered by the network planning algorithm can be regarded as an upper limit for this type of cost; nonetheless, please bear in mind that since the source code of this algorithm is freely available online, it can be modified by anyone, namely by a system designer, in order to incorporate other features such as particular costs' computations according to specific terms of the vendors.

It is worth pointing out that when deciding to add or remove a BBU from the network, the BBU costs play a major role when determining the total costs of the network. More specifically, if the BBU costs were considered negligible, the optimal solution (although unrealistic) would be given by the total number of BBUs equaling the number of RRHs, with colocated placements, as this would mean close to zero distances between the RRHs and the corresponding BBU, thus yielding the minimum possible costs regarding RRH–BBU links. Accordingly, one of the virtues of the network planning algorithm is that the BBU costs are not disregarded when determining the optimal number of BBUs to be deployed for a certain environment.

3.2.2. Network Planning Algorithm Workflow

Figure 3 depicts the flowchart of the network planning algorithm. The first step is to read and store the information contained in the "RRH.dat" file, which is a text file that contains (in each line and separated by commas) the Cartesian coordinates (X, Y) of the RRHs under consideration, along with the required bit rate for each RRH; Table 4 presents the structure of each line regarding the "RRH.dat" file. In addition, the user must provide data regarding the BBUs in the "BBU.dat" text file; once more, a line structure is followed (separated by commas), where the required inputs are the ones indicated in Table 5; these correspond respectively to the maximum number of RRHs supported by a BBU $(RRHs_{max})$, the maximum bit rate supported by a BBU for each RRH–BBU link (B_{max}), the costs of a BBU (Costs_{BBU}), the minimum and maximum number of BBUs to be considered in the analysis $(min\_BBU$ and max_BBU, respectively), and the number of different initializations regarding the network costs optimization procedure (D_{init}).

Afterwards and given the starting number of BBUs (min_BBU), the algorithm computes the positions of the BBUs. This step, combined with the assignment of each RRH to a BBU (where multiple RRHs can be connected to a single BBU), can be regarded as solving a clustering problem. Since smaller RRH–BBU link distances not only increase the likelihood of adopting single-hop wireless communication technologies (which are simpler to install), but also lead to cheaper FO links, as well as lower fronthaul latencies can be achieved, the adopted criterion for clustering is based on minimizing the distances between RRHs and BBUs. Accordingly and by applying the K-means clustering algorithm (as suggested in [31]), the positions of the BBUs are determined. More specifically, the K-means procedure groups the RRHs into K different subsets, where K equals the considered number of BBUs (hence, $K \leq$ RRHs), by minimizing the sum of squared distances between the RRHs belonging to a cluster and the corresponding cluster centroid, i.e., the associated BBU position.

Table 4. Required inputs for the "RRH.dat" file.

RRH.dat	X (m)	Y (m)	B_{min} (Mbps)

Table 5. Required inputs for the "BBU.dat" file.

BBU.dat	$RRHs_{max}$	B_{max} (Mbps)	$Costs_{BBU}$ (¤)	min_BBU	max_BBU	D_{init}

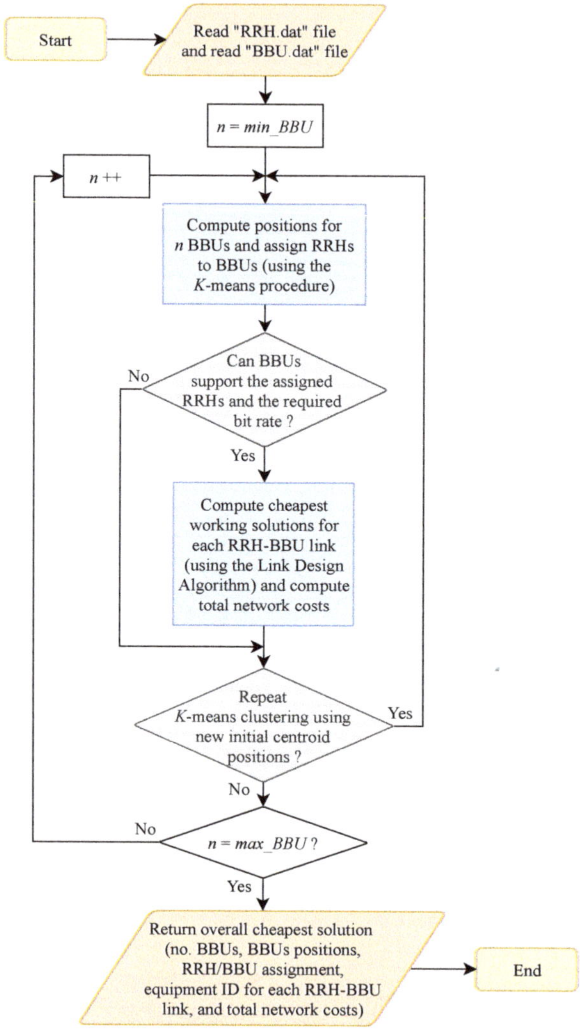

Figure 3. Flowchart of the network planning algorithm.

Next, the algorithm verifies if the BBUs can support the number of RRHs assigned to each one of them, as well as the required bit rate of each RRH–BBU link; if this is the case, then the link design algorithm is used to determine the cheapest working solution satisfying each RRH–BBU link requirement (such as link distance and required bit rate) for the deployment scenario (the scenario characteristics, except link distance and required bit rate, are extracted from the "Scenario.dat" file); cf. Section 3.1. Subsequently, the total costs of the network are computed taking into account the network costs analysis, i.e., by summing up the costs of all RRH–BBU links and the costs of the considered number of BBUs.

Noticing that the K-means approach is an iterative algorithm that relies on an initial random choice of centroid positions, this procedure may yield different clustering results on different runs of the algorithm, which, in turn, may lead to different network costs; note that there are $K^{N(RRHs)}$ ways to partition N RRHs into K clusters. Therefore, the previous two steps (computation of BBUs' positions and determination of the total costs of the network) are repeated multiple times—namely a total of D_{init} times, which is a user-specified value that ideally should be close to $K^{N(RRHs)}$, although this number may be too time consuming to be practical—in order to search for the cheapest network topology regarding the number of BBUs that is under consideration (e.g., min_BBU).

The network planning algorithm then computes again the total costs of the network, but now considering a different number of BBUs, in order to find the cheapest alternative. More specifically, the procedure is repeated, sequentially, from the minimum (min_BBU) to the maximum number of BBUs (max_BBU) defined by the user, and the algorithm will finally return which number of BBUs yields the overall cheapest solution along with other outputs: BBUs positions, RRHs assigned to each BBU, equipment ID for each RRH–BBU link, and total network costs.

4. Fronthaul Design Study Results

In order to study the influence of different environments and weather conditions on the fronthaul design, a set of illustrative equipment was considered: it comprised MRT, FSO, and FO technologies under different functioning conditions, such as maximum supported bit rate, price, and others factors that affected the maximum distance of a viable link (e.g., operating frequency, sensitivity, etc.). The respective test equipment characteristics, which were obtained by combining information from multiple sources [15,32–35], are presented in Table 6.

Table 6. Characteristics of the test equipment.

MRT	ID	B (Mbps)	f (GHz)	P_{Tx} (dBW)	G_{Tx} (dBi)	G_{Rx} (dBi)	A_{equi} (dB)	S_{Rx} (dBW)	N_f (dB)	M (M-QAM)	F.Costs$^{(MRT)}$ (EUR)	V.Costs$^{(MRT)}$ (EUR/$\sqrt{km}$)
	1	380	13	2	28	28	3	−111	4	1024	55,895	11,590
	2	570	42	0	34	34	3	−122	4	1024	85,326	5016
	3	1100	42	0	34	34	3	−122	4	1024	95,736	9680
FSO	ID	B (Mbps)	λ (nm)	P_{Tx} (dBW)	G_{Tx} (dBi)	G_{Rx} (dBi)	A_{equi} (dB)	S_{Rx} (dBW)	F.Costs$^{(FSO)}$ (EUR)			
	4	100	785	−12.2	30	0	3	−60	29,903			
	5	1000	1550	−27	30	0	3	−60	32,353			
	6	1000	1550	−7	30	0	3	−60	48,186			
	7	10,000	1550	−7	30	0	3	−48	86,654			
FO	ID	B (Mbps)	BxD (Mbps·km)	Tx_{min} (dBW)	Rx_{min} (dBW)	L (dB)	F_L (dB/km)		F.Costs$^{(FO)}$ (EUR)			V.Costs$^{(FO)}$ (EUR/km)
	8	10,000	80,000	−27.5	−43	3	0.25		35,135			18,865

The study performed herein is divided into two parts: first, the link design algorithm is used to evaluate which technologies are the preferable cost-effective choices for individual fronthaul segments under different weather conditions, lengths of the links, and bit rate requirements; secondly, by using the network planning algorithm, an assessment is performed regarding the impact of the RRH density on the selection of cost-effective communication technologies for C-RANs.

4.1. Individual Links Assessment

In order to study the performance of MRT, FSO, and FO systems, four different scenarios were considered by taking into account the ranges of values associated with rain and fog intensities [25, 28]; these test scenarios are henceforth denoted as S(r_l/f_l), S(r_h/f_l), S(r_l/f_h), and S(r_h/f_h) and their respective characteristics are given in Table 7; for the sake of clarity, r_l and r_h refer to low and high rain intensity, respectively, whereas f_l and f_h refer to low and high fog intensity, respectively. In addition, the minimum accepted link margins $Ml_{min}^{(MRT)}$, $Ml_{min}^{(FSO)}$, and $Ml_{min}^{(FO)}$ were all set equal to 3 dB [10,32].

Table 7. Characteristics of the test scenarios.

Scenario	U_{max} (%)	T (°C)	$R_i^{(0.01\%)}$ (mm/h)	H (%)	h_a (m)	h_{obs} (m)	$\overline{N_{fog}}$ (days/year)	$\overline{D}$ (h)	Rain Intensity	Fog Intensity
S(r_l/f_l)	0.1	21	15	60	30	−15	10	2	Low	Low
S(r_h/f_l)			135				10		High	Low
S(r_l/f_h)			15				360		Low	High
S(r_h/f_h)			135				360		High	High

Following the previous test scenario settings, the link design algorithm was run for each scenario considering a set of distances ranging from 0 to 20 km and a required bit rate ranging from 0 to 1.5 Gbps, in order to assess how the most cost-effective solution varied for different link requirements and for different weather conditions. Figures 4–7 present the outputs of the link design algorithm (i.e., the cheapest working solution for the considered distance and required bit rate ranges) for the four test scenarios.

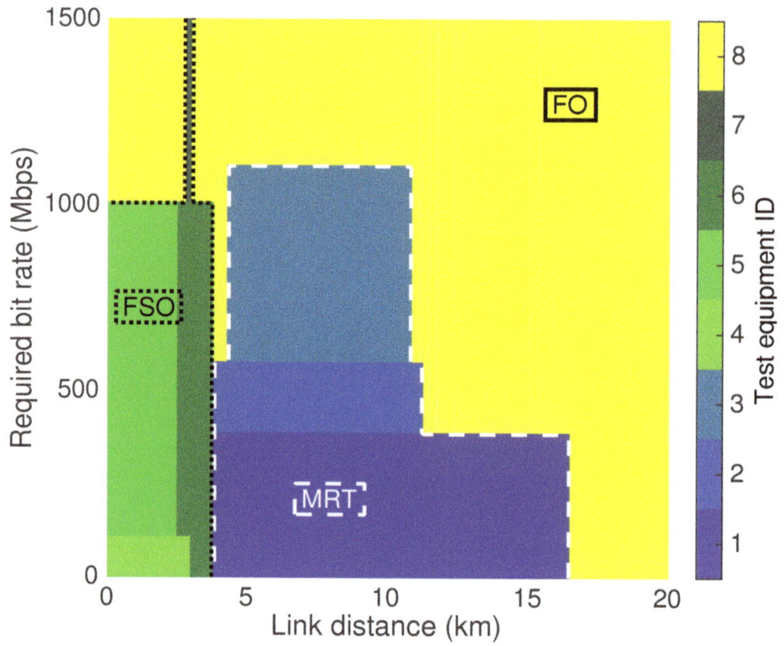

Figure 4. Cost-effective solutions for S(r_l/f_l).

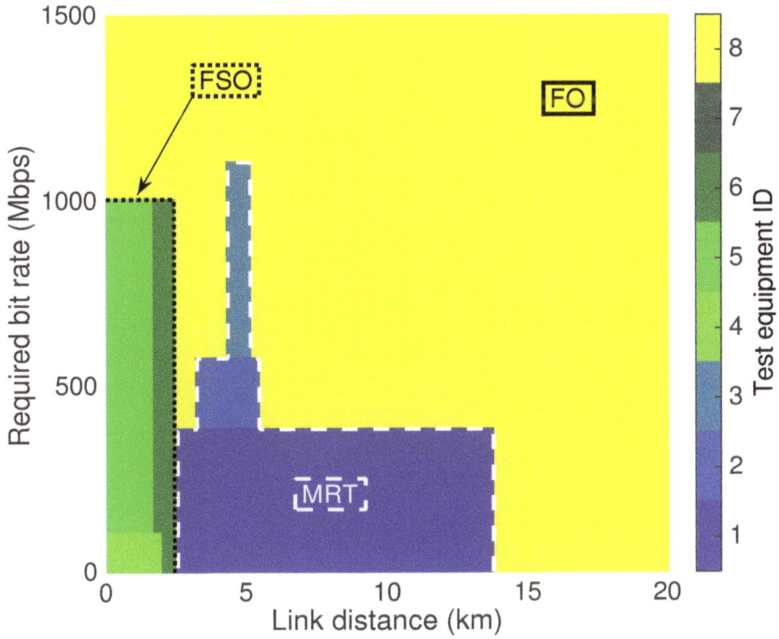

Figure 5. Cost-effective solutions for S(r_h/f_l).

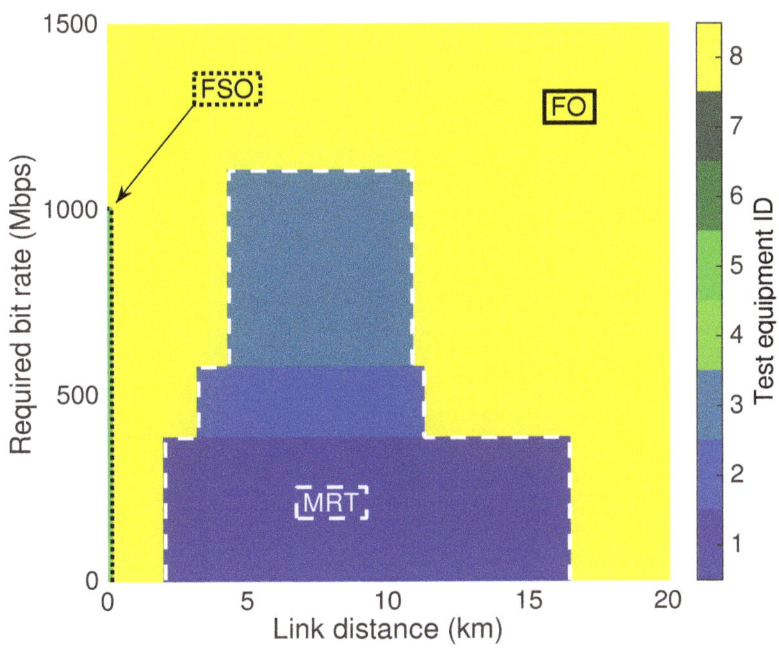

Figure 6. Cost-effective solutions for S(r_l/f_h).

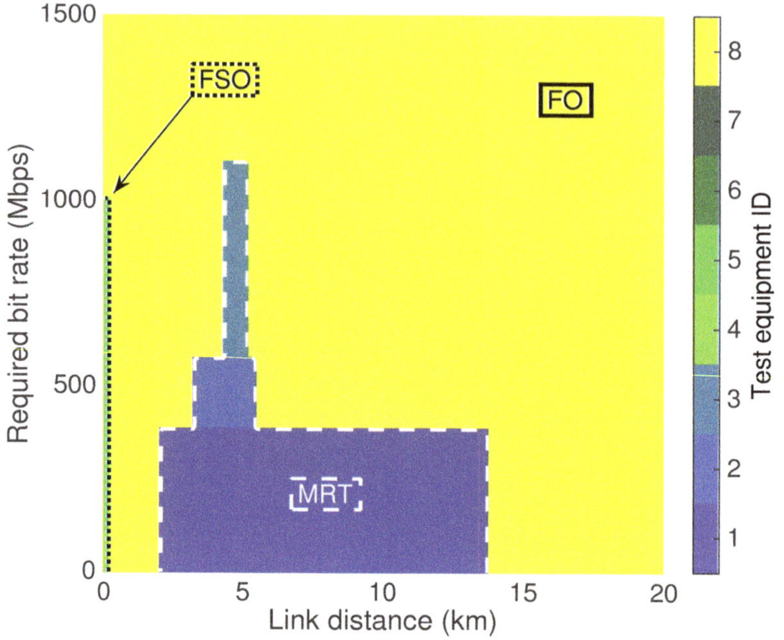

Figure 7. Cost-effective solutions for S(r_h/f_h).

Due to the low rain and low fog intensities considered in S(r_l/f_l), this scenario has very good wireless propagation conditions for both MRT and FSO technologies. Accordingly and as can be observed in Figure 4, an FSO system should be adopted for these weather conditions when the required bit rate is low to medium (lower than 1 Gbps) and the link distance is small (below 4 km) or when the required bit rate is higher (1 to 1.5 Gbps, at least) and the link distance is within a narrow range of small distances (2.5 to 3 km); for the medium bit rate requirement range (570 Mbps to 1.1 Gbps) and for medium link distances (4.5 to 11 km), MRT systems are the best choice; the MRT technology also yields the best results for a wider range of link distances (namely from 4 to 11.5 km and from 4 to 16.5 km), but for lower bit rate requirements (below 570 Mbps and below 400 Mbps, respectively); for all the remaining cases, FO systems are the best choice. Some conclusions can be drawn from these results: first, there is a trade-off between bit rate and distance ranges regarding the cost-effectiveness of MRT and FSO solutions; secondly, the feasibility of a wireless link does not necessarily mean that it is a cost-effective solution; for instance, although FSO links are practicable regarding small distances and high bit rate requirements (notably greater than 1 Gbps), FO systems are a better cost-effective choice in these cases if the link distance is closer to zero (an equivalent example could be given regarding MRT versus FO systems and with respect to small to medium link distances and medium bit rate requirements); nevertheless, FSO solutions can be used as a fallback in these cases, namely if the FO technology cannot be used, e.g., due to the orography of the terrain or the necessity of a quick installation following a natural disaster.

Considering now S(r_h/f_l), which is a scenario characterized by high rain and low fog intensities, Figure 5 shows that both wireless technologies decreased their cost-effective performance in terms of maximum link distance when compared to S(r_l/f_l); this performance decrease was more pronounced regarding MRT systems, especially in the case of medium bit rate requirements (they had a reduction of 6 km regarding the maximum feasible link length). From here, it could be concluded that although the presence of rain makes it harder to select one of the wireless technologies considered herein, if the fog intensity is low, then FSO systems are still a good cost-effective option if the required bit rate is low to medium (lower than 1 Gbps) and the link distance is small (below 2.5 km); on the other hand, even in the case of high rain intensity, MRT solutions should not be disregarded, since they are cost-effective for medium link distances (2.5 to 14 km) and small bit rate requirements (lower than 400 Mbps).

Turning now the attention to the scenario characterized by low rain and high fog intensities, S(r_l/f_h), Figure 6 shows that FSO systems were practically excluded from the cost-effective solutions due to harsh visibility conditions; in particular, FSO links were only feasible for very small distances (up to 200 m). With respect to MRT systems, their cost-effectiveness was similar to the one observed for S(r_l/f_l) (cf. Figure 4). Another noteworthy result is that, when compared to the previously analyzed scenarios, it was the FO technology (and not the MRT one) that replaced FSO as the cost-effective solution for small link distances (below 2 km) and low to medium bit rate requirements (lower than 1 Gbps); this result reinforces the previous conclusion that the feasibility of a wireless link does not necessarily mean that it is a cost-effective solution, namely when considering small link distances.

With respect to S(r_h/f_h), which is a scenario characterized by high rain and high fog intensities, Figure 7 shows that the cost-effectiveness of MRT systems was similar to the one observed for S(r_h/f_l) (cf. Figure 5), whereas the cost-effectiveness of FSO systems was similar to the one observed for S(r_l/f_h) (cf. Figure 6), thus yielding analogous conclusions, respectively.

Combining all the findings gathered so far, a system designer, when planning an RRH–BBU link, can somewhat foresee, without making computations, which technology is more likely to be adopted; this is useful to decide, especially in the first stage of link planning, which equipment should be thoroughly surveyed on the market, thus saving time and money. A summary of the mapping between the pair link distance/bit rate requirement and the respective cost-effective solutions is given in Figure 8, which led to the following rules of thumb:

- For short link distances (under 4 km) and small to medium bit rate requirements (lower than 1 Gbps), an FSO system is the best option, followed by an FO one if the former is unfeasible, notably in foggy scenarios (MRT systems can also be used as fallback);
- For short link distances (under 3 km), but for a higher bit rate requirement (greater than 1 Gbps), FO systems should be adopted, followed by FSO ones;
- For medium to high link distances (5 to 11 km) and small to medium bit rate requirements, as well as some high link distances (14 to 16 km) and lower bit rate requirements (lower than 400 Mbps), the first choice corresponds to the MRT technology, followed by FO if the former is unfeasible, notably in rainy scenarios;
- For medium to high link distances (4 to 14 km) and lower bit rate requirements, as well as some medium link distances (4 to 5 km) and medium bit rate requirements (lower than 1 Gbps), not only MRT systems are the best option, but they also show a high resilience even regarding scenarios with high rain intensity;
- For all the remaining cases, FO systems are the only option, which means that if these systems cannot be deployed, then the link becomes unfeasible (unless wireless relay stations are considered).

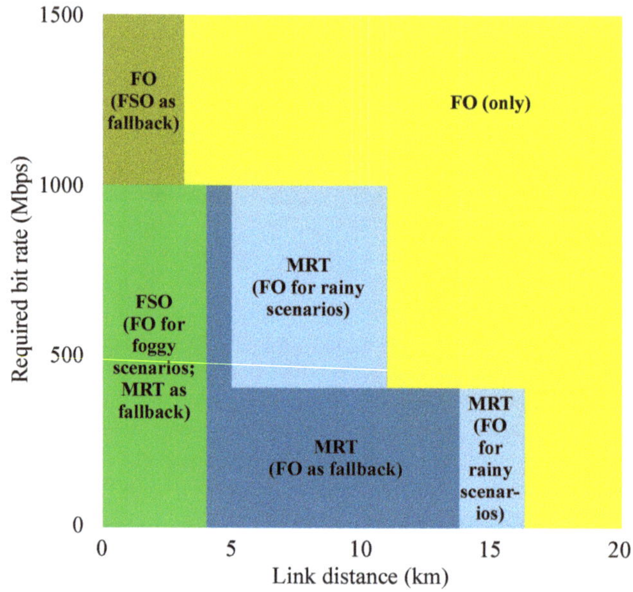

Figure 8. Summary of the cost-effective solutions for individual link deployment.

4.2. Network Assessment

In order to assess the impact of the RRH density on the selection of cost-effective communication technologies for C-RANs, two distinct environments were considered, namely a rural and an urban environment [36], which had different RRH densities; the characteristics of these test environments are presented in Table 8. In addition, the costs of one BBU was assumed to be equal to 167,000 euros [33,37], as well as it was assumed that each BBU had enough capacity to support as many RRH–BBU links as needed; furthermore, for each of these links, it was assumed that the BBUs could support the required bit rate, which was set equal to 1 Gbps for all cases. Moreover, the *K*-means clustering procedure was repeated enough times so that each solution using a certain number of BBUs converged to a minimum in terms of network costs.

Table 8. Characteristics of the test environments for network assessment.

Environment	RRH Density	Test Area	Total No. of RRHs
Rural	0.015 RRHs/km^2	20 km × 20 km	6
Urban	0.3 RRHs/km^2		120

Following the previous settings, the network planning algorithm was run multiple times for each environment and considering, in a separate manner, the four aforementioned test scenarios (cf. Table 7), in order to evaluate the impact of different weather conditions on the fronthaul design and costs.

4.2.1. Rural Environment

Figure 9 presents, for each of the considered test scenarios, the obtained results for the rural environment, namely the average total network costs along with the average technology usage percentage; the colocated percentage corresponds to the average percentage of RRHs that were colocated with the placed BBUs; thus, the respective RRH–BBU links did not make use of MRT, FSO, or FO individual links. Additionally, the average number of BBUs required for each test scenario is also indicated in this figure.

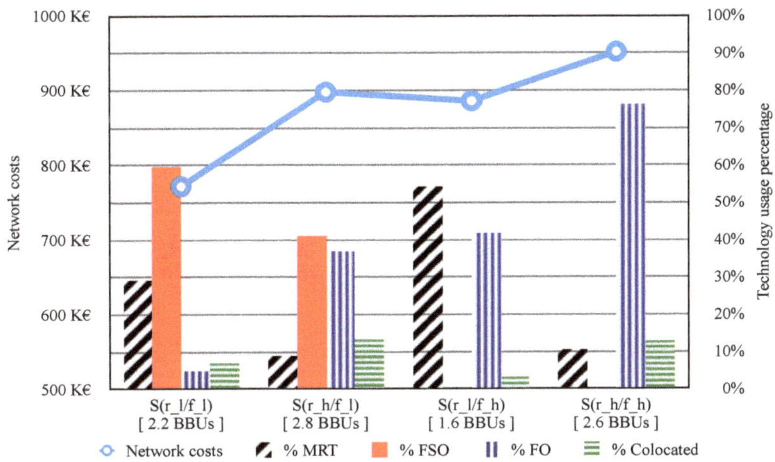

Figure 9. Network costs and technology usage for the rural environment.

The results showed that even though the average number of BBUs required for a cost-effective network design was similar in all cases (ranging from 1.6 to 2.8), the distribution of technology usage regarding the RRH–BBU links clearly varied according to the considered scenario. Nevertheless, it is important to notice that, in all cases, cost-effective solutions were obtained (on average) when some BBUs were colocated with RRHs. Moreover, the FO technology was always adopted for RRH–BBUs links, and its usage percentage increased as the weather conditions deteriorated.

As expected, the scenario characterized by low rain and low fog intensities, S(r_l/f_l), yielded the lowest average network costs, whereas the high rain and high fog intensities of S(r_h/f_h) led to the highest average network costs. However, these scenarios did not give rise to the lowest or highest average number of required BBUs, respectively, concerning all four test scenarios. This shows that the number of required BBUs on its own is not sufficient to have an estimate regarding the final network costs in rural environments.

Another noteworthy result is that regardless of the rain intensity, MRT systems were always adopted for RRH–BBUs links, albeit with lower usage percentages when the rain intensity was high. On the contrary, an equivalent behavior was not observed by the FSO technology; more specifically, although this technology was the most used when the fog intensity was low, FSO systems were not adopted in the scenarios with high fog intensity. This particular outcome in the rural environment enables to highlight an advantage of considering the FSO technology when planning fronthaul networks: if the visibility conditions are favorable, then the use of FSO systems entails significant savings in the network cost; for instance, there was a cost reduction of 6% in rainy scenarios (cf. S(r_h/f_h) vs. S(r_h/f_l)), whereas a cost reduction of 13% was verified under non-rainy scenarios (cf. S(r_l/f_h) vs. S(r_l/f_l)). Furthermore and besides the cost savings, the use of FSO systems leads to an increase of the number of BBUs required for a cost-effective network, which may represent an advantage in future network expansions, especially in rural environments, as new RRHs have a higher likelihood of being served with lower costs by the existing BBUs.

4.2.2. Urban Environment

Figure 10 presents, for each of the considered test scenarios, the obtained results for the urban environment. As can be observed, the distribution of technology usage regarding the RRH–BBU links can be categorized into two groups, according to the visibility conditions: (1) low fog intensity scenarios, where the FSO technology usage is largely dominant (roughly 90–98%) and FO systems account for the remaining RRH–BBU links, with a usage percentage that increases along with the increase of the rain intensity (particularly because the performance of FSO systems is also impaired by the presence of rain, as seen in Section 4.1); and (2) high fog intensity scenarios, in which the distribution of technology usage is practically unaffected by the rain intensity and almost all RRH–BBUs links (about 99%) are served by FO systems. Still regarding the latter group, it is important to notice that even in the presence of harsh visibility conditions in the respective scenarios, FSO systems were adopted for the remaining RRH–BBU links (i.e., approximately for 1% of the cases); this occurred because, unlike the rural environment, some of the RRH–BBU link lengths were always very small (notably less than 200 m) due to the higher RRH density of the urban environment.

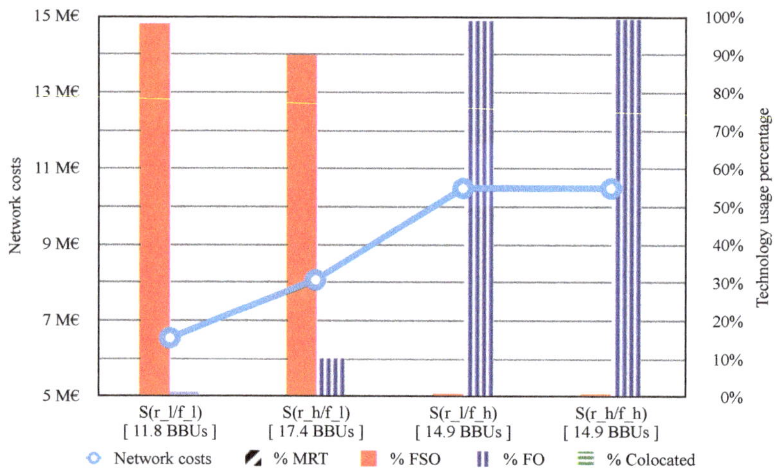

Figure 10. Network costs and technology usage for the urban environment.

Turning the attention now to the network costs, the results showed that the visibility conditions also have a major impact on the urban fronthaul costs: not only high fog intensities led to more expensive networks, but also cheaper ones were obtained when there was low fog intensity; in particular, the possibility of exploiting FSO systems when the visibility conditions were favorable led to a cost reduction of 23% in rainy scenarios (cf. S(r_h/f_h) vs. S(r_h/f_l)), whereas a cost reduction of 38% was verified under non-rainy scenarios (cf. S(r_l/f_h) vs. S(r_l/f_l)). It is also worth to point out that, regardless of the rain intensity, the network costs were very similar whenever the fog intensity was high (cf. S(r_l/f_h) vs. S(r_h/f_h)), mainly because MRT systems were not adopted; in any case, although it is unnoticeable in Figure 10, the costs were slightly cheaper in S(r_h/f_l) than in S(r_l/f_l)) owing to a non-negligible performance increase of FSO systems when the rain intensity decreased, which in turn slightly increased the FSO technology usage.

With respect to the average number of required BBUs, although the lowest value was obtained for the scenario that yielded the lowest average network costs (i.e., S(r_l/f_l)), the opposite was not true, as the highest number of required BBUs occurred for the second cheapest scenario (i.e., S(r_h/f_l)). This shows that the number of required BBUs on its own is not sufficient to have an estimate regarding the final network costs also in urban environments. Finally, notice that although colocated BBUs with RRHs were never adopted as cost-effective solutions regarding the urban environment results, in practice, it may compensate economical to change the placement of a BBU (which was initially suggested by preliminary cost-effectiveness studies) in order for it to match the position of a nearby RRH (for instance, if the initial distance between them is below 20 m), thus saving, e.g., site rental costs.

5. Conclusions

This paper addressed the use of MRT, FSO, and FO systems in the C-RAN fronthaul segment of a mobile network. In particular, a methodology was proposed to evaluate and compare the performance of MRT, FSO, and FO technologies, thus enabling determining the most cost-effective solution for each RRH–BBU link and computing the required number of BBUs and where they should be positioned in order to minimize the overall network costs. In addition, a study was carried out regarding the fronthaul design; more specifically, a cost-effectiveness comparison of the aforementioned communication technologies was performed for individual fronthaul segments under different weather conditions, link lengths, and bit rate requirements; an assessment was also performed regarding the impact of the RRH density on the selection of cost-effective communication technologies for C-RANs.

With respect to individual links, the study results showed how sensitive the wireless communication systems addressed herein (MRT and FSO) are to weather conditions. Moreover, the effects of rain and fog were mapped into a cost-effective link solution chart that is dependent on link length and the required bit rate; in this manner, a system designer can be aware beforehand, and without making computations, about which technologies suit better for the intended link and that deserve to be thoroughly surveyed on the market.

Concerning the analysis performed for different RRH densities, it was seen that for few and sparsely distributed RRHs (rural environment), the technology usage distribution to set up a cost-effective fronthaul significantly changed when the weather conditions varied. On the other hand, when considering an area with a high RRH density (urban environment), it was concluded that one technology was extremely dominant when setting up a cost-effective network; additionally, it was concluded that this technology varied between FSO systems, regarding scenarios with good visibility conditions, and FO systems, namely when FSO systems were strongly impaired by the presence of fog.

One of the main findings of this work was that, regardless of the environment type, the performance of FSO systems is a key factor in terms of the total costs of the fronthaul network. More specifically, the possibility of exploiting the FSO technology when there is low fog intensity entails significant savings in the network costs. Accordingly, this shows how it is of utmost importance for a project manager to collect information about the visibility conditions regarding the implementation scenario (e.g., frequency and duration of fog events), in order to get the most out of FSO systems and, consequently, lowering the fronthaul segment costs.

Lastly, it is worth mentioning that any real implementation of an RRH–BBU link requires additional considerations other than the ones addressed herein (such as site availability, rental contract terms, reuse of previously acquired technology, etc.). In spite of that, the provided methodology (namely the supplied software tool) along with the conclusions of this work can serve as important guidelines for fronthaul network designers. For instance, after adjusting the inputs of the aforementioned tool according to the specific situation being analyzed, a project manager will have a better notion of not only which equipment can be disregarded in the first place, thus saving time, but also which situations require further evaluations for a successful network deployment—e.g., if an FSO system is being considered, then it should be investigated whether the Sun or the presence of trees will not impair this system—which also enables to obtain a more productive and time-efficient fronthaul design.

Author Contributions: Conceptualization, I.S., N.S., and A.R.; methodology, I.S. and N.S.; software, I.S. and N.S.; validation, I.S. and N.S.; formal analysis, I.S. and N.S.; investigation, I.S. and N.S.; resources, M.P.Q. and A.R.; data curation, I.S. and N.S.; writing, original draft preparation, I.S. and N.S.; writing, review and editing, I.S., M.P.Q., and A.R.; visualization, I.S., M.P.Q., and A.R.; supervision, M.P.Q. and A.R.; project administration, M.P.Q. and A.R.; funding acquisition, M.P.Q. and A.R. All authors read and agreed to the published version of the manuscript.

Funding: This work was funded by Instituto de Telecomunicações and by FCT/MCTES through national funds and when applicable co-funded EU funds under the projects UIDB/EEA/50008/2020 and OCTHOPUS.

Conflicts of Interest: The authors declare no conflict of interest.

References

1. ITU. *Measuring Digital Development: Facts and Figures 2019*; International Telecommunication Union: Geneva, Switzerland, 2019.
2. Ericsson. *Ericsson Mobility Report (November 2019)*; Ericsson: Stockholm, Sweden, 2019.
3. Openet Telecom. *Closing the Mobile Data Revenue Gap*; White Paper; Openet Telecom: Dublin, Ireland, 2010.
4. Checko, A.; Christiansen, H.L.; Yan, Y.; Scolari, L.; Kardaras, G.; Berger, M.S.; Dittmann, L. Cloud RAN for Mobile Networks—A Technology Overview. *IEEE Commun. Surv. Tutor.* **2015**, *17*, 405–426. [CrossRef]
5. Peng, M.; Wang, C.; Lau, V.; Poor, H.V. Fronthaul-constrained cloud radio access networks: Insights and challenges. *IEEE Wirel. Commun.* **2015**, *22*, 152–160. [CrossRef]
6. Chanclou, P.; Pizzinat, A.; Le Clech, F.; Reedeker, T.L.; Lagadec, Y.; Saliou, F.; Le Guyader, V.; Guillo, L.; Deniel, Q.; Gosselin, S.; et al. Optical Fiber Solution for Mobile Fronthaul to Achieve Cloud Radio Access Network. In Proceedings of the 22nd Annual Future Network & Mobile Summit (FuNeMS 2013), Lisboa, Portugal, 3–5 July 2013; pp. 1–11.
7. Hailu, D.H.; Gebrehaweria, B.G.; Kebede, S.H.; Lema, G.G.; Tesfamariam, G.T. Mobile fronthaul transport options in C-RAN and emerging research directions: A comprehensive study. *Opt. Switch. Netw.* **2018**, *30*, 40–52. [CrossRef]
8. Schulz, D.; Jungnickel, V.; Alexakis, C.; Schlosser, M.; Hilt, J.; Paraskevopoulos, A.; Grobe, L.; Farkas, P.; Freund, R. Robust Optical Wireless Link for the Backhaul and Fronthaul of Small Radio Cells. *J. Light. Technol.* **2016**, *34*, 1523–1532. [CrossRef]
9. Song, S.; Liu, Y.; Guo, L.; Song, Q. Optimized relaying and scheduling in cooperative Free Space Optical fronthaul/backhaul of 5G. *Opt. Switch. Netw.* **2018**, *30*, 62–70. [CrossRef]
10. Mitsolidou, C.; Vagionas, C.; Mesodiakaki, A.; Maniotis, P.; Kalfas, G.; GHRoeloffzen, C.; van Dijk, P.W.L.; Oldenbeuving, R.M.; Miliou, A.; Pleros, N. A 5G C-RAN Optical Fronthaul Architecture for Hotspot Areas Using OFDM-Based Analog IFoF Waveforms. *Appl. Sci.* **2019**, *9*, 4059. [CrossRef]
11. Zakrzewski, Z. D-RoF and A-RoF Interfaces in an All-Optical Fronthaul of 5G Mobile Systems. *Appl. Sci.* **2020**, *10*, 1212. [CrossRef]
12. Manning, T. *Microwave Radio Transmission Design Guide*, 2nd ed.; Artech House: Norwood, MA, USA, 2009.
13. Salema, C. *Microwave Radio Links: From Theory to Design*; John Wiley & Sons: Hoboken, NJ, USA, 2003.
14. Naveh, T. *Mobile Backhaul: Fiber vs. Microwave*; Ceragon White Paper; Ceragon Networks: Tel Aviv, Israel, 2009.
15. ANACOM. *Administrative Rule*; no. 157/2017, of 10 May; Fees due to ANACOM; Portuguese Official Gazette: Lisbon, Portugal, 2017; pp. 2228–2229. (In Portuguese)
16. Lynch, S. *Fixed Link Licence Fee Algorithm in Force from 2nd June 2005*; Ofcom (Fixed Service Unit): London, UK, 2005.
17. Malik, A.; Singh, P. Free Space Optics: Current Applications and Future Challenges. *Int. J. Opt.* **2015**, *2015*, 945483. [CrossRef]
18. Muhammad, S.S.; Flecker, B.; Leitgeb, E.; Gebhart, M. Characterization of fog attenuation in terrestrial free space optical links. *Opt. Eng.* **2007**, *46*, 066001. [CrossRef]
19. NTT. One Petabit per Second Fiber Transmission over a Record Distance of 200 km. *NTT Tech. Rev.* **2017**, *15*, 1–3.
20. Yeganeh, H.; Vaezpour, E. Fronthaul network design for radio access network virtualization from a CAPEX/OPEX perspective. *Ann. Telecommun.* **2016**, *71*, 665–676. [CrossRef]
21. Alimi, I.A.; Teixeira, A.L.; Monteiro, P.P. Toward an Efficient C-RAN Optical Fronthaul for the Future Networks: A Tutorial on Technologies, Requirements, Challenges, and Solutions. *IEEE Commun. Surv. Tutor.* **2018**, *20*, 708–769. [CrossRef]
22. Alzenad, M.; Shakir, M.Z.; Yanikomeroglu, H.; Alouini, M. FSO-Based Vertical Backhaul/Fronthaul Framework for 5G+ Wireless Networks. *IEEE Commun. Mag.* **2018**, *56*, 218–224. [CrossRef]
23. Yaacoub, E.; Alouini, M. A Key 6G Challenge and Opportunity—Connecting the Base of the Pyramid: A Survey on Rural Connectivity. *Proc. IEEE* **2020**, *108*, 533–582. [CrossRef]
24. Sousa, I.; Sousa, N.; Queluz, M.P.; Rodrigues, A. Fronthaul Design for Wireless Networks: A Software Tool. 2020. Available online: https://github.com/ivo-sousa/fronthaul_network_design (accessed on 12 June 2020).
25. ITU. *P.837: Characteristics of Precipitation for Propagation Modelling*; ITU-R Recommendations; ITU: Geneva, Switzerland, 2017.

26. ITU. *P.1814: Prediction Methods Required for the Design of Terrestrial Free-Space Optical Links*; ITU-R Recommendations; ITU: Geneva, Switzerland, 2007.
27. Forin, D.M.; Incerti, G.; Tosi Beleffi, G.M.; Teixeira, A.L.J.; Costa, L.N.; De Brito Andre, P.S.; Geiger, B.; Leitgeb, E.; Nadeem, F. Free Space Optical Technologies. In *Trends in Telecommunications Technologies*; Bouras, C., Ed.; IntechOpen: London, UK, 2010; Chapter 13, pp. 257–296.
28. Sousa, I.; Queluz, M.P.; Rodrigues, A. An Efficient Visibility Prediction Framework for Free-Space Optical Systems. *Wirel. Pers. Commun.* **2017**, *96*, 3483–3498. [CrossRef]
29. Majumdar, A.K. *Advanced Free Space Optics (FSO): A Systems Approach*; Springer: New York, NY, USA, 2015.
30. CISCO. *Fiber Types in Gigabit Optical Communications*; White Paper; CISCO: San Jose, CA, USA, 2008.
31. Mahapatra, B.; Turuk, A.K.; Patra, S.K.; Kumar, R. Optimal Placement of Centralized BBU (C-BBU) for Fronthaul and Backhaul Optimization in Cloud-RAN Network. In Proceedings of the 18th Annual International Conference on Information Technology (ICIT 2017), Bhubaneswar, India, 21–23 December 2017; pp. 107–112.
32. Rodrigues, M.; Sousa, I.; Sebastião, P.; Rodrigues, A. Backhaul Planning of Mobile Networks using Microwaves and Free Space Optics. In Proceedings of the 10th Conference on Telecommunications (ConfTele 2015), Aveiro, Portugal, 17–18 September 2015; pp. 1–4.
33. Optcore Standard Optical Transceiver and Accessories Guide. Available online: https://www.optcore.net/product-list/standard-optical-transceiver-guide/ (accessed on 10 April 2020).
34. McGarty, T.P. *Fiber to the Home—Capital Costs and the Viability of Verizon's*; FIOS: Keller, TX, USA, 2006.
35. ERSE. *Tariffs and Prices for Electricity and Other Services in 2020*; Directive no. 3/2020, of 17 February; Portuguese Official Gazette: Lisbon, Portugal, 2020; pp. 142–213. (In Portuguese)
36. Chiaraviglio, L.; Cuomo, F.; Maisto, M.; Gigli, A.; Lorincz, J.; Zhou, Y.; Zhao, Z.; Qi, C.; Zhang, H. What is the Best Spatial Distribution to Model Base Station Density? A Deep Dive into Two European Mobile Networks. *IEEE Access* **2016**, *4*, 1434–1443. [CrossRef]
37. De Andrade, M.; Tornatore, M.; Pattavina, A.; Hamidian, A.; Grobe, K. Cost models for Baseband Unit (BBU) Hotelling: from Local to Cloud. In Proceedings of the IEEE 4th International Conference on Cloud Networking (CloudNet 2015), Niagara Falls, ON, Canada, 5–7 October 2015; pp. 201–204.

applied sciences

MDPI

Article

Comparative Analysis of 5G Mobile Communication Network Architectures

Woosik Lee [1], Eun Suk Suh [2,*], Woo Young Kwak [2] and Hoon Han [2]

1 Korea Telecom, Seongnam-si13606, Korea; woosik.lee@kt.com
2 Graduate School of Engineering Practice, Institute of Engineering Research, Seoul National University, Seoul 08826, Korea; wooykwak@snu.ac.kr (W.Y.K.); hahnhoon@snu.ac.kr (H.H.)
* Correspondence: essuh@snu.ac.kr; Tel.: +82-2-880-7175

Received: 6 March 2020; Accepted: 2 April 2020; Published: 4 April 2020

Abstract: Mobile communication technology is evolving from 4G to 5G. Compared to previous generations, 5G has the capability to implement latency-critical services, such as autonomous driving, real-time AI on handheld devices and remote drone control. Multi-access Edge Computing is one of the key technologies of 5G in guaranteeing ultra-low latency aimed to support latency critical services by distributing centralized computing resources to networks edges closer to users. However, due to its high granularity of computing resources, Multi-access Edge Computing has an architectural vulnerability in that it can lead to the overloading of regional computing resources, a phenomenon called regional traffic explosion. This paper proposes an improved communication architecture called Hybrid Cloud Computing, which combines the advantages of both Centralized Cloud Computing and Multi-access Edge Computing. The performance of the proposed network architecture is evaluated by utilizing a discrete-event simulation model. Finally, the results, advantages, and disadvantages of various network architectures are discussed.

Keywords: 5G mobile communication network; multi-access edge computing; centralized cloud computing; hybrid cloud computing

1. Introduction

With the emergence of the smartphone, the amount of data traffic has been growing exponentially since 2008. Based on user demand and the introduction of new technologies, mobile communication technology is currently evolving from fourth generation (4G) to fifth generation (5G). The 5G mobile communication technology (5G herein) can be characterized by three major features: enhanced Mobile Broadband (eMBB), massive Machine Type Communication (mMTC), and Ultra Reliable Low Latency Communication (URLLC). The main concept of the eMBB is to extend frequency resources to a millimeter Wave (mmWave) above 6 GHz. mMTC is defined as a network capacity to simultaneously accommodate millions of devices in an area of 1 km^2. Finally, URLLC is a feature for guaranteeing end-to-end latency within 10 ms.

Among the main features of 5G, URLLC can be recognized as the most distinguished feature that differentiates 5G from previous generations of mobile communication technologies. Prior to the discussion of the URLLC, the mobile network architecture needs to be defined. A mobile network can be sub-divided into radio access network and core network. The radio access network is the network link assigned on radio frequency resources and the core network is the network link assigned to wire-line resources. For 5G standard, significant efforts have been made to reduce the latency of radio access networks, including the use of mini-slots and shortened Transmission Time Interval [1–5]. However, reducing the core network latency is more difficult because of the physical distance limitation (~200 km/ms).

To address the challenges in reducing core network latency, a new network architecture called Multi-access Edge Computing (MEC) architecture has been proposed [6–8]. The main concept of MEC is to reduce the physical distance between user entity (UE) and application server. Network functions are virtualized on clouds, in which application servers are collocated; MEC is an architecture that allocates computing resources closer to UE. Figure 1 shows the current 4G architecture and the new MEC architecture. The 4G architecture shown in this paper will be referred to as the Centralized Cloud Computing (CCC) architecture [9,10], as tasks are processed at the central cloud.

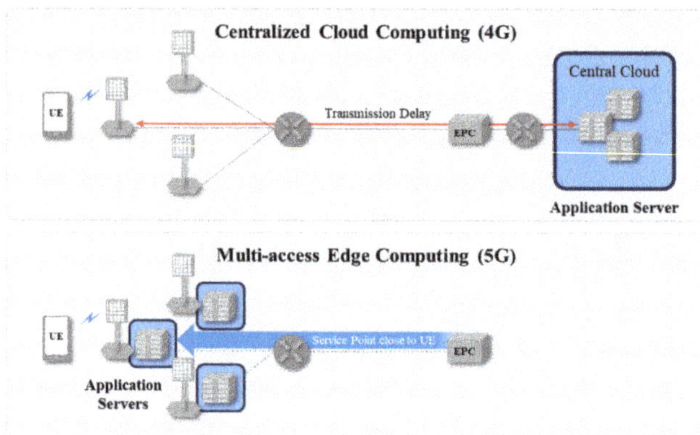

Figure 1. Centralized Cloud Computing (CCC) and Multi-access Edge Computing (MEC) architectures.

In the CCC architecture, the application server is located at the central cloud, causing a delay in the core network if the physical distance between the central cloud and the UE is long. The MEC architecture can reduce transmission delay by placing application server and computing resource (edge cloud) near the target region. However, a network based on the MEC architecture can experience significant delay under certain disruptive situations. Figure 2 shows two of these situations and illustrates the different responses of the CCC and MEC architectures.

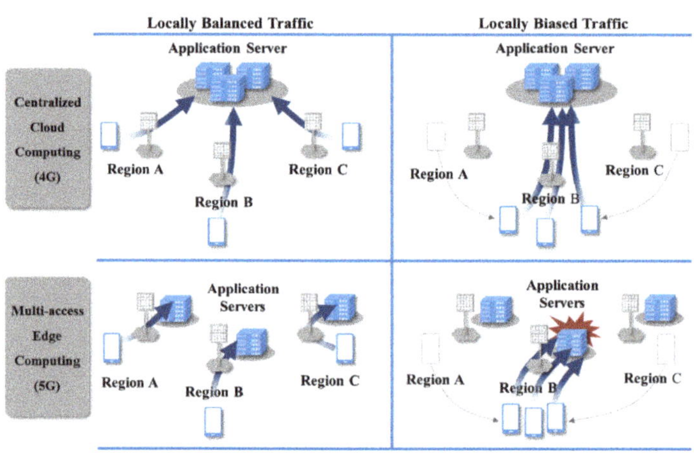

Figure 2. CCC and MEC architecture responses to different data traffic situations.

The left side of the figure shows both the CCC and MEC-based networks operating under normal data traffic conditions. All the data traffic in the CCC-based network is transmitted to the central application server, where the central cloud is located. In the MEC-based network, the regional data traffic is transmitted to respective regional application server to be processed at the edge cloud. The right side of the figure depicts a situation where a data traffic spike occurs in a particular region. Events where large crowds gather, such as the New Year's Eve celebration in Times Square, can cause unusually high data traffic in a specific region. In the current CCC-based network, no noticeable delay occurs as the data traffic is handled by the central cloud, which has sufficient computing resources. However, the MEC-based network has a high probability of experiencing significant delay in that specific region. This is due to the fact that the computing resources allocated to an application server for that particular region may not be sufficient to handle the congested data traffic in the allocated time. Although MEC-based networks can significantly decrease delay time through computer resource distribution and forward deployment of application servers, they are vulnerable to regional data traffic spikes, if insufficient computing resources are allocated in that region.

In summary, the CCC and MEC architectures have their advantages and disadvantages. The CCC architecture can utilize centralized computing resources for all users' traffic, thus being more robust to regional data traffic spike. However, the transmission delay in the CCC architecture is directly proportional to the physical distance between feature elements, resulting in greater time delay for data traffic generated far away from the central cloud. Conversely, the MEC architecture, by locating feature elements in close proximity, can reduce transmission delay. However, it is vulnerable to regional data traffic spikes due to the limited computing resources allocated to each application server.

In this study, we propose a new hybrid network architecture, appropriately named Hybrid Cloud Computing (HCC) architecture, which combines the advantages of both CCC and MEC network architectures while minimizing their risks. This newly proposed architecture, along with the standard CCC and MEC architectures, is analyzed in terms of the data traffic processing capability using a discrete event-based simulation model. The comparative analysis results for the three network architectures are presented in this paper. The following are the contributions of this study: (1) Establishing a simulation model of a communication network architecture that incorporates an offloading decision algorithm considering the queuing status in the system, which is a major factor influencing the latency in actual network environment, and (2) Assessing the feasibility of various communication network architectures using this model, including the newly proposed HCC architecture, with actual field data to reflect real-life situations.

2. Previous Works

Mobile cloud computing technology is the foundation of the MEC architecture. The technology attempts to overcome the limited computing resources of the mobile terminals of subscribers and to reduce power consumption by offloading the workload to a cloud located in a remote location. Research related to this area include work by Kemp et al. [11], who constructed a prototype of and image processing application that incorporated an offloading methodology, where the terminal workload is offloaded to a server. Cuervo et al. [10] proposed a mobile cloud computing related framework named MAUI. Using the proposed framework, they defined a system to classify the offloading code, client-server connection, and offloading decision function based on optimization using linear programming and profiling function to improve the decision-making accuracy. Chun et al. [9] proposed the CloneCloud framework, which enables more effective offloading through virtualization-based synchronization of program execution environment between the user terminal and server. More recent work focuses on establishing the application offloading process for augmented reality through mobile virtual reality [12].

The offloading methodology is another closely related topic that has received a lot of attention. Research on this subject is divided into two different areas, namely the single layer offloading methodology and multiple layer offloading methodology. The former focuses on developing methods to improve the efficiency of decision making for offloading between user terminal and edge cloud, while the latter focuses on solving edge cloud overload problem from a more macroscopic (system level) perspective.

There are two different approaches to the single layer offloading methodology. The first approach involves the optimization of computing resources. Taking this perspective, Huang et al. [13] used the Lyapunov optimization technique to solve the offloading decision problem for the purpose of minimizing terminal power consumption, subject to a dependency between the off-loadable modules. Recent work to implement this approach was published by Chen and Hao [14], who proposed a method to minimize the delay time through efficient allocation of the computing resources required for offloading in the virtualized network environment. Additionally, Hao et al. [15] proposed a new concept called task caching, which solved the terminal power consumption minimization problem through mixed integer optimization, while considering MEC, storage resources, and execution time.

The second approach to the single layer offloading methodology attempts to coordinate and optimize both computing and communication resources. In the work by Liu et al. [16], the authors modeled the terminal task buffer queuing state, task processing state, and offloading transmission state based on a Markov decision process and proposed a solution to the delay time optimization problem using a one-dimensional searching algorithm. Hong et al. [17] proposed a method to obtain an optimal trade-off condition between delay time and power consumption using the finite-state Markov chain model. In recent work by You et al. [18], the authors proposed an algorithm to solve the computing and communication resource optimization problems using an offloading decision model, which is divided into computing resource model based on time division access and communication resource model based on orthogonal frequency division access.

The multiple layer offloading methodology focuses on solving the MEC architecture overload from a more macroscopic perspective. Srinivasan and Agrawal [19] proposed a new architecture called the mobile-central office re-architected as a data center (M-CORD) to address the MEC architecture's overload problem. The M-CORD is a network architecture that utilizes the service orchestration platform called XOS to link the distributed and virtualized base stations to the core network, while allocating centrally located computing resources to each individual edge cloud. Through utilization of this architecture, sharing of computing resource between individual edge clouds is possible, and the edge cloud overload problem can be addressed. Lin et al. [20] defined the MEC architecture as a multiple layer system consisting of terminal edge cloud, and core layers. They proposed a methodology for optimizing computing resources and minimizing delay time through best allocation of computing resources and traffic to nodes in each layer. Kiani and Ansari [21] divided an MEC architecture into three different layers and implemented an auction-based optimization algorithm to obtain minimum cost for the network.

Recently, new methodologies have been proposed for optimizing offloading tasks between clouds. Table 1 lists the recent works in this area.

Table 1. Published works on offloading optimization between clouds.

Previous Research	Multi-Tiered	Offloading	Optimization Objectives	Optimization Methodology	Queuing Status Sharing
Hou et al. [22]	No	Horizontal	Latency	Heuristic	No
Ren et al. [8]	Yes	Vertical	Latency	Convex	No
Ahn et al. [23]	Yes	Vertical	Cost	Heuristic	No
Zhao et al. [24]	Yes	Vertical	Utilization	Convex	No
Zhang et al. [25]	Yes	Vertical	Profit	Convex	No
Zhang et al. [26]	Yes	Vertical	Profit	Convex	No
Ruan et al. [27]	Yes	Vertical	Latency	Heuristic	No
Thai et al. [28]	Yes	Vertical and Horizontal	Cost	Approximation	No

Hou et al. [22] proposed a model for horizontal offloading between autonomous vehicles and road infrastructure, and optimized for latency using a heuristic algorithm. Ren et al. [8] assumed a vertical offloading model for edge and cloud, and optimized for latency using convex optimization. Ahn et al. [23] proposed a cooperation model between edge and cloud for video analytics based on Internet of Things (IoT). They defined the optimization as a mixed integer problem and applied the heuristic algorithm to optimize the cost required for cloud operation. Zhao et al. [24] employed a vehicle network comprising edge and cloud, and tried to optimize cloud utilization using convex optimization. Zhang et al. [25,26] used whole-sale and buy-back models between edge and cloud to share computing resources, and optimized the profit from the edge's perspective. Ruan et al. [27] assumed an energy management infrastructure as a cloud model comprising three tiers, and optimized for latency using joint optimization between Stackelberg and Lyapunov-based pricing and energy demand. Thai et al. [28] proposed vertical and horizontal offloading models between edge to cloud and edge to edge, and used an approximation algorithm to minimize the cost.

Surveying previous works revealed several research gaps that can be addressed. Several papers proposed new frameworks for offloading between user terminals and the cloud and they established foundations for future MEC architecture related research. However, the limitation of their works is that their main direction was focused on the design of the system's functional elements, and they did not propose ways to optimize offloading under various operating conditions. Moreover, the single layer offloading methodology is another research area on which the majority of research has been published. However, the single layer offloading methodology exhibits limitations in addressing the MEC architecture's issues of computing resource shortage, work overload and quality degradation resulting from local traffic spike. The studies on multiple layer offloading methodology focus optimal allocation of computer resources and traffic to multiple terminals, edge clouds, and cloud nodes at the upper layer. Numerous works published in this area were focused on solving the high complexity optimization problem at the center. Although many scholars have studies offloading optimization, as listed in Table 1, no study has incorporated queuing status sharing between clouds in the offloading decision making framework, which has a significant impact on the performance in an actual cloud network operating environment. In this research, by incorporating queuing status sharing between clouds into the simulation model, we estimate realistic latency values to make offloading decision. This is used to assess the newly proposed HCC architecture.

3. Proposed Architectures

3.1. Overview

In this study, a new communication architecture that can address the architectural problems encountered by current CCC and MEC architectures, is proposed. Figure 3 illustrates the three architectures analyzed in this study: (1) CCC architecture, which is the current 4G architecture; (2) newly proposed HCC architecture; (3) MEC architecture, which is the architecture proposed for 5G systems.

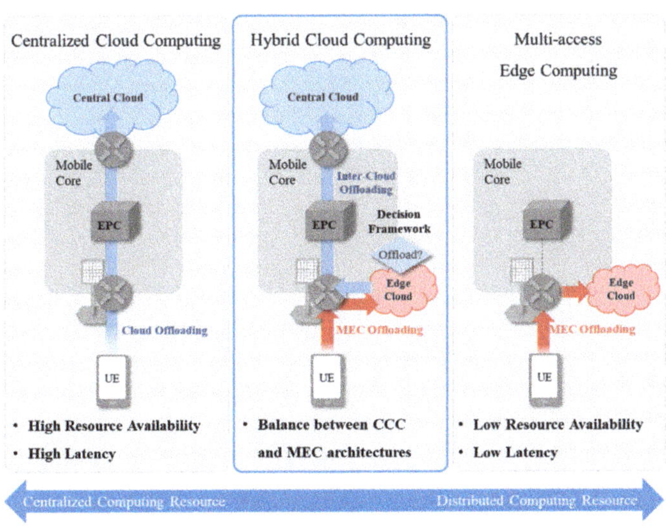

Figure 3. Three network communication architectures proposed for the analysis.

3.2. CCC Architecture

The CCC architecture is the current architecture used in 4G systems. In this architecture, the service request of the user passes through the local level base station and backhaul network and is then delivered to core functions, such as a serving gateway or a public data network gateway. The request then goes through the internet before finally arriving at the central cloud to complete the service process. In the actual mobile communication network, the physical transmission distance for a related service could be several hundred kilometers, causing high latency. However, the CCC architecture features the benefit of centralized pooled computing resources as shown in Figure 2.

3.3. HCC Architecture

To incorporate the advantages of both CCC and MEC architectures, a new network architecture, named HCC architecture, is proposed in this study. As illustrated in Figure 3, the HCC architecture has computing resource at the central cloud, and at application servers (edge clouds) for the regions they service. The normal volumes of local data traffic are handled by the application server and edge cloud for that region, resulting in short delay time. However, when a data traffic spike occurs and the edge cloud at the local application server is tied up with previous tasks, the extra data traffic is offloaded to the computing resource located at the central cloud, rather than waiting in queue at the local application server. The local application server makes an offloading decision for every task to achieve lower latency at that moment. To estimate the latency, real time statistical data is gathered at the respective application server, including the status of computing resources and process queue for the edge and central clouds. This can alleviate potential data traffic spike problem experienced by a completely distributed MEC architecture. The HCC architecture provides a more balanced approach to handling data traffic compared to CCC architecture, in which every task must be processed at the central cloud regardless, and MEC architecture, in which every task must be processed at local edge clouds. The HCC architecture proposed in this study is based on the MEC standard reference architecture established by the European Telecommunication Standard Institute (ETSI) [6]. The architecture and associated offloading decision-making framework can be realized in the form of a mobile edge service in the standard reference architecture. Moreover, the information between clouds, required for the offloading decision-making framework, is exchanged through an Mp3 interface in the standard reference architecture. The proposed HCC architecture is a high-level design based on technologies that satisfy standard requirements. Further, the functional elements of the system proposed in this paper are mapped to lower modules in the standard reference architecture, and therefore, can be easily implemented in a real network environment.

3.4. MEC Architecture

In a 5G system based on network virtualization, the logical functions to process the user traffic at the base station and core instrument are integrated into a user plane function and is distributed to serviced regions. Furthermore, considering the service-based architecture, which aims to fuse the network and application, the 5G system supports a communication path between the user plane function and the application. The MEC architecture takes advantage of this feature by placing an application function, which was located on the central cloud originally, to a local cloud at individual application servers and connects it with the user plane function. The MEC architecture reduces the delay time significantly by moving the application function closer to the UE. However, due to the distributed allocation of computing resources, the available computing resource per location is limited, making the architecture vulnerable to regional traffic spikes as shown in Figure 2.

4. Simulation Model

4.1. Simulation Model Overview

To assess the capabilities of the different architectures, a simulation model of the network architecture was constructed. Figure 4 shows the communication network modeled on the left, and the overall structure of the corresponding simulation model on the right.

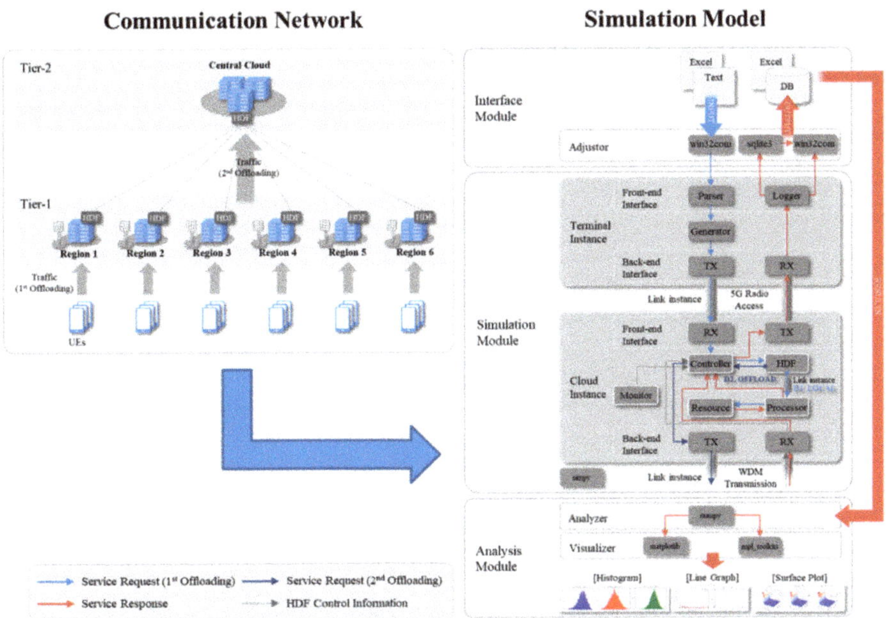

Figure 4. Network architecture simulation model.

The communication network shown on the left in the figure has a multi-tiered structure. The central cloud is located in Tier-2, while six edge clouds are located in Tier-1. The central cloud is connected to the edge cloud in a star topology structure. The various UEs communicate continuously with the edge clouds. The edge clouds and the central cloud are connected through an IP network based on optical transmission technology. The network simulation model was constructed so that it represented the CCC, HCC, or MEC architectures depending on the computing resource allocation between the central cloud and edge clouds. Thus allocating 100% of the computing resources to the central cloud represented the CCC architecture, while allocating 100% of computing resources to the edge clouds represented the MEC architecture. The HCC architecture was represented by any architecture in which the computing resources were allocated to both the central and edge clouds.

The simulation model was developed in the Microsoft Windows 10 (64 bit) environment on a laptop computer equipped with Intel Core i5-6200U processor with 8 GB of RAM. The main module used to develop simulation model was Python-based SimPy (version 3.0.10), which is a discrete-event simulation framework. The additional modules used were NumPy for data analysis and Matplotlib for graphic output.

As shown on the right side of Figure 4, the simulation model is divided into interface, simulation, and analysis modules. The interface module permits the user to input simulation parameter values, collect simulation results, and produce status information. The simulation module receives information from the interface module and simulates the communication network response.

The simulation module can be decomposed into a cloud module and a link module. The cloud module can be realized as an edge cloud or a central cloud, where the central cloud can be placed in a higher tier. Moreover, for each individual cloud, an arbitrary amount of computing resources can be allocated, and the offloading decision-making framework and policy to execute it can be inputted. The link module is used to realize network links between clouds, and by adjusting the physical parameters, such as the latency characteristics and variance, it can simulate wired and wireless links.

The analysis module receives the simulation results from the interface module and visualizes it appropriately to suit the objective of the analysis. By displaying the results in different types of graphs, it allows efficient analysis and interpretation of the simulation results.

There are two newly designed modules incorporated into the network architecture simulation model to specifically simulate and optimize the HCC architecture. The first module is the HCC offloading decision module, and the second one is the HCC computing resource allocation module.

4.2. HCC Offloading Decision Module

The HCC offloading decision module is specifically designed for the HCC architecture. The role of the HCC offloading decision module is to execute offloading decisions for individual tasks entering the edge cloud to minimize end-to-end latency from user perspectives. Since the CCC and MEC architectures do not need to make task offloading decisions, the HCC offloading decision module is utilized when the network model is simulating the HCC architecture. When a specific task enters the edge cloud, it compares the time it takes to process the task directly at the edge cloud (t_{EC}) to the time it takes to process by offloading it to the central cloud (t_{CC}). Here, t_{EC} can be expressed as

$$t_{EC} = t_{EC}^{queue} + t_{EC}^{process} \tag{1}$$

where t_{EC}^{queue} is delay time in the queue and $t_{EC}^{process}$ is the actual task processing time. Additionally, the task processing time through the central cloud, t_{CC}, can be expressed as

$$t_{CC} = t_{CC}^{queue} + t_{CC}^{process} + 2t_{EC}^{transport} \tag{2}$$

where t_{CC}^{queue} is delay time in the queue, $t_{CC}^{process}$ is the actual task processing time, and $2t_{EC}^{transport}$ is the round trip time for the task from the edge cloud to central cloud and back.

Equations (1) and (2) are further expanded by introducing new parameters: w, R_{EC}, and R_{CC}. Parameter w is the actual work load of an individual task. R_{RC} and R_{CC} are the computing resources for the edge cloud and the central cloud, respectively, expressed in CPU cycle frequency in units of megahertz (MHz). With these parameters, the time required to process a specific task with work load w for the edge cloud and the central cloud can be expressed, respectively, as follows:

$$t_{EC} = t_{EC}^{queue} + \frac{w}{R_{EC}} \tag{3}$$

$$t_{CC} = t_{CC}^{queue} + \frac{w}{R_{CC}} + 2t_{EC}^{transport} \tag{4}$$

In addition to the parameters defined, the decision variable x need to be defined. If we set $x = 0$, then it is assumed that $t_{EC} \leq t_{CC}$ and the task is processed at the edge cloud. If we set $x = 1$ then the task is offloaded to the central cloud ($t_{EC} > t_{CC}$). The overall latency (t_{HC}), due to the task offloading decision, can be expressed as

$$t_{HC} = (1 - x)t_{EC} + xt_{CC} \tag{5}$$

4.3. HCC Architecture Computing Resource Allocation Module

The second simulation module is used to determine optimal computing resource allocation between individual edge clouds and the central cloud within the HCC architecture. The module is constructed on the basis of queueing theory.

Assume that λ is the arrival rate for tasks into the HCC network, and μ is the service rate. Then, by using the Little's Law, the queuing time (t^{queue}) for each edge cloud or central cloud can be expressed as

$$t^{queue} = \frac{\lambda/\mu}{\mu-\lambda} \tag{6}$$

Here, if we assume a simple network in which the edge cloud and central cloud are connected in a 1:1 manner, then the queuing time for the edge cloud and the central cloud are expressed, respectively, as follows:

$$t_{EC}^{queue} = \frac{\lambda_{EC}/R_{EC}}{R_{EC}-\lambda_{EC}} \tag{7}$$

$$t_{CC}^{queue} = \frac{\lambda_{CC}/R_{CC}}{R_{CC}-\lambda_{CC}} \tag{8}$$

The arrival rate for edge cloud (λ_{EC}) and central cloud (λ_{CC}) can alternatively be expressed as the ratio of tasks offloaded from edge cloud to the central cloud (p), which can then be stated as $\lambda_{EC} = \lambda(1-p)$ and $\lambda_{CC} = \lambda p$.

To prevent queuing time divergence for a specific edge cloud or central cloud, the HCC offloading decision must be made to minimize the difference, Δ, between t_{EC}^{queue} and t_{CC}^{queue}, which can then be expressed as

$$\Delta = \left| \frac{\lambda(1-p)/R_{EC}}{R_{EC}-\lambda(1-p)} - \frac{\lambda_p/R_{CC}}{R_{CC}-\lambda_p} \right| \tag{9}$$

Taking the assumption one step further, a more realistic network structure is considered, where edge clouds and central cloud are connected in N:1 star topology configuration. Here, Δ can be expressed as

$$\Delta = \left| \frac{1}{N} \sum_{i=1}^{N} \frac{\lambda_i(1-p_i)/R_{EC,i}}{R_{EC,i}-\lambda_i(1-p_i)} - \frac{\sum_{i=1}^{N} \lambda_{p,i}/R_{CC}}{R_{CC}-\sum_{i=1}^{N} \lambda_p} \right| \tag{10}$$

In the simulation model, $R_{EC,i}$ and R_{CC} are set as the design variables, and t_{HC} and Δ are measured through the simulation. Additionally, variable λ is represented by the data traffic and μ is represented by the computing resource.

5. Simulation Assumptions and Architecture Robustness Assessment

5.1. Simulation Assumptions

Assumptions for the CCC, HCC, and MEC architecture simulations were made for fair comparative analysis. They are listed below.

Network structure: As shown in Figure 4, the network consists of one central cloud facility, connected to six edge clouds in cluster that is in a star shaped topology. For the simulation, it was assumed that the distance between the city where edge clouds are located and the central cloud facility was 200 km for all three architectures. For the radio access network path, it was assumed that the latency for all three architectures are same.

Edge cloud connectivity: For the simulation case study, one of the key simplifying assumption was the implementation of the simplified network policy, which assumes that network will only use dedicated resource allocated to the region. Another assumption made was that each edge cloud interact with central cloud only, not with other edge clouds. Although the standard from ETSI takes into account that edge clouds can connect to nearby edge clouds for mobile application purposes, the case study presented in this study primarily focuses on computational offloading from edge clouds to the central cloud, which is not related to the task transfer from an edge cloud to another edge cloud, thus justifying the assumption made.

Latency in radio access network and core network: For communication network, latency occurs in radio access network and core network paths. In the simulation, it was assumed that all three architectures had equal amount of latency for both paths for fair comparison. For the radio access network path, the assumed latency for the simulation was set at 3 ms. For the core network path, the amount of latency was set so it corresponds to 200 km distance equivalent latency between the central cloud and edge clouds for all architectures.

Measurement units: For the simulation, the task is defined as a job that a mobile device requests to the cloud, and the workload is defined as the amount of computing resource required to process the task. The cloud computing resource is expressed as the CPU cycle frequency in units of MHz. The workload, which is the amount of CPU cycles required to complete a given task, is expressed in 1000 cycle. The processing time, which is the time required to complete a given task, can be obtained by dividing the task workload by the available computing resource. It is expressed in units of ms. Data traffic, which is the number of tasks generated per time unit, is expressed as tasks per second.

Application and key performance index for network architectures: For the case study, the network was used to transmit data to and from personal mobile communication devices. The performance measure used for this architectural analysis was the percentage of tasks that exceeded end-to-end processing time of 10 ms. This is based on the latency guideline proposed by Lee et al. [29]. The threshold for out-of-specification percentage was set to 5% after consulting Korea Telecom's internal subject matter experts regarding the out-of-specification deviation for this particular application.

5.2. Architecture Robustness Assessment

Using the simulation model constructed, CCC, HCC, and MEC architectures are simulated. For the initial architecture assessment, we tested the robustness of each architecture with respect to the increase in task data traffic.

For the robustness assessment, the total computer resource for each architecture was set at 60,000 MHz. For CCC architecture, all computer resources are placed in the central cloud. For MEC architecture, six edge clouds were assigned 10,000 MHz each. For the HCC architecture, the trend analysis was performed using the HCC architecture computing resource allocation module introduced in the previous section. Figure 5 shows the analysis results in the form of a log-scale. The x-axis represents the percentage of computing resources allocated to the central cloud, and the y-axis represents the Δ value in Equation (10) for the HCC architecture with $N = 6$.

For this particular HCC architecture, the trend analysis revealed that the optimum central cloud to edge clouds distribution ratio was 25:75, meaning 25% of the total computing resource (15,000 MHz) was allocated to the central cloud and 75% of the resources (45,000 MHz) were allocated in equal amounts to six edge clouds. It should be noted that the trend analysis for determining the optimal resource allocation ratio between the central cloud and edge clouds need to be conducted for different situations, since the optimal ratio may be different for each specific situation.

The task data traffic experienced by individual region was varied from 1800 tasks per second to 10,800 tasks per second. For each architecture, the percentage of tasks with out-of-specification percentage was measured as the function of task traffic. Figure 6 shows the results of the robustness assessment for all three architectures under increasing data traffic.

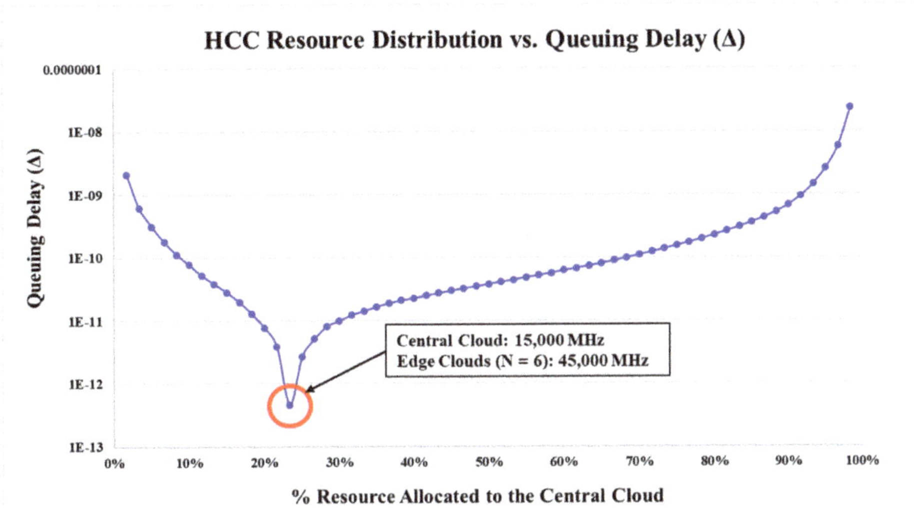

Figure 5. Trend analysis results for computing resource allocation in the Hybrid Cloud Computing (HCC) architecture.

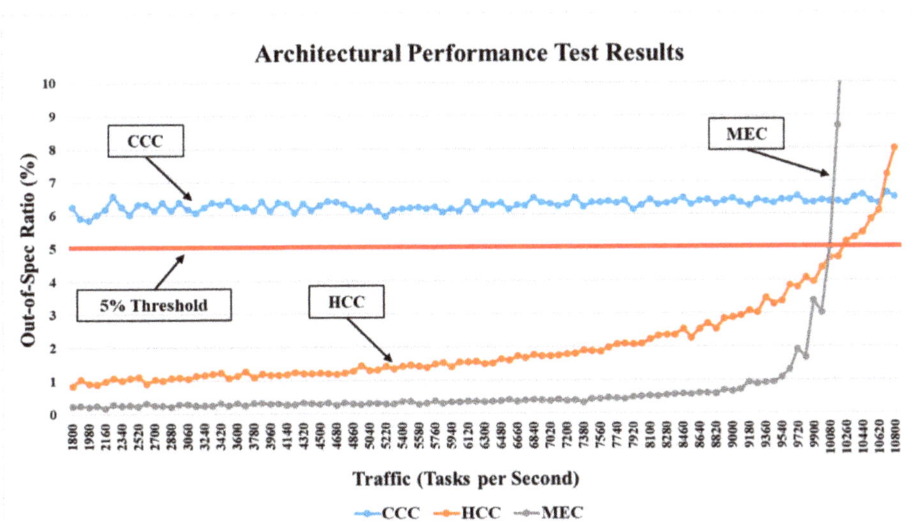

Figure 6. Robustness assessment for CCC, HCC, and MEC architectures.

The results show that the CCC architecture does not always satisfy the 5% out-of-specification ratio; nevertheless, it is also most robust among the three architectures. The HCC architecture shows more sensitivity to the data traffic increase than the CCC architecture. However, for most data traffic ranges simulated, the HCC architecture processed data traffics well within the specification. The MEC architecture had the best performance in terms of the percentage of out-of-specification ratio. However, as the data traffic reaches more than 9000 tasks per second, the MEC architecture's performance deteriorates rapidly, which shows that there is an upper limit to the data traffic range in which MEC is robust. This initial assessment revealed key characteristics of each architecture.

6. Simulation Using Actual Field Data

6.1. Simulated Scenarios

Using the network simulation model developed in this work, the CCC, MEC, and HCC architectures were modeled and their robustness to task traffic increase was assessed. To evaluate their capabilities in a realistic situation, two different daily data traffic scenarios were generated. The first scenario simulated normal daily traffic patterns for six regions of the same city in Korea, while the second scenario assumed data traffic spike in certain regions of the city. Figures 7 and 8 show the data traffic patterns for the two scenarios.

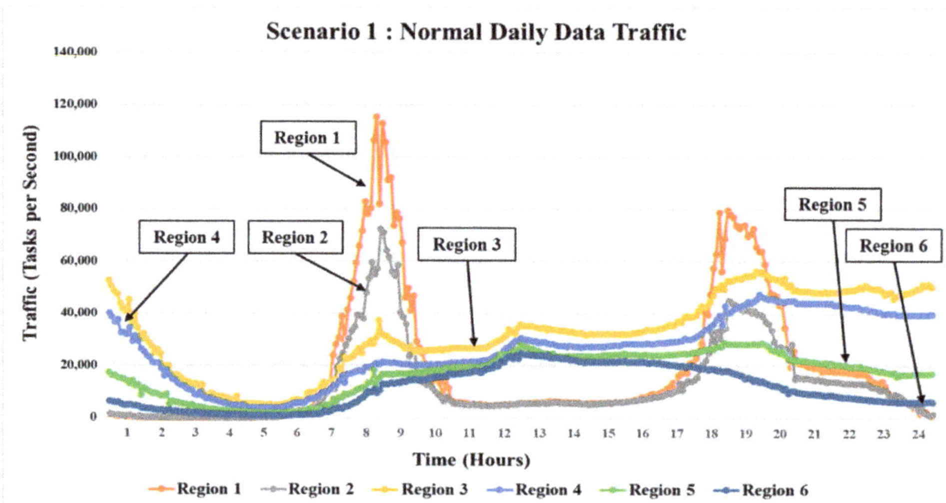

Figure 7. Twenty-four hour data traffic patterns for Scenario 1: normal daily traffic conditions.

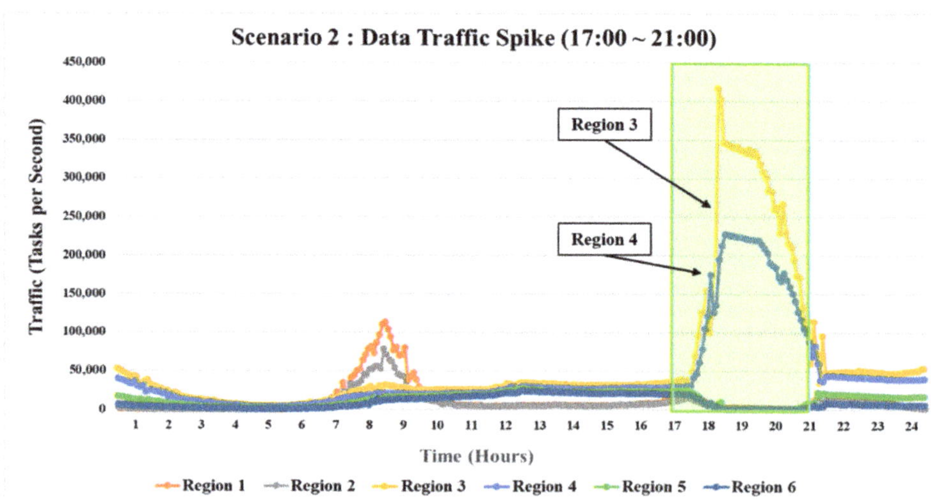

Figure 8. Twenty-four hour data traffic patterns for Scenario 2: regional data traffic spike.

Scenario 1 (normal daily data traffic): In this scenario, six data traffic patterns shown in Figure 7 were used as inputs into six base stations. All the data traffic patterns shown are actual normal daily data traffic patterns acquired during the same 24-h period from the six different regions. Data for each region are gathered at regular intervals, and are grouped by an hour segment to determine the amount of data traffic per hour. In the figure, data traffic pattern for each region are labeled for clarification purposes. Unfortunately, further detail regarding these regions were not available due to confidentiality issues. Each data traffic pattern shows a different peak traffic hour. This is attributed to the type of region where data was acquired. Certain regions were residential areas, with data traffic peaks before and after work hours. Some were commercial areas, with data traffic peaks after normal work hours. Other regions were business districts, with peak traffic hours occurring during normal business hours. Using these data, responses for the three network architectures were obtained through the simulation.

Scenario 2 (regional data traffic spike): In this scenario, a crowd gathering event in two of the six regions in the city was simulated. The six data traffic patterns in Figure 8 shows how the city's data traffic pattern was disrupted. During the evening hours (17:00~21:00 h), the data traffic for four regions were decreased, while the other two regions were loaded with data traffic from the four regions. As with Scenario 1, the responses for the three network architectures were obtained through the simulation using these inputs.

6.2. Simulation Assumptions

Total amount of computing resources for the clouds: For the simulation, all three network architectures were assigned the same amount of total computing resources. This was done to assess proposed architectures under same set of conditions for fair comparison. The total amount of computing resources was determined as follows. Initially, the amount of computing resources necessary to handle the data traffic during the peak time was determined for each of the six regions. Next, the total amount of computing resources was set to the sum of the individual resources determined for each region, which totaled to 52,200 MHz. Therefore, for all three network architectures, the total amount of computing resources was set to 52,200 MHz. Table 2 shows computing resource allocation for each network architecture.

Table 2. Computing resource allocation for three architectures.

	Computing Resource Allocation (MHz)		
Regions	**CCC**	**HCC**	**MEC**
Central Cloud	52,200	7830	-
Region 1	-	8271	9730
Region 2	-	7735	9100
Region 3	-	7353	8650
Region 4	-	7072	8320
Region 5	-	7013	8250
Region 6	-	6928	8150
Total Resources	52,200	52,200	52,200

Computing resource allocation for CCC architecture: For the CCC architecture, the entire 52,200 MHz of computing resources were allocated to the central cloud.

Computing resource allocation for HCC architecture: For the HCC architecture, the computing resources were allocated to both the central cloud and edge clouds. The key issue was determining the allocation ratio between the central cloud and edge clouds. This was done by performing a trend analysis using the simulation model. Computing resources were distributed to six edge clouds, and the amount allocated to each edge cloud was the same as that of the MEC architecture. For each edge cloud, the computing resources were then decreased in steps of 1% from the original resource amount, and allocated to the central cloud. The simulation was then conducted to assess the performance of the HCC architecture. The results of the trend analysis revealed that the optimal allocation ratio between the central cloud and edge clouds for this particular task traffic was 15% allocation to the central cloud and 85% allocation to edge clouds. Therefore, out of 52,200 MHz of computing resources, approximately 7830 MHz was allocated to the central cloud, and the rest of the resources were allocated to the six edge clouds in their respective regions, as listed in Table 2. One thing to note is that for HCC, the amount of computing resource allocated to the central cloud is 7830 MHz, which is not too much different from other edge clouds allocation amounts. In real situation, there would be more amount of resource allocated to the central cloud. However, for this case study, since all three network architectures compared must have same amount of computing resources, this was determined to be reasonable.

Computing resource allocation for MEC architecture: For the MEC architecture, the entire amount of computing resources was allocated to the edge clouds located in the six base stations. The resource allocated to each edge cloud corresponded to the computing resources required to handle peak task traffic for Scenario 1.

7. Simulation Results and Discussion

7.1. Simulation Results

The response of the three network architectures to data traffic patterns were expressed as the percentage of tasks that exceeded a processing time of 10 ms during the 24-h period. Figures 9 and 10 show the simulation results for the three architectures for Scenarios 1 and 2. The response of each architecture is shown with appropriate labels.

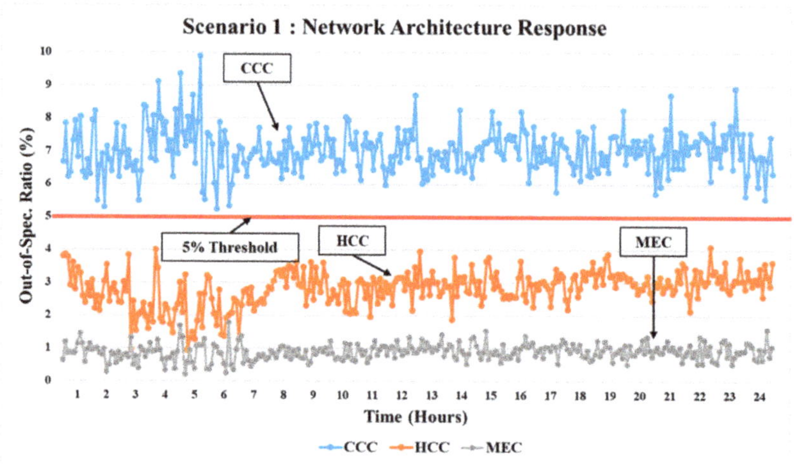

Figure 9. Individual network architecture's response to data traffic shown in Scenario 1.

The results obtained for Scenario 1 demonstrated each architecture's response to normal daily traffic. For the CCC architecture, the percentage of tasks that required more than 10 ms of processing time was always greater than the 5% specification threshold. For the HCC architecture, the percentage of tasks that required more than 10 ms was less than the threshold limit at all times. The MEC architecture performed the best, with the lowest percentage of tasks (~1%) that required more than 10 ms.

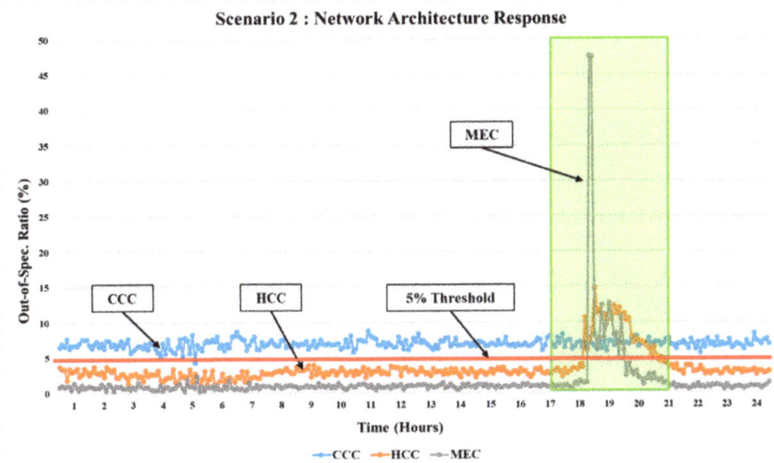

Figure 10. Individual network architecture's response to data traffic shown in Scenario 2.

Scenario 2, which simulated crowd gathering at two regions, demonstrated each network architecture's behavior when the data spike occurred. The CCC architecture demonstrated robust response as expected. This was because, as long as the total amount of data traffic at a given time was the same, the processing time for data traffic remained largely unchanged. However, the percentage of delayed tasks was above the threshold, which is similar to the results of Scenario 1. Conversely, the MEC architecture was significantly sensitive to the data traffic spike. As shown in the Scenario 2 results in Figure 10, the percentage of delayed tasks during the data traffic spike rose to nearly 50%, demonstrating that the architecture is vulnerable to such disturbances. Finally, the HCC architecture's response also demonstrated its sensitivity to the data traffic spike, but in a considerably more inhibited manner than the MEC architecture. The peak percentage of delayed tasks was approximately 15%, which was significantly lower than that of MEC architecture.

The results of the simulation-based analysis demonstrated the advantages and disadvantages for three network architectures. The CCC architecture was robust to data traffic spikes. However, due to the physical distance between the central cloud and base stations, the number of delayed tasks was always above the permitted threshold. The MEC architecture performed reasonably well under normal condition. However, it was considerably vulnerable if a data traffic spike occurred. The HCC architecture offers a compromise between the CCC and MEC architectures, as it meets the specification threshold for delayed task, but still impedes the detrimental effect of data traffic spikes.

7.2. Discussion

Simulation results provided valuable insights for newly proposed HCC architecture in terms of how they behave under different situations, and how these insights gained be used when the HCC architecture is actually implemented. More discussions on simulation results, issues for improvement, and limitations are presented.

In the simulation environment and conditions assumed in the case study, HCC architecture processed tasks well within the allowed specification under the normal condition, although it was slower than processing times by the MEC architecture. This has to do with the restriction of the total resource assigned for allocation between the central cloud and edge clouds of the HCC architecture. For fair comparative analysis of three network architectures, the total computing resource was fixed at 52,200 MHz for each architecture. Since the total computing resource amount was the sum of computing resource required to process peak hour data traffic for each of six regions, task processing time by the MEC architecture was the best among three architectures compared. HCC architecture, on the other hand, was forced to allocate some of these regional computing resources to the central cloud due to the restriction imposed by simulation assumption. This resulted in somewhat higher average processing time than the MEC architecture, due to some tasks forced to be offloaded to the central cloud during busy hours.

Another reason for higher processing time has to do with HCC offloading decision module located in each application server. When the HCC offloading decision module makes a decision for processing the task at the edge cloud or offload to the central cloud, it checks the queue at the central cloud at that instant. However, since all six application servers in the simulation model are making this decision, only checking the status at the central cloud but not the other application servers, this might generate more unexpected delay at the central cloud than expected, due to the random arrival of tasks from several application servers.

Insights gained from the simulation model and case study can serve as valuable guidelines for the actual HCC architecture development and deployment. The first way to further reduce HCC architecture's task processing time is to allocate the computing resource required to process the peak hour traffic to each application server, and allocate additional computing resource to the central cloud. This, in other words, means that the HCC architecture with amount of computing resources equal to that of MEC architecture at each cloud, and then add extra computing resource at the central cloud. With this implementation option, the performance of HCC architecture under normal data traffic condition should be equal to that of the MEC architecture. During the data traffic spike, as shown in Scenario 2 of the case study, further reduction of task processing delay than results shown in Figure 10 can be expected. However, the marginal overhead required for the HCC architecture over the MEC architecture is the extra investment for the central cloud, along with connective infrastructure between the central clouds and edge clouds.

The second approach to implement the HCC architecture in the field is to differentiate offloading paths for latency critical tasks and non-critical tasks. In reality, latency critical tasks, such as tasks from autonomous vehicles, and non-critical tasks, such as web-browsing tasks from mobile devices, arrive in mixed batches. The offloading decision module determines the priority of the task, and makes decision to either process it locally or to offload to the central cloud. In this case, the computing resource required for each edge cloud would correspond to the amount of latency critical tasks for peak hour, not the total amount of tasks. This may result in less amount of computing resources required than the first option previously mentioned. Although these two implementation options for the HCC architecture require more marginal overhead compared to the MEC architecture, they offer noticeable advantages over the MEC architecture.

8. Conclusions and Future Work

In this study, three communication network architectures were analyzed for their response capability under normal and disruptive data traffic conditions. The current CCC architecture, the MEC architecture, and the newly proposed HCC architecture were modeled using a Python-based simulation software. The behaviors of these architectures were assessed using actual data traffic patterns obtained from the field. The results were used to quantify the advantages and disadvantages of each architecture, and they demonstrated the potential benefit of the new HCC architecture. Insights gained from the case study were discussed, along with how these insights can be used to benefit the actual deployment of the HCC architecture in the field.

There is scope to further analyze the HCC architecture. The architecture can be assessed in terms of other key performance indices, such as the latency, cost, utilization, and profit, which are critical performance criteria for 5G network architectures. Additionally, for the HCC architecture, determining the optimal allocation between the central and edge clouds is an important research topic. What are some other realistic data traffic scenarios that need to be considered? How can computing resources be optimally allocated among the central cloud and edge clouds based on possible scenarios? Is there a "golden allocation ratio" that can include the majority of possible scenarios? Is there any variation of the HCC architecture that can further enhance its performance? Will there be any performance improvement by integrating machine learning algorithms into offloading decision framework? All of these questions present valuable opportunities for future research.

Author Contributions: Conceptualization, W.L. and E.S.S.; Methodology, W.L., E.S.S, W.Y.K. and H.H.; Software, W.L.; Validation, W.L., E.S.S., W.Y.K. and H.H.; Formal analysis, W.L.; Investigation, W.L. and E.S.S.; Resources, E.S.S.; Data curation, W.L.; Writing—original draft preparation, W.L. and E.S.S.; Writing—review and editing, W.L. and E.S.S.; Visualization, W.L. and E.S.S.; Supervision, E.S.S., W.Y.K. and H.H.; Project administration, E.S.S.; Funding acquisition, E.S.S. All authors have read and agreed to the published version of the manuscript.

Funding: This work was supported by the National Research Foundation of Korea (NRF) grant funded by the Korea government (MSIT) (No. NRF-2016R1D1A1A09916273).

Conflicts of Interest: The authors declare no conflict of interest.

Nomenclature

4G	Fourth Generation
5G	Fifth Generation
AI	Artificial Intelligence
CCC	Centralized Cloud Computing
eMBB	Enhanced Mobile Broadband
ESTI	European Telecommunication Standard Institute
HCC	Hybrid Cloud Computing
IoT	Internet of Things
MEC	Multi-access Edge Computing
mMTC	Massive Machine Type Communication
mmWave	Millimeter Wave
ms	Millisecond
URLLC	Ultra Reliable Low Latency Communication
UE	User Entity

References

1. Bag, T.; Garg, S.; Shaik, Z.; Mitschele-Thiel, A. Multi-Numerology Based Resource Allocation for Reducing Average Scheduling Latencies for 5G NR Wireless Networks. In Proceedings of the 2019 European Conference on Networks and Communications (EuCNC), Valencia, Spain, 18–21 June 2019; pp. 597–602.
2. Li, J.; Sahlin, H.; Wikstrom, G. Uplink PHY Design with Shortened TTI for Latency Reduction. In Proceedings of the 2017 IEEE Wireless Communications and Networking Conference (WCNC), San Francisco, CA, USA, 19–22 March 2017; pp. 1–5.
3. Wang, L.; Gao, Y.; Dong, Y.; Li, Z.; Zhang, Z.; Zhang, X.; Liu, X. Study on Flexible TTI Scheduling for LAA Systems. In Proceedings of the 2018 15th International Symposium on Wireless Communication Systems (ISWCS), Lisbon, Portugal, 28–31 August 2018; pp. 1–5.
4. Xiaotong, S.; Nan, H.; Naizheng, Z. Study on system latency reduction based on Shorten TTI. In Proceedings of the 2016 IEEE 13th International Conference on Signal Processing (ICSP), Chengdu, China, 6–10 November 2016; pp. 1293–1297.
5. Zhang, X. Latency reduction with short processing time and short TTI length. In Proceedings of the 2017 International Symposium on Intelligent Signal Processing and Communication Systems (ISPACS), Xiamen, China, 6–9 November 2017; pp. 545–549.
6. *Multi-access Edge Computing (MEC);* Framework and Reference Architecture, ETSI GS MEC 003 V2.1.1; ETSI Industry Specification Group (ISG): Sophia-Antipolis, France, 2019.
7. Montero, R.S.; Rojas, E.; Carrillo, A.A.; Llorente, I.M. Extending the Cloud to the Network Edge. *Computer* **2017**, *50*, 91–95. [CrossRef]
8. Ren, J.; Yu, G.; He, Y.; Li, G.Y.; Li, Y. Collaborative Cloud and Edge Computing for Latency Minimization. *IEEE Trans. Veh. Technol.* **2019**, *68*, 5031–5044. [CrossRef]
9. Chun, B.-G.; Ihm, S.; Maniatis, P.; Naik, M.; Patti, A. CloneCloud: Elastic execution between mobile device and cloud. In Proceedings of the Sixth Conference on Computer Systems, Salzburg, Austria, 10–13 April 2011; pp. 301–314.

10. Cuervo, E.; Balasubramanian, A.; Cho D-k Wolman, A.; Saroiu, S.; Chandra, R.; Bahl, P. MAUI: Making smartphones last longer with code offload. In Proceedings of the 8th International Conference on Mobile Systems, Applications, and Services, San Francisco, CA, USA, 15–18 June 2010; pp. 49–62.
11. Kemp, R.; Palmer, N.; Kielmann, T.; Seinstra, F.; Drost, N.; Maassen, J.; Bal, H. eyeDentify: Multimedia Cyber Foraging from a Smartphone. In Proceedings of the 2009 11th IEEE International Symposium on Multimedia, San Diego, CA, USA, 14–16 December 2009; pp. 392–399.
12. Wei, X.; Wang, S.; Zhou, A.; Xu, J.; Su, S.; Kumar, S.; Yang, F. MVR: An Architecture for Computation Offloading in Mobile Edge Computing. In Proceedings of the 2017 IEEE International Conference on Edge Computing (EDGE), Honolulu, HI, USA, 25–30 June 2017; pp. 232–235.
13. Huang, N.; Wang, P.; Niyato, D. A Dynamic Offloading Algorithm for Mobile Computing. *IEEE Trans. Wirel. Commun.* **2012**, *11*, 1991–1995. [CrossRef]
14. Chen, M.; Hao, Y. Task Offloading for Mobile Edge Computing in Software Defined Ultra-Dense Network. *IEEE J. Sel. Areas Commun.* **2018**, *36*, 587–597. [CrossRef]
15. Hao, Y.; Chen, M.; Hu, L.; Hossain, M.S.; Ghoneim, A. Energy Efficient Task Caching and Offloading for Mobile Edge Computing. *IEEE Access* **2018**, *6*, 11365–11373. [CrossRef]
16. Liu, J.; Mao, Y.; Zhang, J.; Letaief, K.B. Delay-optimal computation task scheduling for mobile-edge computing systems. In Proceedings of the 2016 IEEE International Symposium on Information Theory (ISIT), Barcelona, Spain, 10–15 July 2016; pp. 1451–1455.
17. Hong, S.-T.; Kim, H. QoE-Aware Computation Offloading to Capture Energy-Latency-Pricing Tradeoff in Mobile Clouds. *IEEE Trans. Mob. Comput.* **2019**, *18*, 2174–2189. [CrossRef]
18. You, C.; Huang, K.; Chae, H.; Kim, B.-H. Energy-Efficient Resource Allocation for Mobile-Edge Computation Offloading. *IEEE Trans. Wirel. Commun.* **2017**, *16*, 1397–1411. [CrossRef]
19. Srinivasan, K.; Agrawal, N.K. A study on M-CORD based architecture in traffic offloading for 5G-enabled multiaccess edge computing networks. In Proceedings of the 2018 IEEE International Conference on Applied System Invention (ICASI), Chiba, Japan, 13–17 April 2018; pp. 303–307.
20. Lin, Y.-D.; Lai, Y.-C.; Huang, J.-X.; Chien, H.-T.; Huang, C.-X. Three-Tier Capacity and Traffic Allocation for Core, Edges, and Devices for Mobile Edge Computing. *IEEE Trans. Netw. Serv. Manag.* **2018**, *15*, 923–933. [CrossRef]
21. Kiani, A.; Ansari, N. Toward Hierarchical Mobile Edge Computing: An Auction-Based Profit Maximization Approach. *IEEE Internet Things J.* **2017**, *4*, 2082–2091. [CrossRef]
22. Hou, X.; Ren, Z.; Wang, J.; Cheng, W.; Ren, Y.; Chen, K.-C.; Zhang, H. Reliable Computation Offloading for Edge Computing-Enabled Software-Defined IoV. *IEEE Internet Things J.* **2020**, *1*. [CrossRef]
23. Ahn, S.; Lee, J.; Kim, T.Y.; Choi, J.K. A Novel Edge-Cloud Interworking Framework in the Video Analytics of the Internet of Things. *IEEE Commun. Lett.* **2020**, *24*, 178–182. [CrossRef]
24. Zhao, J.; Li, Q.; Gong, Y.; Zhang, K. Computation Offloading and Resource Allocation For Cloud Assisted Mobile Edge Computing in Vehicular Networks. *IEEE Trans. Veh. Technol.* **2019**, *68*, 7944–7956. [CrossRef]
25. Zhang, Y.; Lan, X.; Li, Y.; Cai, L.; Pan, J. Efficient Computation Resource Management in Mobile Edge-Cloud Computing. *IEEE Internet Things J.* **2018**, *6*, 3455–3466. [CrossRef]
26. Zhang, Y.; Lan, X.; Ren, J.; Cai, L. Efficient Computing Resource Sharing for Mobile Edge-Cloud Computing Networks. *IEEE/ACM Trans. Netw.* **2020**, 1–14. [CrossRef]
27. Ruan, L.; Yan, Y.; Guo, S.; Wen, F.; Qiu, X. Priority-Based Residential Energy Management With Collaborative Edge and Cloud Computing. *IEEE Trans. Ind. Inform.* **2020**, *16*, 1848–1857. [CrossRef]
28. Thai, M.-T.; Lin, Y.-D.; Lai, Y.-C.; Chien, H.-T. Workload and Capacity Optimization for Cloud-Edge Computing Systems with Vertical and Horizontal Offloading. *IEEE Trans. Netw. Serv. Manag.* **2020**, *17*, 227–238. [CrossRef]
29. Lee, K.; Kim, J.; Park, Y.; Wang, H.; Hong, D. Latency of Cellular-Based V2X: Perspectives on TTI-Proportional Latency and TTI-Independent Latency. *IEEE Access* **2017**, *5*, 15800–15809. [CrossRef]

applied sciences

MDPI

Article

Primary User Traffic Pattern Based Opportunistic Spectrum Handoff in Cognitive Radio Networks

Kaleem Arshid [1], Iftikhar Hussain [2,*], Muhammad Khawar Bashir [3], Shahid Naseem [4], Allah Ditta [4], Natash Ali Mian [2], Misha Zahid [2] and Israr Ali Khan [5]

1 Faculty of Information Technology, Beijing University of Technology, Beijing 100124, China; kaleemarshid@yahoo.com
2 School of Computer and IT, Beaconhouse National University, Lahore 53700, Pakistan; natash.ali@bnu.edu.pk (N.A.M.); mishz@hotmail.com (M.Z.)
3 Department of Statistics and Computer Science, University of Veterinary and Animal Sciences, Lahore 54000, Pakistan; mkbashir@uvas.edu.pk
4 Department of Information Sciences, Division of Sciences and Technology, University of Education, Lahore 54770, Pakistan; shahid.naseem@ue.edu.pk (S.N.); allahditta@ue.edu.pk (A.D.)
5 Department of Mathematics, Namal Institute Mianwali, Talagang Road, Mianwali 42250, Pakistan; Israr.khan@namal.edu.pk
* Correspondence: iftikhar.hussain@bnu.edu.pk; Tel.: +92-42-3810-0156

Received: 2 October 2019; Accepted: 21 February 2020; Published: 2 March 2020

Abstract: Through the expeditious expansion of the wireless network, the unlicensed bandwidth-based devices are growing substantially as compared to the present vacant bandwidth. Cognitive radio networks present a proficient solution to the spectrum shortage diminution hitch by allowing the usage of the vacant part of the spectrum that is not currently in use of the Primary User licensed bandwidth to the secondary user or cognitive radio user. Spectrum management procedure in cognitive radio network comprises of spectrum sharing, sensing and handoff. Spectrum handoff plays a vital role in spectrum management and primarily focuses on single handoff strategies. This paper presents a primary user traffic pattern-based opportunistic spectrum handoff (PUTPOSH) approach to use in the cognitive radio networks. PUTPOSH permits a secondary user to sense the arrival of a primary user and use an opportunistic handoff scheme. The opportunistic handoff scheme firstly detects the arrival of the primary users by energy detection sensing and secondly, it allows a cognitive radio user to decide whether to do handoff or not contingent upon the overall service time to reduce the unused handoffs. The handoffs can either be reactive or proactive based on the arrival rate of the primary user. The simulation results show that the presented PUTPOSH approach (a) minimizes the number of handoffs and the overall service time, and (b) maintains the channel utilization and throughput of the system at a maximal point.

Keywords: spectrum handoff; spectrum sensing; cognitive radio networks; cognitive radio user; primary and secondary user

1. Introduction

The number of devices which make use of the licensed or unlicensed spectrum is growing swiftly, and a spectrum shortage problem arises due to the existing unproductive spectrum allocation strategies. The current approaches are static-based and not capable of adjusting to the increasing demand of the bandwidth requirements [1,2]. The cognitive radio network (CRN) provides a proficient solution to overcome the spectrum shortage problem. CRN permits a secondary user (SU) which is also known as a cognitive radio user (CRU) to utilize the provisionally unused bandwidth of the primary user (PU) to improve the narrow spectrum resources. CRU increases the utilization productivity of the channel without resettling the predefined rules of spectrum allocation [3]. The spectrum management method in CRNs is mainly composed of three major steps. Firstly, spectrum sensing, where CRU or SU get provisional access to the vacant spectrum; spectrum sensing scans and detects the vacant spectrum bands constantly by probing the PU actions. Secondly, spectrum sharing, where many SUs access those vacant spectrums; spectrum sharing must be synchronized to avoid crash between various SUs. Finally, the spectrum handoff (SH)—in which an CRU ought to continue its spectrum access on an available channel (if the related PU originates).

Recent research on CRN focuses mainly on spectrum detection and exchange of information [4–8]. Spectrum transfer is becoming a challenging topic due to different research actions related to the heterogeneous network and also remains less studied [3,9]. Spectrum transfer is considered as a vigorous step in spectrum management due to its position for continuous switching from one available channel to another without compromise to its Quality of Service (QoS) [3]. There are two types of SH, proactive and reactive handoff [10–12]. In the proactive handoff, the target channel is selected with regard to the PU's interarrival pattern. It is used for data transfer prior to any event of the actual handoff process. In the reactive handoff, to resumed the paused transmission a channel is selected by real-time sensing after the handoff event takes place. The CRU can maintain the paused transmission on the recently explored channel [10]. The prevailing work mainly focuses on the single handoff approach which may be proactive or reactive. The handoff method is preselected without considering the PU arrival rate and it can be the reason for channel underutilization and large handoff delay. The handoff approach between reactive or proactive is preselected without caring for the primary user traffic samples in the existing single handoff strategies. The research efforts described in [9–14] are focused on single SH approaches, where the handoff process is previously determined. The existing handoff approaches function in the theory of shifting the channel on PU's arrival and no methods in these practices are to reside and wait until PU completes a successful transmission. Preferably, a CRU should be familiar with the PU traffic pattern using its sensing capabilities and the most appropriate handoff method wherever required.

In this paper, a primary user traffic pattern-based opportunistic spectrum handoff (PUTPOSH) scheme is presented. It permits a CRU to sense the PU's traffic through energy-efficient sensing [15–17] and then adopt a proactive or reactive handoff strategy according to the interarrival rate. PUTPOSH is comprised of three modules: (1) spectrum sensing, (2) spectrum mobility management, and (3) spectrum handoff decision. In the spectrum sensing module, the energy detection technique is used due to its low computational features and working difficulty. The spectrum mobility management obtains the PU interarrival information by spectrum sensing and the dwelling period of a PU is forecasted with respect to the previous waiting behavior in its licensed band. The major contribution of this research is the spectrum handoff decision module. The spectrum handoff decision chooses on a suitable handoff class between reactive and proactive with regards to (w.r.t.) the overall service time of a CRU. The overall service time of a CRU comprises of sensing, processing, waiting and transmission times. Spectrum handoff takes place when the PU appears on its bandwidth which is temporarily engaged by the CRU. Apart from maintaining its transmission CRU must have to sense for a new free channel. When a free channel is sensed, the CRU can continue its transmission on the recently vacant spectrum, packet failure can occur throughout this handoff method [2,18]. Moreover, A preemptive resume priority (PRP) M/M/1 queuing model is used to manage the transmission and incorporated to differentiate the spectrum usage behavior of PUs and SUs. The research aims to develop a PUTPOSH algorithm with various handoff modules that permit a CRU to detect patterns of PU's traffic and select a reactive or proactive scheme accordingly. On the arrival of PU, the CRU intelligently decides whether to do handoff or to wait for the on-going channel depending upon the overall service time (in both cases). A simulation which included PUTPOSH was setup in Matlab to validate and show the performance of the presented approach by comparing the results with some of the existing reactive, proactive, and hybrid schemes in terms of the overall service time, the number of handoffs and the channel utilization.

This paper is organized as follows: the literature review is presented in Section 2. The PUTPOSH approach is described in Section 3. It comprises of three sections (a) the presented approach, (b) the queuing model, and (c) the iming diagram of CRU transmission. Section 4 describes the simulation setup and some of the results. Finally, the conclusion is given in Section 5.

2. Literature Review

The literature presented in this section is generally absorbed on the spectrum sensing and handoff features.

2.1. Spectrum Sensing

Under channel shadowing and fading, the PU signal cannot be noticeable to the CRU. The detection capability of a single CRU is limited, which in turn strongly affects the PU transmission. A spectrum sensing scheme in CRNs with cooperative nature is presented in [19] and is based upon the Amplify and Forward (AF) protocol to reduce the sensing time of CRUs. Authors in [19] considered an infrastructure of the cognitive radio network where a band manager is responsible for communication between the cooperative CRUs. Authors defined two protocols to detect the PU arrival: (1) a non-cooperative protocol, and (2) a totally cooperative protocol. Another cooperative spectrum sensing framework is proposed by Won-Yoel Lee and Ian. F. Akyildiz in [20] which delivers solution for the sensing efficiency and interference avoidance problems. Sensing and transmission cannot be done concurrently in CRNs, therefore, it has to be bounded with the transmission. The basic concept of sensing in CRNs is to provide efficient and opportunistic communication to SUs. Data communication cannot be done which sufficiently reduces the transmission capabilities of SU.

In [21], authors presented a framework based on the theory of multiclass to exploit the attainable throughput in CRN. Through the absence of the PU in the two-class hypothesis, the energy range of sensing signal is separated into quantized areas while during the presence of the PU, the sensing signal is conserved. Authors claim that the obtainable approach increases the throughput by providing a higher amount of transmission actions. In [22], authors formulated and analyzed the amount of spectrum operation with different groups of channels and with different primary and secondary users in synchronized network structures. Authors considered realistic channels for the SUs and each channel is licensed to the PUs. The CRU is supported by the spectrum handoff and is authorized to utilize the channels by sensing outcomes and PU interruptions. In [23], authors studied resource sharing for chunk based multi-carrier with time varied spectrum resources and presented a novel opportunistic capacity model. Authors divided the novel opportunistic capacity model into two modules to reduce the computational complexity and solve them using the Lagrangian dual method. In [24], authors developed a supportive sensing method, which is based on the pairwise secondary user transmitter and receiver. Authors presented the detailed protocol description to illustrate the working of the projected framework. The main objective is to minimize the false alarm rate of the missed detection rate by adaptively correcting the discovered threshold of every sensor.

Energy finding is a simple spectrum sensing method that can be depicted as a Neyman Pearson-like a binary hypothesis testing issue which is developed using the chi-square, gamma/normal statistical distributions [25–27]. Energy detection is usually used for the low computational features complexities [28,29]. Therefore, immense work can be found in the literature discussing energy detection in terms of fading channels, diversity systems, additive white Gaussian noise and collaborative detection [18,30–37]. However, the energy detector is not capable of differentiating between the PU signal and the noise with a lower value of Signal to Noise Ratio (SNR) [38]. The solution to this problem is presented in the sensing techniques that are based on Eigenvalue based methods [39–42]. Using Eigenvalue based method, an energy detection scheme is proposed in [43] that can efficiently perform under frequency selective channels and noise uncertainty. The method proposed in [44] is a low complexity spectrum sensing method that depends upon the exploitation of the sub-band energy variations. In [45], authors proposed a twofold threshold energy recognition scheme. Two thresholds are used in double threshold schemes instead of one. The region between the two thresholds is considered as the uncertain region and CRU performs sensing again.

2.2. Spectrum Handoff

2.2.1. Proactive Handoff

In [46], authors anticipated a proactive handoff approach which considered Short Time Backup Channel (STBC), the BC for CRU communication is selected prior to the PU arrival. The handoff judgment in STBC is dependent upon the QoS of the ongoing channel. In this scheme, a backup channel achieves the usage of the bandwidth improvement for a short period whereas, in the complete backup approaches, the BC is reserved continually with the ongoing channel. In the STBC the lowest stage of the handoff wait can be attained as compared to without backup schemes. The STBC chooses the marked channel in the handoff prior to the occurrence of the trigger event. A proactive Fuzzy Logic (FL) based Spectrum Handoff approach is proposed in [47] and presented the major principles of FL to manage troubles professionally. Two FL controllers are used; first controller determines the space among CRU and PU. It examines the control of CRU communication (without causing any influence on the communication of the nearest PU). The second controller measures the waiting of the CRU in the ongoing channel. A handoff is started if the QoS of the CRU is not suitable and the high intervention caused by the CRU on the nearest PU's communication. In [14], authors projected a methodology established upon an increased probability Spectrum Handoff having Cumulative Probability (SHCP) to decrease the handoffs to achieve an improved quality of service. In this method, the PU decides whether to wait or stay on its ongoing operating channel or do a handoff. It depends upon an algorithm of probability estimation however maintaining a backup reserve channel. In [48], authors anticipated a proactive handoff scheme based on a probabilistic and predictive approach. It is slightly mandatory due to the indefinite behavior of PUs. It was planned to decrease the latency and loss of facts, and figure out through spectrum handoff. Proactive handoff gives an excessive option to the SUs to continue their half-finished broadcast on the target station. In [49], authors presented a distributed CRN scheme based on the multi-armed bandit method. Authors examined the blind spectrum choice problem of SUs by considering a fixed handoff stay whose detecting aptitude of Cognitive Radio (CR) is inadequate and the channel statistics are not known in advance. In this scenario, SUs have made the choice of either (i) waiting at ongoing spectrum with low accessibility or (ii) handing off to another spectrum with higher accessibility.

2.2.2. Reactive Handoff

Authors in [18], projected a reactive handoff approach named as dynamic frequency hopping communities (DFHC). DFHC is believed to enhance QoS demands of CRUs while giving suitable and reliable sensing of the spectrum for assuring the PU defense. In this research, a wireless regional area network cell or a CRU examined the accessibility of the new objective channel while transmitting on an ongoing channel. To avoid intervention with the PUs, the CRU maintains its communication on the selected channel and opens sensing for other channels. Authors in [50], proposed a reactive handoff approach which uses an M/G/1 queuing model [18] to examine the channel utilization of CRN. In this approach, a user can concurrently use various vacant channels for transmitting CR. In [21], the authors proposed an M/G/1 queuing model to resolve the spectrum handoff problem. The interarrivals to the M/G/1 queue are modeled and shared various vacant channels, for instance, the overall interarrivals from all CRUs. The handoff delay occurs at the maximum level when the PU arrived, hence the handoff judgment is completed. In [51], author presented a framework comprises of probabilistic algorithms, and some other novel approaches including CRU clustering and PRP M/G/1 queuing to attain better competence in spectrum handoff in a CRN. Authors claim that the proposed strategy performs better than existing schemes in terms of accuracy in sensing the right channel, handoff latency, and energy consumption. The authors of [52] projected a reactive handoff scheme for Zigbee in which a CRU can access various channels via incessantly sensing and spectrum handoff. The presented sensing and handoff technique provides better results in a situation where the noise is at a minimal level. However, as an energy detector, it cannot distinguish between noise and PU signal, hence a false alarm can be activated which produces the useless handoffs.

2.2.3. Hybrid Handoff

In order to assure the requirement of the wide bandwidth and to achieve the improved results in terms of QoS and spectrum aggregation (SA), an approach is required which permits a PU or a CRU to concurrently use various spectrum groups [53]. CRU is a clever user who can sense and use vacant spaces by probing the radio atmosphere. This class of sensing builds probably to join the free channels by spectrum aggregation. In [54], the author presented a scheme named hybrid handoff, which is founded on dynamic spectrum aggregation (DSA) to discover the manners of a CRU through handoff. Each time the CRU performs a handoff to BC when PU visits back to its certified channel. According to CRU, all the channels are BCs except the present channel (can be either utilized or idle). When PU arrives at the present channel; on the one hand, if the channel is free, the communication of CRU can be smoothly moved into it while on the other hand if all the channels are assessed to be busy, the CRU will wait till the completion of the PU communication. In [55], authors developed a spectrum-management scheme and estimated the performance in varied spectrum environment (opportunistic and centralized CRNs). Authors considered a determined threshold period for spectrum handoff delay to improve the performance for both opportunistic and negotiated situations with backup channels. In [54], authors describe that the CRU should select the probability of minimum handoff to reduce the handoff delay and the number of handoffs. The main application of this method makes it achievable to give superior QoS. The level of SH gap is minimized whenever the BCs are utilized; it guides towards the utilization of channels which are continuously vacant (most of the time).

Table 1 summarizes the handoff strategies in a comparative manner, which are also discussed in the above subsections. The strategies are compared on the basis of channel backup property, bandwidth utilization and handoff delay. The fuzzy-based analytic hierarchy process (FAHP) is a scheme with a full backup property in which BC is kept (all the time) with the ongoing operating channel. The bandwidth utilization becomes low whenever the handoff delay is at the minimal level. STBC and SHCP schemes have the middle course between full backup and no backup therefore the bandwidth utilization is at an average level while handoff delay is minimal. Since the fuzzy-based scheme has no BCs, the handoff delay can be maximized when the channel underutilization is at its minimal level. DFHC [18], M/G/1 [21] and Zigbee [52] are reactive handoff schemes with no channel backup facility, therefore, the handoff delay is at maximum level and the channel underutilization is at its minimal level. DSA [53] is a hybrid handoff scheme which is considered as a middle course between proactive and reactive handoff schemes. In DSA, the bandwidth utilization is at an average level, however, the handoff delay can be at its maximum level.

Table 1. Comparison of some of the existing handoff strategies with regards to the channel backup, bandwidth utilization and the handoff delay.

Categories	Handoff Approach	Backup Channel	Bandwidth Utilization	Delay of Handoff	Disadvantages
Proactive Handoff Schemes	STBC [46]	Compromise between no and full backup	Average	Minimum	- Bandwidth underutilization
	Fuzzy Based [47]	No backup	Maximum	Can be Maximum	- No on-going sensing mechanism - Increased Handoff delay
	SHCP [14]	Compromise between no and full backup	Average	Minimum	- Increased waiting Time - Bandwidth underutilization
	FAHP [52]	Full backup	Minimum	Minimum	- Bandwidth underutilization - No on-going sensing mechanism
Reactive Handoff Schemes	DFHC [18]	No Backup	Maximum	Maximum	- Improved waiting time - Improved handoff delay
	M/G/1 Queuing Based [50]	No Backup	Maximum	Maximum	- Handoff processing time is not considered - Improved handoff delay
	Zigbee Scheme [52]	No Backup	Maximum	Maximum	- False alarm probability - Cannot work Under multi-path fading or shadowing conditions
Hybrid Handoff Schemes	Hybrid Scheme [53]	No Backup	Average	Can be maximum	- Improved waiting time - Improved handoff delay

3. PUTPOSH Model

3.1. Overview of the Model

The PUTPOSH approach contains the dual handoff modules i.e., proactive and reactive. The PUTPOSH scheme permits CRU to detect patterns of PU's traffic and select a reactive or proactive scheme accordingly. On the arrival of PU, CRU may intelligently decide whether to do handoff or wait for an ongoing channel by considering the overall service time into account.

The PUTPOSH approach has the following properties:

1. The cognitive radio networks (CRNs) is assumed to be a time divided system and every CRU performs spectrum sensing in the first half of every time slot. The actual transmission of the CRU is performed in the second part of the time slot where the target channel is sensed as idle.
2. When the target channel is consumed by a PU, the CRU will perform the actual mobility management function by considering the waiting on an up-to-date channel or shifting its communication to the new channel.
3. CRNs comprise of autonomous channels, where every channel has high and low precedence queues. Every high priority queue has served one PU while the lowest precedence queue served several SUs on the first-come-first-serve (FCFS) basis.
4. A handoff decision protocol is divided into the transmission and sensing time slots. The interarrival of PU is noticed by the cognitive radio user (CRU) in its current channel. It ought to spend the initial part of every timeslot in monitoring the free channels while the communication is completed in the second part of the time slot.
5. When the multiple unused channels are assessed, the CRUs will choose any vacant channel for its next transmission and this random decision is according to the uniform distribution. Furthermore,

if no other channel is found to be free, the CRU will wait and stay on its current channel until the free channel is available.

3.2. Framework

Figure 1 shows the proposed primary user traffic pattern based opportunistic spectrum handoff framework. The proposed design contains three different parts; (i) spectrum sensing, (ii) spectrum mobility management, and (iii) spectrum handoff decision. These are connected in Figure 1 and also described in the following subsections respectively.

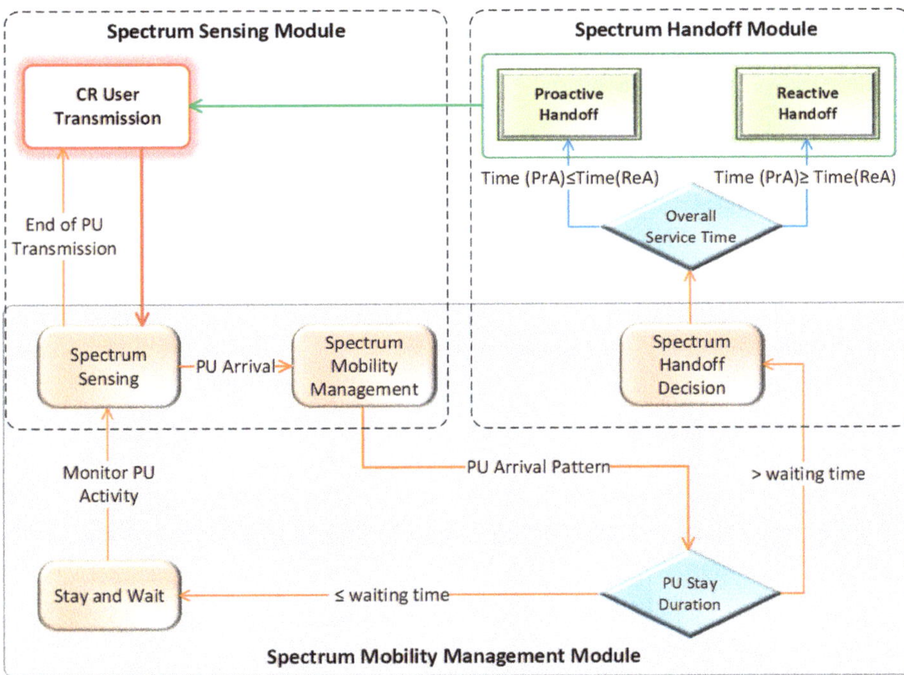

Figure 1. The proposed primary user traffic pattern based opportunistic spectrum handoff framework.

3.2.1. Spectrum Sensing

In the spectrum sensing module (Figure 1), the PU movement is continually observed by the CRU during transmission on the currently occupied channel. The spectrum sensing part can be achieved by integrating one of the existing sensing methods, i.e., matched filter (MF) [15], cooperative sensing (CS) [16,42] or energy detection (ED) [17,43,44]. In the presented PUTPOSH model, the energy detection technique is used in the spectrum sensing part for the reason of its working difficulty and for the low computational features [45,46,52]. It is considered as a generalized approach when it is compared with the MF and CS. In the ED approach, the recipients do not need any information on the PU's signal [18,53] and the signal is observed by comparing the given threshold value with the energy detector's output. The threshold value is fixed but depends on the earlier arrival of the PU. The apparent energy stage is increased when PU arrives on its licensed band. When an energy level's threshold is specified, the apparent energy stage is tested multiple times either in the presence and in the absence of the PU. As the power of the arriving signal increases than the given threshold, the handoff process is initiated to execute handoff actions.

3.2.2. Spectrum Mobility Management

The spectrum mobility management obtains the interarrival information of the PU by spectrum sensing. The dwelling period of a PU is forecasted at the assigned channel with respect to the previous waiting behavior. The stopover stay of the PU at the assigned channel depends upon its communication requirements and can be of any time period. Therefore the dwelling period is assumed and based upon the random distribution method [18,21]. The decision of either to wait and stay at the ongoing path or to execute the handoff event is based on the current result. On the one hand, the CRU does not make the handoff decision when the PU continued communication for shorter intervals of time. The CRU may carry its communication and can wait for the ongoing spectrum band. On the other hand, the CRU chooses to make a handoff decision when PU continued for a longer period at the previous licensed band (see Spectrum mobility management module in Figure 1). To allocate vacant channels, the waiting time is considered as the overall time of a CRU which can be the reason for delay in the priority queue. It can be resolved by the M/G/1 queuing equation for the network model [43,44].

3.2.3. Spectrum Handoff Decision

The spectrum handoff decision chooses on a suitable handoff class between reactive and proactive w.r.t. the lowest overall service time of the CRU. The overall service time of the CRU is based on the following times: sensing time, processing time, waiting time and transmission time. According to Figure 1 (in the spectrum handoff module), ReA time and PrA time represent the overall service time of the reactive and proactive handoff judgments respectively. In this paper, the ReA and PrA time values are assumed to find the threshold and also to compare different approaches. The overall minimal service time is implemented for the spectrum handoff decision. Subsequently, the CRU can maintain its transmission on the recently selected channel.

The Spectrum handoff module in Figure 1 shows that the proposed PUTPOSH strategy begins with the recognition of the PU interarrival by spectrum sensing and forecasts PU dwelled period in the assigned channel. Hence in this step, the CRU either dwells on the recent channel or moves into the SH decision stage.

3.3. The Queuing Model

A preemptive resume priority (PRP) M/M/1 queuing model is used to differentiate the usage of the spectrum behavior of PUs and SUs (see Figure 2).

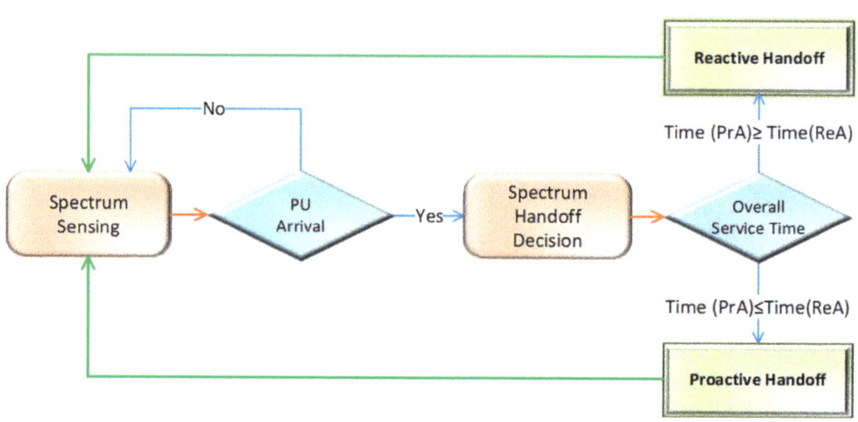

Figure 2. Cognitive radio user behavior on the arrival of primary user (PU).

The characteristics of the proposed queuing model are as follows:

1. Two kinds of users can be served by each channel, i.e., high priority PUs and low priority CRUs.
2. The PUs has preemptive priority to interrupt the communication of CRUs. Once the CRU is interrupted, it leaves the existing channel and starts sensing further available channel to continue its transmission.
3. A CRU may experience several interruptions during its transmission.
4. In the case of multiple users with similar priority; access to the requested channel is served on the basis of first come first serve.

In Figure 3, the presented PUTPOSH approach contains two channels and two queues for PU and CRUs. The PU reserved its position in a high priority queue while the CRU is located in a lower priority queue. When the transmission is interrupted by PU, the CRU may wait and stay with the ongoing operations stage or may transfer its communication to one of the new channels. The outstanding transmission of the CRU is placed at the head of the lower priority queue. The remaining transmission of CRU is positioned at the tail of the lower precedence queue [10–12]. In any case, the channel is available for transmission and CRU will continue its communication.

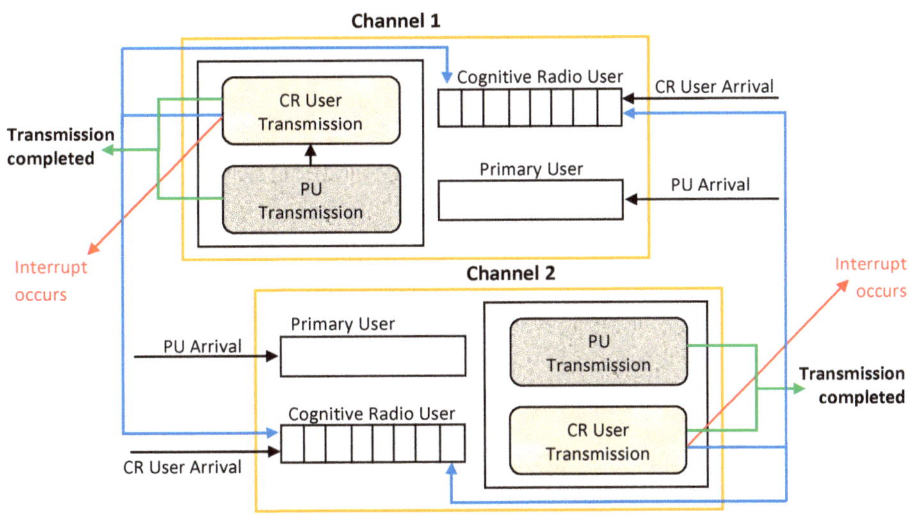

Figure 3. Queuing behavior of cognitive radio user (CRUs) and PUs (two channels and two queues).

3.4. Timing Diagram of the CRU Transmission in the Handoff Scheme

Figure 4 delineates the timing diagram of a CRU transmission in the presented adaptive hybrid handoff scheme. CRU starts its transmission on channel 1 and after some intervals of time, the CRU proactively senses the arrival of the PU. The CRU compares the predicted PU waiting time and also the handoff delay on the same channel. Based on the comparison decision, the CRU reacts to perform a handoff on channel 2. During transmission on channel 2, the CRU senses the PU arrival and again makes the similarities of the overall service time (in case of stay and wait on the recent channel and performing the handoff).

Moreover, CRU predicts the stopover of PU for a short duration and may wait and stay until PU completes its data transmission. As soon as PU resumes the channel, the CRU restarts its transmission on channel 2. After some time intervals, CRU again detects the PU arrival for a longer duration at channel 2 and performs the handoff at channel 1, another free channel. In the last part, the CRU performs a handoff to channel 3 to complete its transmission and may resumes the channel again.

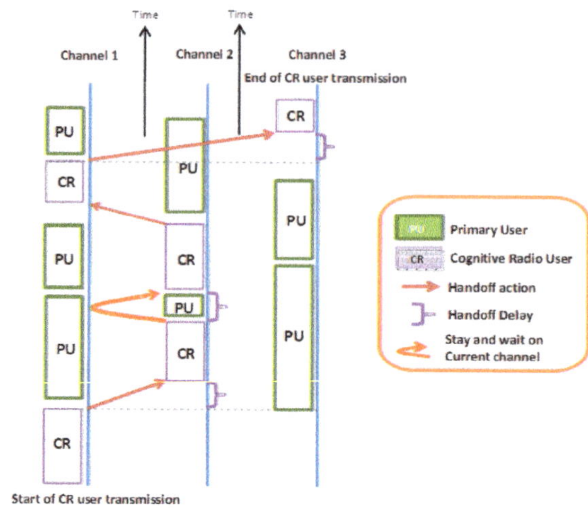

Figure 4. Time stream of the CRU transmission in the handoff scheme.

4. Simulation Setup and Performance Evaluation

4.1. Simulation Setup

The proposed PUTPOSH and some of the existing related approaches were implemented and simulated using Matlab. The existing approaches: (1) reactive handoff—DFHC [18]—(2) proactive handoff—fuzzy logic-based spectrum handoff (FLSH) [47]—and (3) hybrid handoff—DSA [53]—are selected to simulate with PUTPOSH. Two diverse Poisson procedures are used to produce PUs and SUs [20] in the CRN at constant time. The overall service time and the inter-arrival time are measured as non-integer instances of users. The data transfer of the low priority CRUs can be interrupted by the high priority PUs. To avoid the collision of the same priority CRUs during the channel access, the first come first serve scheduling strategy is used.

The parameters utilized to achieve simulation outcomes are listed in Table 2. The simulation was executed multiple times (generally about 1000 times) and the frequent values are taken to plot different graphs. We presume the packet length of SUs and PUs was 10 bytes for conducting the experiments. The overall service time of CRU included the waiting time, the processing time of the channel, the data transmission time and the sensing time respectively. The dealing out time of the channel is supposed to be 0.05 msec [10,12]. The arrival rates of PU and CRUs were followed by the Poisson processes. For simplicity, the arrival rate value of CRU was fixed to 0.1 and compared with the parameters at dissimilar values i.e., 0.02, 0.03, 0.04, 0.05, 0.06, 0.07, and 0.08 of PU interarrival rates. This disparity aspect helped us understand the behavior of a CRU in the proposed PUTPOSH scheme with various interarrival rates of the PU. Moreover, the overall service time of CRU and PU was taken as 0.5 and 0.4, respectively, and the order of the PU was considered a superior precedence than SU. Therefore, the overall service time of CRUs depends on the arrival rate of the PU and different arrival rates of PUs may effect the overall service time.

The simulation starts with the recognition of a PU by a CRU at a momentarily engaged licensed band. The CRU may make a decision to ensure a handoff to the next available channel or to stay and wait at the ongoing channel. The decision of transferring communication to a new vacant channel or may waiting in an ongoing channel depends firmly on the service time. When handoff was achieved, the overall service time for reactive and proactive decision was considered. With the overall least service point, CRU makes the decision to execute reactive or proactive handoff.

Table 2. Experimental settings: parameters with their values and units.

Parameters	of	Values	Units
Packet Length	Primary User	10	Bytes
Packet Length	Cognitive Radio User	10	Bytes
Processing Time	Channel	0.5	MSec
Arrival Rate	Primary User	0.02, 0.03, 0.04, 0.05, 0.06, 0.07, 0.08	Arrival per slot
Arrival Rate	Cognitive Radio User	0.01	Arrival per slot
Service Rate	Primary User	0.5	Slots per arrival
Service Rate	Cognitive Radio User	0.4	Slots per arrival

4.2. *Performance Evaluation*

The efficiency of the PUTPOSH approach is observed by considering the following attributes: overall service time, throughput, number of handoffs and channel utilization. To achieve comparison results, the proposed handoff technique is compared with the DFHC (reactive handoff), FLSH (proactive handoff), and DSA (hybrid handoff) approaches.

4.2.1. PU Detection through Sensing Module

The probability distribution of PU detection through the sensing module of PUTPOSH is shown in Figure 5. The energy recognition method is used to sense the PU arrival. The detection prospect depends on the signal-to-noise ratio (SNR). To measure the efficiency of any strategy, the detection probability is considered an importamt factor. However, the existing techniques in CR did not deliberate the detection probability. In this experiment, the detection probability is calculated to make sure that PUTPOSH works well with the changes in SNR. According to Figure 5, the PU recognition probability remains lofty with the higher values of the SNR (i.e., values larger than 0.8). This is due to the fact that energy recognition works proficiently well with the superior standards of SNR [37,53]. CRU acquires the link to deliver data for a particular communication is recognized as the overall time. The best state for CRU is to attain the minimum time for data delivery with minimum delay. PUTPOSH intelligently switches between reactive and proactive handoff decision in order to reduce the delivery time of data and consequently the threshold value of PU interarrival is extracted. With the lower value of SNR, the energy detector is unable to distinguish between PU signal and the noise. The solution to this problem is the use of sensing techniques that are based on eigenvalue centered methods. Eigenvalue-centered methods are simple and based on the low-complexity spectrum methods that are capable to distinguish between PU signal and the noise with lower SNR. The eigenvalue centered energy detection scheme performs in a decent way alongside the low-frequency selective channels and with the noise uncertainty. With the incorporation of the eigenvalue-based method, the performance of PUTPOSH is well-organized and efficient.

4.2.2. Discovering Threshold Value

The overall service time is basically the total time required by CRU to transmit the entire data at the destination. The perfect situation for CRU to broadcast information with a smaller amount of holdup to attain the least amount of overall service time. In order to reduce rapidly the overall service time, the presented system logically switches between reactive and proactive handoff judgments. In this way, the threshold value is discovered for the interarrival of the PU. The overall service time of the reactive and proactive decisions w.r.t. the handoff is illustrated in Figure 6. The total service time of a proactive handoff is lower than the reactive handoff up to 0.05 and the comparison is performed on the basis of PU interarrival rate. In the proactive method, the handoff process is performed prior to trigger the handoff event. Over the lower PU arrival rate, the proactive handoff shows better performance than the counterpart. When the speed of PU flows over 0.05, the reactive handoff method shows better results because the handoff process is completed after the handoff event. Therefore, threshold value 0.05 is extracted by PUTPOSH when it switches between proactive or reactive handoff.

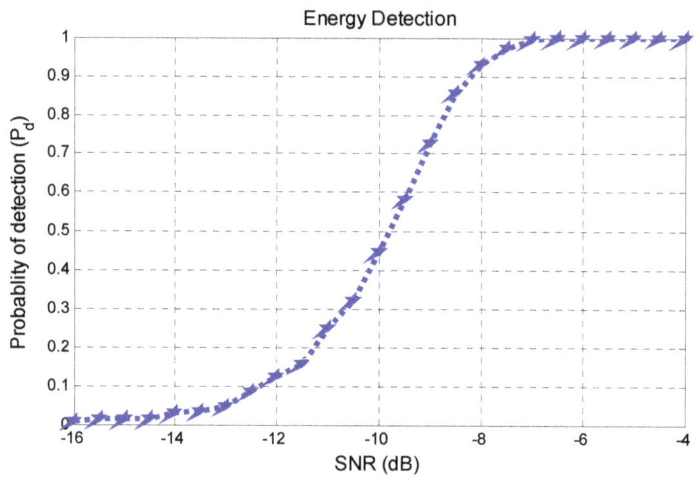

Figure 5. Distribution probability of PU detection over the signal to noise ratio.

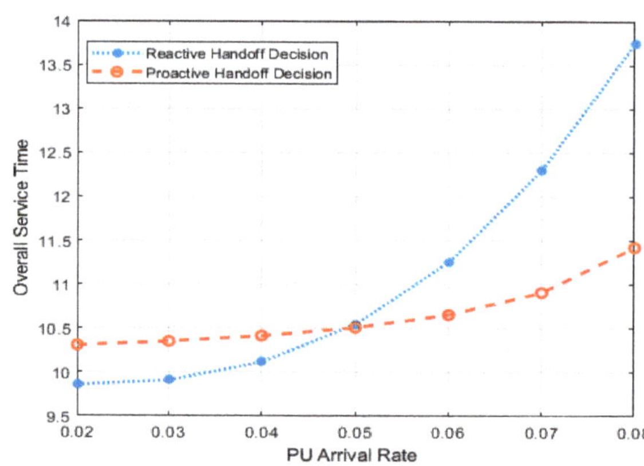

Figure 6. Threshold value detection to toggle among reactive and proactive handoff decisions.

4.2.3. Comparison between PUTPOSH and Existing Handoff Schemes

Overall Service Time

When the arrival rate of PU fits up to the threshold value, the PUTPOSH strategy switches to the reactive handoff and allows a superior rate to the proactive handoff. Similarly, when the threshold value remains or becomes higher than the arrival rate of PU, the PUTPOSH strategy allows CRU to move to the proactive handoff. In this case, PUTPOSH produces 12.80% better results than the reactive handoff (shown in Figure 7). Hence PUTPOSH, as a hybrid approach, uses the advantages of both the proactive and the reactive strategies whenever required.

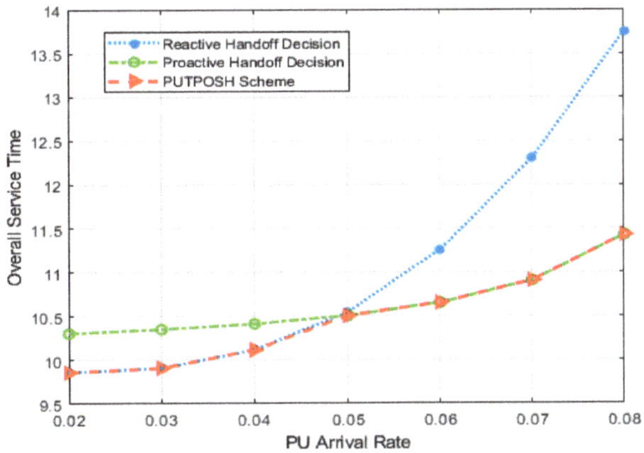

Figure 7. Comparison of primary user traffic pattern-based opportunistic spectrum handoff (PUTPOSH) with reactive and proactive handoffs w.r.t. the overall service time.

Number of Handoffs

Figure 8 represents the number of handoffs executed by the CRU. The existing handoff strategies, as compared to PUTPOSH, executed an additional quantity of handoffs; during each phase, the PU is detected and the handoff is executed by proactive and reactive approaches. On the interarrival of PU, the existing approaches worked by frequently changing between channels and had no mechanism of stay and wait to execute the transmission. The PUTPOSH scheme perceptively chooses whether to execute the handoff or not, conditional to the prediction of the overall service time. The intelligent decision to handle handoff operation helps PUTPOSH to achieve 14.28% better results than the existing DFHC (reactive handoff), FLSH (proactive handoff), and DSA (hybrid handoff) approaches. However, with the help of this way, the infertile handoffs actions are evaded.

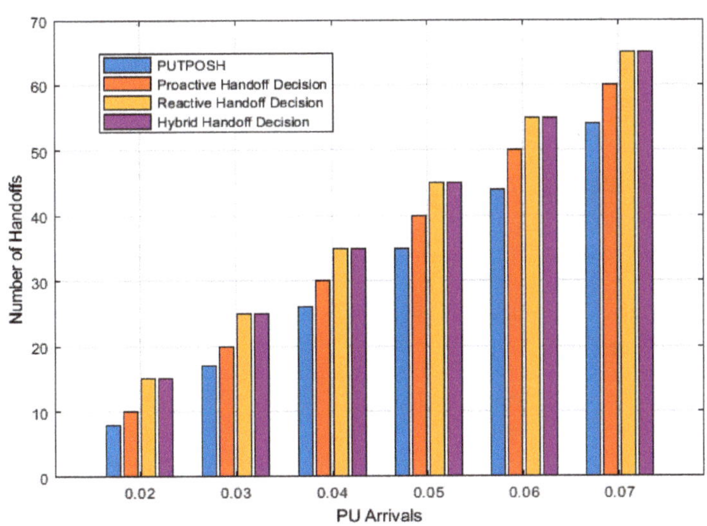

Figure 8. Comparison of PUTPOSH with the existing handoff schemes w.r.t. the number of handoffs.

4.2.4. Comparison of PUTPOSH with Stationary Spectrum Access and Hybrid Approaches

Channel Utilization

The PUTPOSH scheme is compared with the stationary spectrum access method in terms of channel consumption (see Figure 9). A fixed spectrum access approach is categorized by the spectrum access method where only licensed operators are permitted to exploit the channel. CRU cannot utilize the spectrum resourcefully, even when PU is lacking. The PUTPOSH scheme, being a lively spectrum access pattern, depicts extraordinary results in terms of channel consumption because it permits CRU to exploit the vacant spectrum in the absence of PU [2,18]. The solitary PU is permitted to exploit the band in the static spectrum access, and is effected by the underutilization of the channel. The comparison (in Figure 9) reveals that the PUTPOSH scheme produces 20% better results than the static spectrum access (for both reactive and proactive) approaches.

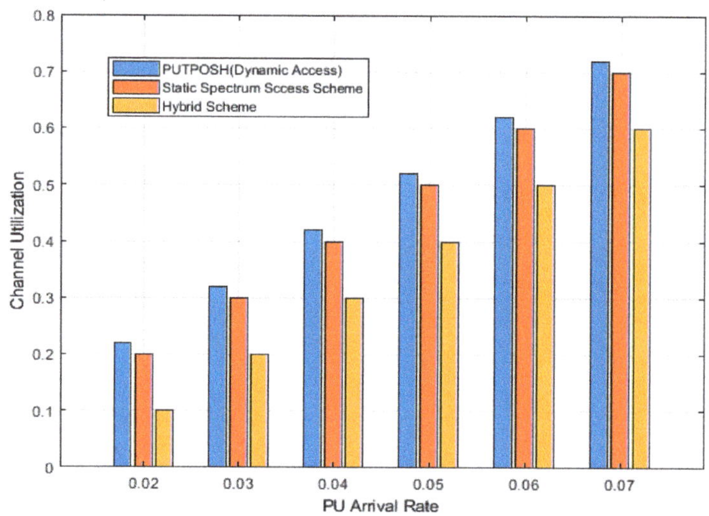

Figure 9. Comparison of PUTPOSH with stationary (reactive and proactive) spectrum access, and hybrid schemes w.r.t. the channel utilization.

4.2.5. Throughput Analysis of PUTPOSH as a Function of Primary User Arrivals

Figure 10 shows the throughput analysis of the PUTPOSH strategy. The throughput of primary, secondary and overall system is calculated in terms of the arrival rate of PU. The CRU has limited or opportunistic access to the channels and the lowest throughput is achieved against every PU arrival. At the channel, PU has priority access and yields greater throughput. The overall system throughput represents the combined throughput of a CRN system that can be achieved through PUTPOSH. With the PUTPOSH approach, the growing number of PU interarrivals due to the condensed total service period and the effective channel exploitation.

Table 3 summarizes the quantitative analysis of the existing and PUTPOSH schemes in a comparative manner. The comparison is based on the overall service time, the number of handoffs and the channel utilization, which are already discussed in the above subsections.

Table 3. Comparison of PUTPOSH and existing schemes based on overall service time, number of handoffs and channel utilization.

Comparison Parameters	PU Arrival Rate (Per Slot)	Reactive Handoff	Proactive Handoff	Hybrid Handoff	Proposed (Handoff)	Results (%) (Proposed is Better)
Overall Service Time	0.03 0.07	9.9 12.3	10.4 10.9	9.9 10.9	9.9 10.9	Has 12.6% superior rate than reactive,
Number of Handoffs	4	45	45	40	35	28.57% than proactive and reactive and 14.28% than hybrid schemes.
Channel Utilization	4	0.4 (static)		0.5 (Proposed—dynamic)		20% than existing

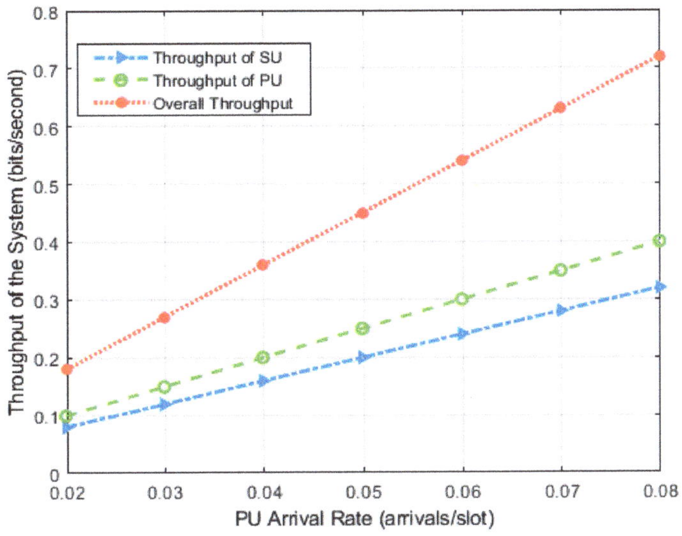

Figure 10. Throughput analysis of the PUTPOSH approach w.r.t. the function of PU arrivals.

4.2.6. Handoff Delay of PUTPOSH and Hybrid (DSA) Schemes

A DSA scheme applies proactive spectrum sensing and reactive handoff action jointly; firstly, the target channel selection is prepared beforehand or during CRU data transmission and secondly, the spectrum handoff is performed after the handoff triggering event. In PUTPOSH, a hybrid solution is referred to as "adaptive", if the decisions of channel selection and handoff are made by continuously monitoring the arrival and departure patterns of the PU. When PU moves quite regularly, a CRU may adapt to a reactive handoff strategy. While in the case of rare PU movements, a proactive handoff solution is preferred by the corresponding CRU. Figure 11 shows the comparison of the PUTPOSH scheme with a hybrid approach (DSA) in terms of handoff delay. The graph in Figure 11 shows that the DSA has more handoff delay than the PUTPOSH scheme because sensing is achieved prior to the arrival of the PU whereas the handoff action is performed after the PU's arrival. Therefore the cumulative handoff delay of the DSA remains high as compared to PUTPOSH. The queuing procedure in PUTPOSH adequately reduces the handoff delay because of its adoptive environment, and thus the cumulative delay becomes lower than the DSA.

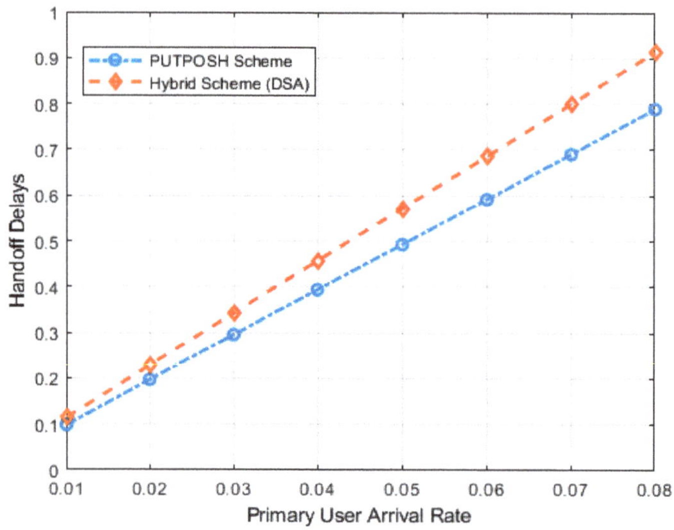

Figure 11. Comparison of proposed handoff and a hybrid approach (dynamic spectrum aggregation (DSA)) schemes in terms of handoff delay.

5. Conclusions

Spectrum Handoff is a vital part of the spectrum management process in CRNs. It is the method of moving the ongoing communication of a CRU to a vacant channel at the arrival of PU without a glitch. In this research, a new handoff (PUTPOSH) approach has presented—on the one hand, it maximizes the channel utilization and the throughput while on the other hand, it minimizes the overall service time and the number of handoffs in CRNs. In PUTPOSH, firstly, an energy detection sensing scheme was employed for the announcement of a PU in its licensed spectrum band. Secondly, a PUTPOSH pattern was projected, where a CRU intelligently switched between reactive and proactive handoffs depending upon the overall service time. Hence the presented approach has achieved the aids of both proactive and reactive schemes. The comparison of the PUTPOSH with the existing proactive, reactive and hybrid handoff strategies has also conducted. The results show that the PUTPOSH approach gets good performance in terms of overall service time—having a superior rate than the proactive and 12.8% robust than the reactive handoff. In terms of the number of handoffs, the PUTPOSH gives 14.28% better results than proactive (FLSH), reactive (DFHC) and hybrid (DSA) approaches. The PUTPOSH scheme produces 20% better results than the static spectrum access and throughput in CRNs. PUTPOSH can professionally reduce the overall service time and the number of fruitless handoffs of a CRU while keeping the channel busy and the system throughput at the highest level.

This work can be extended in a number of means; (1) a situation can be considered where a CRU can transfer its respective data on numerous accessible primary channels concurrently, (2) a management tool amongst the CRUs retrieving the primary spectrum holes can be developed, (3) CRU can interconnect to exchange the information regarding the movement of the PU and their stopover as well on a conforming channel to make the communication more reliable, and (4) spectrum handoff requires frequent spectrum sensing and channel information that takes a significant amount of power. Hence the energy-efficient spectrum sensing and mobility are still open challenges in CRNs.

Author Contributions: Conceptualization, K.A.; Data curation, M.K.B., N.A.M. and M.Z.; Formal analysis, I.H. and I.A.K.; Funding acquisition, M.K.B., S.N., A.D. and N.A.M.; Methodology, K.A., I.H. and I.A.K.; Resources, M.Z.; Software, K.A.; Supervision, I.H. and A.D.; Validation, M.K.B., S.N., A.D. and I.A.K.; Visualization, N.A.M.; Writing—original draft, K.A.; Writing—review & editing, I.H., S.N., M.Z. and I.A.K. All authors have read and agreed to the published version of the manuscript.

Funding: This research received no external funding.

Conflicts of Interest: The authors declare no conflict of interest.

References

1. Ghasemi, A.; Sousa, E.S. Asymptotic Performance of Collaborative Spectrum Sensing under Correlated Log-Normal Shadowing. *IEEE Commun. Lett.* **2007**, *11*, 34–36. [CrossRef]
2. Akyildiz, I.F. Spectrum Management in Cognitive Radio Networks. In Proceedings of the 2008 IEEE International Conference on Wireless and Mobile Computing, Networking and Communications, Avignon, France, 12–14 October 2008. [CrossRef]
3. Christian, I.; Moh, S.; Chung, I.; Lee, J. Spectrum Mobility in Cognitive Radio Networks. *IEEE Commun. Mag.* **2012**, *50*, 114–121. [CrossRef]
4. Sun, H.; Nallanathan, A.; Wang, C.; Chen, Y. Wideband Spectrum Sensing for Cognitive Radio Networks: A Survey. *IEEE Wirel. Commun.* **2013**, *20*, 74–81. [CrossRef]
5. Herath, S.P.; Rajatheva, N.; Tellambura, C. Energy Detection of Unknown Signals in Fading and Diversity Reception. *IEEE Trans. Commun.* **2011**, *59*, 2443–2453. [CrossRef]
6. Zhu, Q.; Wu, Y.; Tsang, D.H.K.; Peng, H. Cooperative Spectrum Sharing in Cognitive Radio Networks with Proactive Primary System. In Proceedings of the 2013 IEEE/CIC International Conference on Communications in China-Workshops (CIC/ICCC), Xi'an, China, 12–14 August 2013; pp. 82–87. [CrossRef]
7. Lin, Y. Jamming-Aware Randomized Spectrum Sharing in Cognitive Radio Communication Networks. In *Applied Science, Materials Science and Information Technologies in Industry*; Applied Mechanics and Materials; Trans Tech Publications Ltd.: Stafa-Zurich, Switzerland, 2014; Volume 513, pp. 834–840. [CrossRef]
8. Lee, W.; Akyildiz, I.F. Spectrum-Aware Mobility Management in Cognitive Radio Cellular Networks. *IEEE Trans. Mob. Comput.* **2012**, *11*, 529–542. [CrossRef]
9. Akyildiz, I.F.; Lee, W.-Y.; Vuran, M.C.; Mohanty, S. NeXt Generation/Dynamic Spectrum Access/Cognitive Radio Wireless Networks: A Survey. *Comput. Netw.* **2006**, *50*, 2127–2159. [CrossRef]
10. Wang, C.W.; Wang, L.C.; Adachi, F. Modeling and Analysis for Proactive-Decision Spectrum Handoff in Cognitive Radio Networks. In Proceedings of the 2009 IEEE International Conference on Communications, Dresden, Germany, 14–18 June 2009; pp. 1–6. [CrossRef]
11. Wang, C.; Wang, L. Analysis of Reactive Spectrum Handoff in Cognitive Radio Networks. *IEEE J. Sel. Areas Commun.* **2012**, *30*, 2016–2028. [CrossRef]
12. Wang, L.; Wang, C.; Chang, C. Modeling and Analysis for Spectrum Handoffs in Cognitive Radio Networks. *IEEE Trans. Mob. Comput.* **2012**, *11*, 1499–1513. [CrossRef]
13. Cacciapuoti, A.S.; Akyildiz, I.F.; Paura, L. Optimal Primary-User Mobility Aware Spectrum Sensing Design for Cognitive Radio Networks. *IEEE J. Sel. Areas Commun.* **2013**, *31*, 2161–2172. [CrossRef]
14. Yi, P.; Yong, Z. A Novel Spectrum Handoff Method Based on Spectrum Reservation. *TELKOMNIKA Indones. J. Electr. Eng.* **2014**, *12*, 653–660. [CrossRef]
15. Pandya, P.; Durvesh, A.; Parekh, N. Energy Detection Based Spectrum Sensing for Cognitive Radio Network. In Proceedings of the 2015 Fifth International Conference on Communication Systems and Network Technologies, Gwalior, India, 4–6 April 2015; pp. 201–206. [CrossRef]
16. Chatterjee, S.; Maity, S.P.; Acharya, T. On Optimal Threshold Selection in Cooperative Spectrum Sensing for Cognitive Radio Networks: An Energy Detection Approach Using Fuzzy Entropy Maximization. *Wirel. Pers. Commun.* **2015**, *84*, 1605–1625. [CrossRef]
17. Swetha, N.; Sastry, P.N.; Rao, Y.R. Analysis of Spectrum Sensing Based on Energy Detection Method in Cognitive Radio Networks. In Proceedings of the 2014 International Conference on IT Convergence and Security (ICITCS), Beijing, China, 28–30 October 2014; pp. 1–4. [CrossRef]
18. Hu, W.; Willkomm, D.; Abusubaih, M.; Gross, J.; Vlantis, G.; Gerla, M.; Wolisz, A. Cognitive Radios for Dynamic Spectrum Access-Dynamic Frequency Hopping Communities for Efficient IEEE 802.22 Operation. *IEEE Commun. Mag.* **2007**, *45*, 80–87. [CrossRef]
19. Cacciapuoti, A.S.; Caleffi, M.; Paura, L. On the Impact of Primary Traffic Correlation in TV White Space. *Ad Hoc Netw.* **2016**, *37*, 133–139. [CrossRef]
20. Lee, W.; Akyildiz, I.F. Optimal Spectrum Sensing Framework for Cognitive Radio Networks. *IEEE Trans. Wirel. Commun.* **2008**, *7*, 3845–3857. [CrossRef]

21. Jan, S.U.; Vu, V.-H.; Koo, I. Throughput Maximization Using an SVM for Multi-Class Hypothesis-Based Spectrum Sensing in Cognitive Radio. *Appl. Sci.* **2018**, *8*, 421. [CrossRef]
22. Khalid, W.; Yu, H. Sum Utilization of Spectrum with Spectrum Handoff and Imperfect Sensing in Interweave Multi-Channel Cognitive Radio Networks. *Sustainability* **2018**, *10*, 1764. [CrossRef]
23. Huang, J.; Zeng, X.; Jian, X.; Tan, X.; Zhang, Q. Opportunistic Capacity-Based Resource Allocation for Chunk-Based Multi-Carrier Cognitive Radio Sensor Networks. *Sensors* **2017**, *17*, 175. [CrossRef]
24. Liu, P.; Qi, W.; Yuan, E.; Wei, L.; Zhao, Y. Full-Duplex Cooperative Sensing for Spectrum-Heterogeneous Cognitive Radio Networks. *Sensors* **2017**, *17*, 1773. [CrossRef]
25. Wang, W.; Zhang, Z.; Huang, A. Spectrum Aggregation: Overview and Challenges. *Netw. Protoc. Algorithms* **2010**, *2*, 184–196. [CrossRef]
26. Mishra, S.M.; Sahai, A.; Brodersen, R.W. Cooperative Sensing among Cognitive Radios. In Proceedings of the 2006 IEEE International Conference on Communications, Istanbul, Turkey, 11–15 June 2006; Volume 4, pp. 1658–1663. [CrossRef]
27. Jiang, C.; Beaulieu, N.C.; Zhang, L.; Ren, Y.; Peng, M.; Chen, H. Cognitive Radio Networks with Asynchronous Spectrum Sensing and Access. *IEEE Netw.* **2015**, *29*, 88–95. [CrossRef]
28. Li, S.; Zheng, Z.; Ekici, E.; Shroff, N. Maximizing System Throughput by Cooperative Sensing in Cognitive Radio Networks. *IEEE/ACM Trans. Netw.* **2014**, *22*, 1245–1256. [CrossRef]
29. Haykin, S.; Moher, M. *Modern Wireless Communication*; Prentice-Hall, Inc.: Upper Saddle River, NJ, USA, 2004.
30. Joseph, S.M.; Vijayakumar, P. Energy Efficient Spectrum Sensing and Accessing Scheme for Zigbee Cognitive Networks. *Int. J. Online Eng.* **2014**, *10*, 59–62. [CrossRef]
31. Yin, C.; Tan, X.; Ma, L. A Hybrid Handoff Strategy Based on Dynamic Spectrum Aggregation in Cognitive Radio System. In Proceedings of the IEEE 2013 Tencon-Spring, Sydney, Australia, 17–19 April 2013; pp. 213–217. [CrossRef]
32. Ruttik, K.; Koufos, K.; J'antti, R. Detection of Unknown Signals in a Fading Environment. *IEEE Commun. Lett.* **2009**, *13*, 498–500. [CrossRef]
33. Sofotasios, P.C.; Rebeiz, E.; Zhang, L.; Tsiftsis, T.A.; Cabric, D.; Freear, S. Energy Detection Based Spectrum Sensing Over $\kappa-\mu$ and $\kappa-\mu$ Extreme Fading Channels. *IEEE Trans. Veh. Technol.* **2013**, *62*, 1031–1040. [CrossRef]
34. Atapattu, S.; Tellambura, C.; Jiang, H. MGF Based Analysis of Area under the ROC Curve in Energy Detection. *IEEE Commun. Lett.* **2011**, *15*, 1301–1303. [CrossRef]
35. Wu, Y.; Hu, F.; Zhu, Y.; Kumar, S. Optimal Spectrum Handoff Control for CRN Based on Hybrid Priority Queuing and Multi-Teacher Apprentice Learning. *IEEE Trans. Veh. Technol.* **2017**, *66*, 2630–2642. [CrossRef]
36. Hall, M.J.; Chamberlain, R.D. Using M/G/l Queueing Models with Vacations to Analyze Virtualized Logic Computations. In Proceedings of the 2015 33rd IEEE International Conference on Computer Design (ICCD), New York, NY, USA, 18–21 October 2015; pp. 78–85. [CrossRef]
37. Ghasemi, A.; Sousa, E.S. Spectrum Sensing in Cognitive Radio Networks: Requirements, Challenges and Design Trade-Offs. *IEEE Commun. Mag.* **2008**, *46*, 32–39. [CrossRef]
38. Wu, Y.; Yang, Q.; Liu, X.; Kwak, K.S. Delay-Constrained Optimal Transmission With Proactive Spectrum Handoff in Cognitive Radio Networks. *IEEE Trans. Commun.* **2016**, *64*, 2767–2779. [CrossRef]
39. Tandra, R.; Sahai, A. SNR Walls for Signal Detection. *IEEE J. Sel. Top. Signal Process.* **2008**, *2*, 4–17. [CrossRef]
40. Zeng, Y.; Liang, Y.C. Eigenvalue-Based Spectrum Sensing Algorithms for Cognitive Radio. *IEEE Trans. Commun.* **2009**, *57*, 1784–1793. [CrossRef]
41. Zeng, Y.; Liang, Y. Spectrum-Sensing Algorithms for Cognitive Radio Based on Statistical Covariances. *IEEE Trans. Veh. Technol.* **2009**, *58*, 1804–1815. [CrossRef]
42. Dikmese, S.; Renfors, M. Performance Analysis of Eigenvalue Based Spectrum Sensing under Frequency Selective Channels. In Proceedings of the 2012 7th International ICST Conference on Cognitive Radio Oriented Wireless Networks and Communications (CROWNCOM), Stockholm, Sweden, 18 October 2012; pp. 356–361. [CrossRef]
43. Dikmese, S.; Sofotasios, P.C.; Renfors, M.; Valkama, M. Subband Energy Based Reduced Complexity Spectrum Sensing Under Noise Uncertainty and Frequency-Selective Spectral Characteristics. *IEEE Trans. Signal Process.* **2016**, *64*, 131–145. [CrossRef]

44. Dikmese, S.; Wong, J.L.; Gokceoglu, A.; Guzzon, E.; Valkama, M.; Renfors, M. Reducing Computational Complexity of Eigenvalue Based Spectrum Sensing for Cognitive Radio. In Proceedings of the 8th International Conference on Cognitive Radio Oriented Wireless Networks, Washington, DC, USA, 8–10 July 2013; pp. 61–67. [CrossRef]
45. Verma, P.; Singh, B. Throughput Maximization by Alternative Use of Single and Double Thresholds Based Energy Detection Method. *Optik* **2016**, *127*, 1635–1638. [CrossRef]
46. Lertsinsrubtavee, A.; Malouch, N.; Fdida, S. Spectrum Handoff Strategies for Multiple Channels Cognitive Radio Network. In *Proceedings of the ACM CoNEXT Student Workshop*; CoNEXT '10 Student Workshop; Association for Computing Machinery: New York, NY, USA, 2010. [CrossRef]
47. Alhammadi, A.; Roslee, M.; Alias, M.Y. Fuzzy Logic Based Negotiation Approach for Spectrum Handoff in Cognitive Radio Network. In Proceedings of the 2016 IEEE 3rd International Symposium on Telecommunication Technologies (ISTT), Coimbatore, India, 29–31 March 2016; pp. 120–124. [CrossRef]
48. Yawada, P.S.; Dong, M.T. Intelligent Process of Spectrum Handoff/Mobility in Cognitive Radio Networks. *J. Electr. Comput. Eng.* **2019**, *2019*. [CrossRef]
49. Chen, Y.; Zhou, H.; Kong, R.; Zhu, L.; Mao, H. Decentralized Blind Spectrum Selection in Cognitive Radio Networks Considering Handoff Cost. *Future Internet* **2017**, *9*, 10. [CrossRef]
50. Akyildiz, I.F.; Wang, X.; Wang, W. Wireless Mesh Networks: A Survey. *Comput. Netw.* **2005**, *47*, 445–487. [CrossRef]
51. Aggarwal, M.; Velmurugans, T.; Karuppiah, M.; Hassan, M.M.; Almogren, A.; Ismail, W.N. Probability-Based Centralized Device for Spectrum Handoff in Cognitive Radio Networks. *IEEE Access* **2019**, *7*, 26731–26739. [CrossRef]
52. Maity, S.P.; Chatterjee, S.; Acharya, T. On Optimal Fuzzy C-Means Clustering for Energy Efficient Cooperative Spectrum Sensing in Cognitive Radio Networks. *Digit. Signal Process.* **2016**, *49*, 104–115. [CrossRef]
53. Lertsinsrubtavee, A.; Malouch, N.; Fdida, S. Spectrum Handoff Strategy Using Cumulative Probability in Cognitive Radio Networks. In Proceedings of the 2011 3rd International Congress on Ultra Modern Telecommunications and Control Systems and Workshops (ICUMT), Budapest, Hungary, 5–7 October 2011; pp. 1–7.
54. Gittins, J.; Glazebrook, K.; Weber, R. Multi-Population Random Sampling (Theory). In *Multi-Armed Bandit Allocation Indices*; John Wiley & Sons, Ltd.: Hoboken, NJ, USA, 2011; pp. 173–211. [CrossRef]
55. Hoque, S.; Arif, W. Performance Analysis of Spectrum Handoff under Heterogeneous Spectrum Environment in Ad Hoc and Centralized CR Networks. *Ad Hoc Netw.* **2019**, *91*, 101877. [CrossRef]

applied sciences

MDPI

Article

Velocity-Aware Handover Self-Optimization Management for Next Generation Networks

Abdulraqeb Alhammadi [1], Mardeni Roslee [1,*], Mohamad Yusoff Alias [1], Ibraheem Shayea [2] and Abdullah Alquhali [1]

1 Faculty of Engineering, Multimedia University, Selangor 63100, Malaysia; abdulraqeb.alhammadi@gmail.com (A.A.); yusoff@mmu.edu.my (M.Y.A.); alhadi1abd@gmail.com (A.A.)
2 Department of Electronics and Communication Engineering, Istanbul Technical University, Istanbul 34467, Turkey; ibr.shayea@gmail.com
* Correspondence: mardeni.roslee@mmu.edu.my

Received: 9 December 2019; Accepted: 20 January 2020; Published: 17 February 2020

Abstract: The fifth generation (5G) network is an upcoming standard for wireless communications that coexists with the current 4G network to increase the throughput. The deployment of ultra-dense small cells (UDSC) over a macro-cell layer yields multi-tier networks, which are known as heterogeneous networks (HetNets). HetNets play a key role in the cellular network to provide services to numerous users. However, the number of handovers (HOs) and radio link failure (RLF) greatly increase due to the increase in the UDSC in the network. Therefore, mobility management becomes a very important function in a self-organizing network to improve the system performance. In this paper, we propose a velocity-based self-optimization algorithm to adjust the HO control parameters in 4G/5G networks. The proposed algorithm utilizes the user's received power and speed to adjust the HO margin and the time to trigger during the user's mobility in the network. Simulation results demonstrate that the proposed algorithm achieves a remarkable reduction in the rate of ping-pong HOs and RLF compared with other existing algorithms, thereby outperforming such algorithms by an average of more than 70% for all HO performance metrics.

Keywords: handover; HetNets; mobility management; self-optimization

1. Introduction

Heterogeneous networks (HetNets) have gained considerable attention in the past few years. These networks consist of different types of cells, such as macro, pico, and femto, which are introduced to meet user demand. As stated by the third generation partnership project (3GPP), the main role of the HetNets is to improve the network performance in terms of boosting capacity and coverage [1]. In next generation cellular networks, an ultra-dense small cell, which comprises numerous small cells overlapped with macro cells, is introduced to boost coverage and improve user experience [2]. Non-standalone (NSA) and standalone (SA) are two road maps of transition from 4G to 5G mobile networks [3]. In the former, the 5G networks are deployed along with the existing 4G core network. Thus, the first stage of providing the 5G service starts with NSA. Then, 5G SA is implemented after the 5G coverage is completely established. Implementing a 5G network that overlaps with the 4G network leads to mobility issues when the users move from one base station (BS) to another [4,5]. Mobility management in HetNets is complex due to several types of inter-frequency technologies involved, which requires the user to perform the handover (HO) process while moving from one BS to another. Although the mobility robustness optimization (MRO) function can improve the user mobility experience in 4G networks, further improvement is necessary to address and resolve the issues of the HO probability (HOP), HO ping-pong (HOPP), and HO failure (HOF). The improvement can be implemented by optimizing the values of the HO control parameters (HCPs) according to user experience. Mobility management should be properly addressed to avoid service degradation due to high rates of HO, HOPP, and radio link failure (RLF). Furthermore, this issue should be resolved to ensure that the 5G of cellular networks can provide a seamless communication during user movement among different deployment scenarios [6,7]. Several functions are introduced in self-optimization networks, such as MRO and load balancing optimization [8]. Both functions perform optimization to achieve different objectives during user's mobility and aim to dynamically optimize the HCP values to handle different HO problems. For example, the MRO function automatically adjusts the HCP values to maintain system quality and performs automatic optimization for HCPs with minimal human intervention. The HCP consists of two main parameters: HO margin (HOM) and time to trigger (TTT). By adjusting these parameters to the proper values during user movements in cell coverage, the rates of HOPP and HOF are minimized, consequently improving the service quality.

Numerous studies have proposed different algorithms to address and solve the HO issue in HetNets. Ni et al. [9] have optimized HCPs based on user equipment (UE) velocity and HO types using mobility state estimation (MSE). In this work, the authors only updated the TTT based on the user velocity with limited updating values of TTT. This method did not fully optimize the HO performance since the gap between these updating values is very big and only three fixed values are selected in accordance with UE velocity. Similarly, Tiwari and Deshmukh [10] have presented an HO decision strategy and an MSE scheme to avoid unnecessary HOs (UHOs) and service failures in HetNets. The proposed model utilizes the number of HOs and sojourn time measurements to estimate the UE velocity. The simulation results show that the proposed MSE model reduces the number of UHOs and service failures. However, the performance with respect to other HO performance metrics, such as HOPP, RLF, and delay, have not been discussed. Researchers [11] have also introduced an adaptive algorithm that selects different values of HOM and load balancing for each UE in HetNets. The HO decision in the proposed algorithm depends on the signal-to-interference-plus-noise-ratio (SINR) rather than on the received signal strength indicator, which is then used to calculate the actual level of the HOM. Shayea et al. [12–14] have proposed an HO optimization technique based on a weighted function for carrier aggregation. The proposed algorithm automatically adjusts the values of the HOM in accordance with three functions: SINR, traffic load, and velocity. The simulation results have demonstrated that the proposed algorithm enhances the system performance in terms of spectral efficiency at cell edge and outage probability.

Three types of HCPs are considered in this work: TTT, measurement interval, and hysteresis, all of which are adjusted according to the number of HOPP performed in a measurement interval. A three-layer filter technique is applied to improve the HO performance, which requires information about the mobility condition and speed state of UE. Su et al. [15] have adopted a comparison method that compares two parameters, namely, cell boundary crossings and HO execution, to optimize the overall network performance. The HO decision on a target cell is completely dependent on the signal strength measurement. Saeed et al. [16] have developed a model to optimize HOM based on fuzzy logic for HetNets. The fuzzy logic consists of two inputs: call drop rate and load balancing index, both of which adapt the HOM for macro and small cells. In other work [17], the reinforcement learning concept is used to detect an HO in the network. Effective session HOs lead to low drop call rates and also reduce the HOF and HOPP. However, this technique only supports the UE mobility speed up to 120 km/h. In [18–20], several algorithms have been proposed to investigate and evaluate the issue of mobility management in different mobile speed scenarios. Three types of HO are considered to adapt the HCPs: too early, too late, and HO to wrong cells. The results demonstrated that the adapted HCPs reduce the rates of HOPP, HOP, and RLF.

The major contributions of this paper can be summarized as follows. First, we formulate and address the issue of MRO in HetNets by focusing on two performance main metrics: RLF and HOPP, which lead to service interruption. Then, we propose a velocity-based optimization algorithm to continuously adjust the HCPs based on the condition of UE. Finally, we evaluate and compare the performance of the proposed algorithm to other existing related works. The remainder of this paper is organized as follows. Section 2 explains the system model that used for this work. The proposed velocity-based algorithm is presented in Section 3. Section 4 presents the simulation and performance evaluation. The conclusions are provided in Section 5.

2. System Model

We consider a HetNet architecture that comprises several 4G macro cells and a 5G small cell BS where three small cells are uniformly distributed in each macro BS. Each macro BS has a three-sectored hexagonal layout, where each of the sectors operates as an individual cell and the small BSs are omni-directional single-sector cells. Figure 1 displays an example of the HetNets deployment scenario with three small cells placed on one macro cell. R and r represent the radius of macro and small cells, respectively.

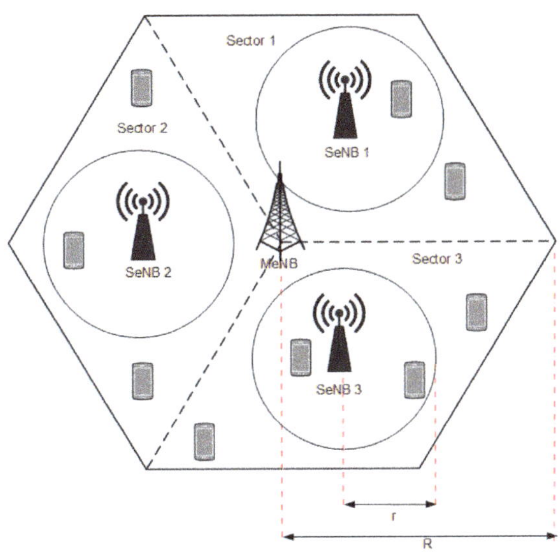

Figure 1. System model for HetNets.

The macro and small BSs operate at low and high frequency bands, receptivity, with reuse frequency factor being equal to one. Table 1 lists the notations used in this paper. The set of macro and small cells are defined as $N_k = 1, ..., N_m$, $N_l = 1, ..., N_n$, respectively. The user set is denoted as N_u where $U = 1, ..., U$, where U is randomly distributed in the network with a random mobility model. The UEs receive requested traffic through either the macro or small cells. At each small and macro cell, a distributed self-organizing network collects HO information and optimizes HCPs. The execution of the HO procedure initializes when a UE moves from a serving cell to a target cell in the same or different network. The serving cell makes an HO decision based on the measurement report (MR) from the UE to begin the HO process to a target cell. The path loss model for different bands in the urban area between a BS, and the user is expressed as [21]:

$$PL_{u,k,l} = 20\log_{10}(\frac{4\pi r_0}{\lambda_l}) + 20\log_{10}(\frac{d_{u,k}}{d_0}) + \chi, \tag{1}$$

where:

$$BS = \begin{cases} small\ cell & if \quad l = 1 \\ macro\ cell & otherwise \end{cases}$$

where d_0 and $d_{u,k,l}$ represent the reference distance and the distance between the user U and BS k, respectively ($d_{u,k,l} \geq d_0$; d_0 is assumed to be 50 m). λ_l is the wavelength at the carrier frequency $f_{c,l}$. χ is a Gaussian random variable with zero mean and variance σ^2.

The maximum quality of service (QoS) requirements limit the interference by reducing the RLF. The performance of each UE u should meet the minimum data rate requirement for QoS satisfaction. For channel modeling, the SINR experienced by UE u is modeled as [22,23]:

$$SINR_{u,k,l} = \frac{p_{u,k,l} g_{u,k,l} b_{ij}}{\sum_{i \in K \backslash \{k\}} \sum_{j \in U \backslash \{u\}} p_{ij} g_{u,k,l} + P_{AWGN}}, \tag{2}$$

where $p_{u,k,l}$ is the received signal power at u. $g_{u,k,l}$ is the channel gain experienced by UE u at k. b_{ij} is the binary association indicator of user u ($b_{ij} = 1$ indicates that user u associates with one BS. Otherwise, $b_{ij} = 0$.). p_{ij} represents the interference of the received signal power by UE u at k. P_{AWGN} is the additive white Gaussian noise power.

Table 1. List of notations.

Notation	Description
N_u	Set of UE $U \in 1, 2, ..., \|U\|$
N_k	Number of macro BSs
N_l	Number of small BSs
V_u	Velocity for UE u
$PL_{u,k,l}$	Path loss model
d_0	Reference distance between UE u and BS
d_u, k, l	Distance between UE u and BS k
λ_l	Wavelength at carrier frequency $f_{c,l}$
χ	Gaussian random variable with zero mean and variance σ^2
$SINR_{u,k,l}$	SINR experienced by UE u at BS k
b_{ij}	Binary association indicator of user u
$g_{u,k,l}$	Channel gain for UE u at BSk
$p_{u,k,l}$	Received signal power by UE u at BS k
P_{AWGN}	AWGN power
θ_{th}	Threshold level for HO decision
$V_{u,t}$	UEu Speed level at time t
V_r	Reference speed (medium speed)for entire simulation
$RSRP_S$	Serving RSRP UE u at BS k
$RSRP_t$	Target RSRP for UE u at BS k
$\Delta\mathcal{HOM}_{u,t}$	Updated HOM for each UE u at time t
$\Delta\mathcal{TTT}_{u,t}$	Updated TTT for each UE u at time t
N_F	Number of failure HO
N_C	Number of successful HO

3. Velocity-Based Optimization Algorithm

3.1. HO Problem

Improper configurations of HCPs increase the rate of RLF, thereby degrading the system performance. The high and low values of HCPs cause too late and too early HOs, respectively. Thus, the HCPs should be frequently adjusted according to the UE's mobility. In this work, the adjustment of the HCPs depends on the UE's speed and the reference signal received power (RSRP), which causes high HOPP and RLF. For example, when a user moves at high speeds, numerous cells (macro or small cells) will be crossed. In this case, the value of the HCPs should be decreased to avoid too late HO. By contrast, when a user moves at low speeds, it experiences a short distance movement and attains a good signal quality; thus, high values of HCPs are required to avoid too early HO. The objective of this work is to minimize the probability of occurrence of RLFs and HOPP during the HO processes by adjusting the HCPs, which can be mathematically expressed as:

$$\underset{\mathcal{TTT},\mathcal{HOM}}{\text{argmin}} \quad P(\mathcal{TTT},\mathcal{HOM}) \tag{3a}$$

$$Subject \quad to: \quad \sum_{j=1}^{N} b_{ij} = 1, \forall_j \tag{3b}$$

$$TTT_{\min} \leq \mathcal{TTT} \leq TTT_{\max} \quad TTT \in (0, ..., 5260ms) \tag{3c}$$

$$HOM_{\min} \leq \mathcal{HOM} \leq HOM_{\max} \quad HOM \in (0, ..., 10dm) \tag{3d}$$

$$v_{\min} \leq v_{u,t} \leq v_{\max} \tag{3e}$$

$$b_{ij} \in 0, 1, \forall_{ij} \tag{3f}$$

where P refers to the resulting probability of HOPP, RLF, and HOF, which control by a proper selection of $\mathcal{TTT}$ and $\mathcal{HOM}$. Constraint (3b) ensures each monitoring user U is associated with one BS k; (3c) and (3d) ensure that the adjusted values of $\mathcal{TTT}$ and $\mathcal{HOM}$ are within the bounded range. (3e) ensures the UE speed at time $v_{u,t}$ within the bounded range $(v_{\min}, v_{\max})$. The last constraint is (3f), which refers to binary constraint that involves exactly two variables to indicate the user association. The procedure HO begins after the serving BS receives the MRs from a UE that makes the HO decision. The UE periodically measures the RSRPs of all serving BSs (every 50 ms) and reports a measurement-triggered HO if a certain condition is satisfied. In this work, all considered events in the 3GPP TR 36.331 are configured for measurement reporting to simulate the network as a realistic environment [24]. The A3 event is triggered when a serving BS becomes worse than the target BS on the basis of a certain margin level. The suitable instance for the UE to report the measurements is immediately after the expiration of the TTT timer. The condition is described as:

$$RSRP_S > RSRP_T + \mathcal{HOM}_{S \to T}, \tag{4}$$

where $RSRP_S$ and $RSRP_T$ are the averaged values RSRP measured for serving BS and target BS respectively. $\mathcal{HOM}_{S \to T}$ is the HOM from serving BS to target BS.

3.2. Proposed Algorithm

To address the HO optimization, we propose a distributed velocity-based optimization algorithm to enhance the system performance. The proposed algorithm is an enhancement of our previous work in [19,22]. It depends on the monitoring of the two considered parameters (i.e., UE speed and RSRP) that exert a high impact on HO performance. The adapted values of HCPs depend on these parameters, which are periodically optimized over time. Algorithm 1 describes the proposed velocity-based optimization algorithm. The initial values of TTT and HOM are set to 100 ms and 2 dB, respectively.

The algorithm starts with the condition selection according to the measurements of $RSRP_S$ and $RSRP_T$ pulse the threshold level θ_{th}, which is assumed to be 2 dB. Then, the $\hat{\mathcal{HOM}}$ and $\hat{\mathcal{TTT}}$ are adjusted in accordance with the three UE speed states: low, medium, and high. We considered four mobile speeds scenarios: 40, 80, 120, and 160 km/h, where the medium speed is assumed to be V_r = 70 to 90 km/h. $\Delta\mathcal{HOM}_t$ and $\Delta\mathcal{TTT}_t$ represent the adaptive HOM and TTT, respectively, where they are updating for each UE in cell to avoid the RLF in each simulation time. α and β are the step levels to adapt the values of TTT and HOM in next simulation time which carry values of 50 ms and 1 dB, respectively. The abovementioned steps are implemented when the HO condition is encountered. Table 2 shows the increase and decrease in the step levels according to mobile speeds. The flowchart of the proposed algorithm is illustrated in Figure 2.

Algorithm 1 Velocity-based optimization algorithm

```
begin
    Set up initial values of TTT and HOM
    if Simulation time t = 1 then
        HO Decision ⟵ false
    else
        if RSRP_S > RSRP_T + θ_th then
            Update HOM and TTT according to condition 1.
            Run Algorithm 2.
            Update HOM and TTT
        else if RSRP_S < RSRP_T + θ_th then
            Update HOM and TTT according to condition 3.
            Run Algorithm 2.
            Update HOM and TTT
        else
            Update HOM and TTT according to condition 2.
            Run Algorithm 2.
            Update HOM and TTT
        end if
        ΔHOM = HOM_{t+1}
        ΔTTT = TTT_{t+1}
        calculate HOF, HOPP, RLF
    end if
end
```

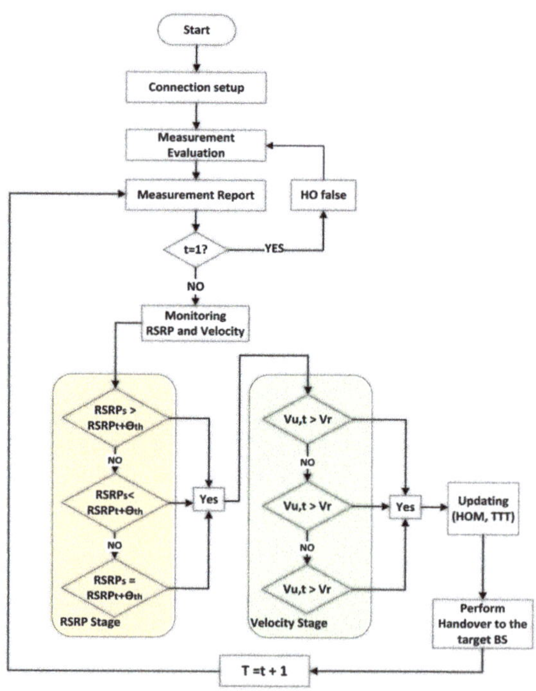

Figure 2. Flowchart of the proposed algorithm.

Table 2. Adjusting the table of HCPs.

Conditions	Speeds	$\Delta\mathcal{HOM}$	$\Delta\mathcal{TTT}$
	$V_{u,t} > V_r$	−Step	−Step
Condition 1	$V_{u,t} = V_r$	+Step	+Step
	$V_{u,t} < V_r$	+Step	+Step
	$V_{u,t} > V_r$	−Step	−Step
Condition 2	$V_{u,t} = V_r$	None	None
	$V_{u,t} < V_r$	+Step	+Step
	$V_{u,t} > V_r$	−Step	−Step
Condition 3	$V_{u,t} = V_r$	−Step	−Step
	$V_{u,t} < V_r$	−Step	−Step

The UE sends the MR to the serving BS during their mobility in the network. Then, the serving BS checks if the serving RSRP of the UE meets one of the above conditions. If the serving RSRP is greater than the target RSRP plus the threshold θ_{th}, then the current $\Delta\mathcal{HOM}_t$ and $\Delta\mathcal{TTT}_t$ are adjusted by adding the current values to the previous values of $\mathcal{HOM}_{t-1}$ and $\mathcal{TTT}_{t-1}$ according to UE speed. After updating the HCPs, the proposed algorithm will check if the HO decision is true or false. Algorithm 2 elaborates the HO trigger and decision process.

Algorithm 2 HO Trigger and Decision

if then $RSRP_{TBS} > RSRP_{SBS} + \Delta\mathcal{HOM}_t$
    if then $Trigger\ \ timer \geq TTT$
        $HO\ \ Decision \longleftarrow True$
        Send HO request
    else if
        then $HO\ \ Decision \longleftarrow false$
        Run Trigger Timer
    end if
end if

Finally, the main HO performance metrics are calculated and evaluated. The averages of HOP, HOPP, and HOF in each simulation for all UEs are calculated as:

$$\overline{HOP} = \frac{\sum_{j=1}^{N_u} HOP}{N_u} \quad \forall\ j^{th}UE, \tag{5}$$

where N_u denotes number of UEs overall simulation time. The average of the HOPP per UE are expressed as follows:

$$\overline{HOPP} = \frac{N_{HOPP}}{N_C + N_F}, \tag{6}$$

where N_{HOPP} represents the number of HOPPs over the entire simulation. N_F and N_C represent the number of failed and successful HO, respectively.

RLF is considered when the UEs lose connectivity with the BS during the HO process. The main source of RLF is the failure during HO initialization, which disrupts or causes the radio link to fail. The average probability of the RLF from all UEs is expressed as:

$$\overline{P(RLF)} = \frac{\sum_{j=1}^{N_u} P(RLF)}{N_u} \quad \forall\ j^{th}UE \tag{7}$$

4. Simulation and Performance Analysis

4.1. Simulation Scenario

The performance of the proposed algorithm is evaluated in a realistic cellular communication environment using MATLAB. The simulation environment consists of a two-tier model with multiple small cells that are uniformly distributed within each macro cell coverage [1,22,25]. The network model enables multi-radio access technologies with multi-technology that allow the UEs to connect to one BS. The UEs are randomly distributed in the network and move in random directions within a coverage area. All UEs move with the same speed in four possible directions in each scenario. We consider a line-of-sight connection where no restrictive assumptions are drawn about any obstacles. Table 3 presents the main simulation parameters. As previously mentioned, four mobile speed scenarios are considered to represent the average vehicle speeds in the urban areas.

Table 3. Simulation parameters.

Parameter	Value	
	4G Macro Cell	**5G Small Cell**
Carrier frequency (GHz)	2.1	28 [26]
Number of BS	61	183
Number of UEs/BS	100	200 [7]
Cell radius (m)	500	200
Cell height (m)	25	15
System bandwidth (MHz)	20	500
Transmit power (dBm)	46	30 [7]
Shadowing standard deviation (dB)	8	10
Simulation area (Km^2)	8×8	
UE height (m)	1.5	
Mobility model	Random direction model	
Noise figure (dB)	9	
Thermal noise density (dBm/Hz)	−174	
TTT (ms)	Adaptive	
HOM (dB)	Adaptive	
HO execution time (ms)	50	

4.2. Performance Evaluation

4.2.1. Fixed Values of HCPs

The HO performance of various fixed HCPs is investigated and validated in this section. The performance represents the average values computed over all UEs in the cells throughout the simulation cycles under various UE speeds. The impact of the different fixed HOM levels on the system performance under different UE speeds are also investigated. The mobility robustness can be investigated using differentiated HOM and TTT settings for all UEs' mobile scenarios as shown Table 4. The performance of the different settings of HCPs is analyzed using two key performance indicators, namely, average HOP and average probability of HOPP.

The effect of user mobility on the system performance is investigated. The average experienced HOP is obtained under different settings of HOM and TTT with respect to different mobile speed scenarios.

Table 4. Fixed HOM and TTT.

Set	HCPs	
	HOM (dB)	**TTT (ms)**
Set1	5	480
Set2	10	2560
Set3	8	1500
Set4	2	512
Set5	10	320
Set6	0	40

Figure 3 shows the average HOP over all mobile speeds and simulation times. In general, the reduction in the HOP leads to a significant reduction in the HOPP and the HOF. This phenomenon occurs because at low HOM levels and TTT intervals, UEs can perform early HO to the target cell. By contrast, high HOM level reduces the possibility of UE HO. The result also shows that the Set2 using a relatively high HOM and long TTT (HOM = 10, TTT = 2560) provides a significant reduction of the average HO probability compared to other settings' overall simulation time. However, the high HOP recorded in Set6 using a low value of HOM and short time of TTT (HOM = 0, TTT = 40).

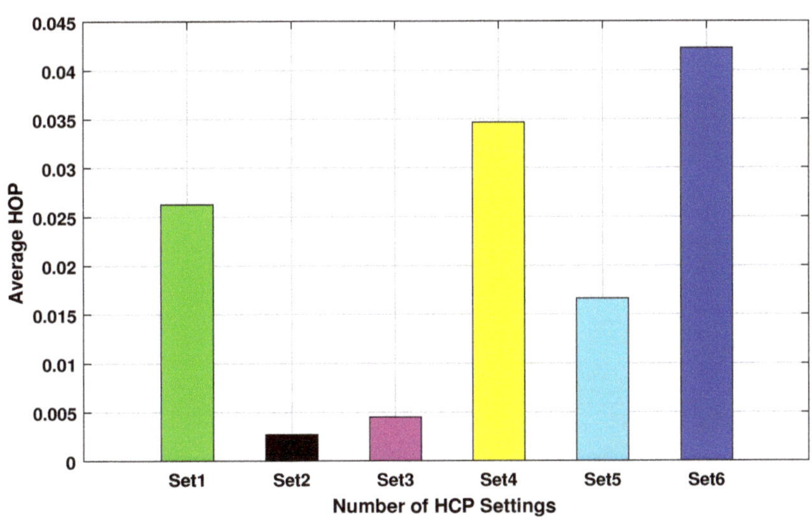

Figure 3. Overall average HOP vs. HCP setting scenarios.

Figures 4 and 5 display the average probability of HOPP versus different mobile speed scenarios. The result shows that the highest achieved rate of HOPP is obtained by Set 6, which can be attributed to the low values of HCPs that cause the UE to bounce between the serving and target BSs. The lowest rate of the HOPP is obtained by Set 2 due to the high values of HCPs.

The behavior of HOP and HOPP with different values of HOMs and TTTs is consistent with the effect of HOM on both parameters. At high TTT intervals, the HOP and HOPP rates decrease and reduce the signaling overhead, whereas at low intervals, both rates increase and raise the signaling overhead. The high UE speed may also cause a high RLF rate because neither the serving BS nor the target BS serves the UE.

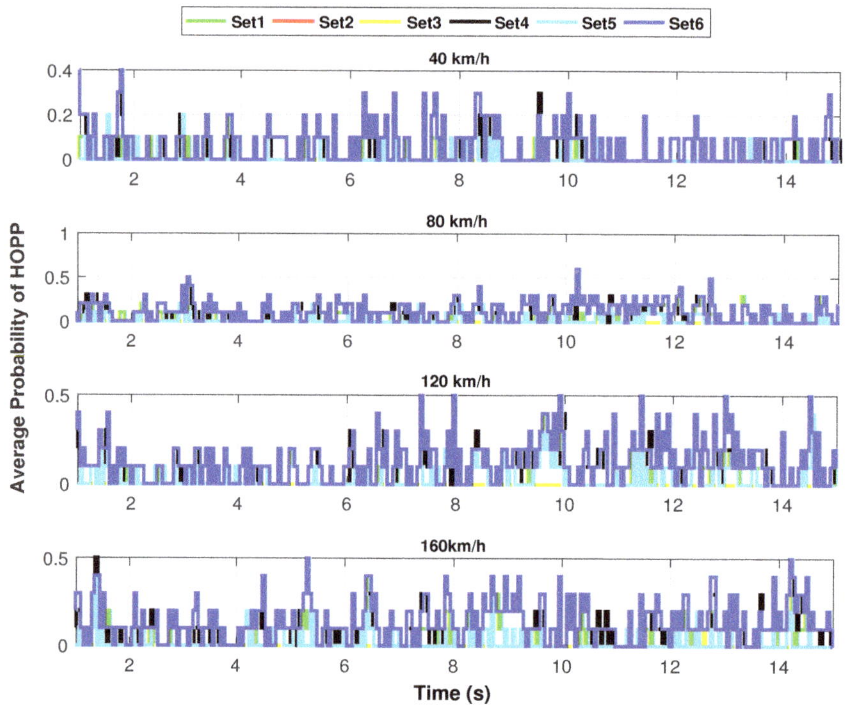

Figure 4. Average probability of HOPP vs. time.

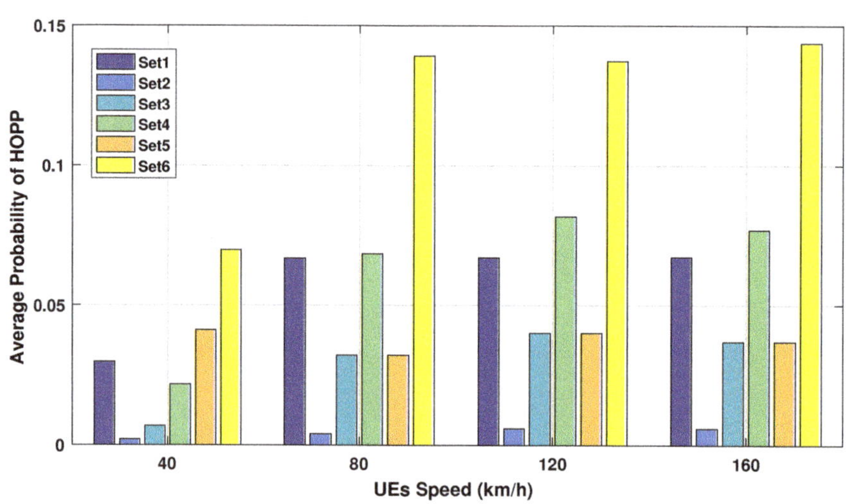

Figure 5. Average probability of HOPP vs. UE speeds.

4.2.2. Proposed Algorithm

To analyze the performance of the proposed algorithm, simulations are performed under different mobile speeds. The proposed algorithm is compared with a dynamic algorithm and with those proposed by Ray et al. [11] and Nie et al. [9] algorithms. Figure 6 depicts the average HOP of all algorithms under different UE speeds. The proposed algorithm significantly reduces the average HOP compared with the three algorithms under all speeds. The overall average HOP obtained by the proposed algorithm is 70% less than that obtained by the other algorithms.

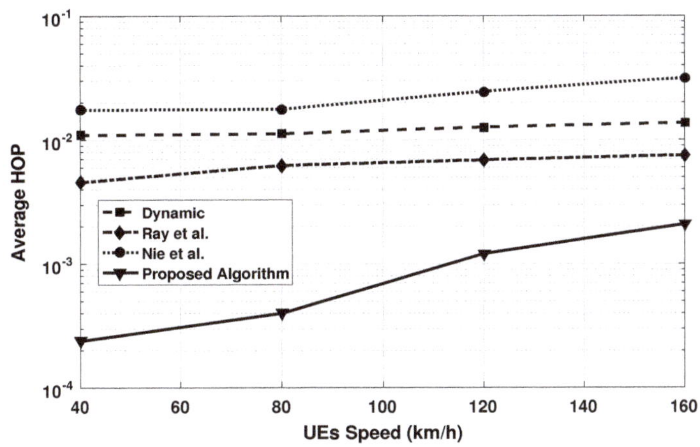

Figure 6. Average HOP with varying UE speeds.

Figure 7 illustrates the average HOPP probability with respect to the UE speeds over the entire simulation time. The simulation results demonstrate that the proposed algorithm obtains a lower HOPP reduction rate than other algorithms under all UE speeds because the other algorithms do not effectively optimize the HCPs according to UE experience, especially when the UEs move near the cell edge. Consequently, the number of UHOs increase due to the high HOPP rate, especially when the UE moves in high speed. Moreover, a high rate of HOP may result in increased HOPP and HOF rates, where a low rate may decrease the HOPP rate.

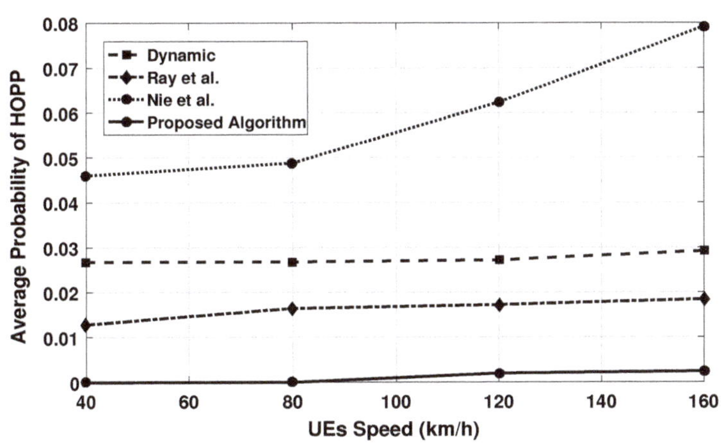

Figure 7. Average probability of HOPP with varying UE speeds.

Figure 8 shows the average probability of RLF with respect to UE mobile speed, which presents the frequent rate of disconnection radio links between BSs and UE mobility. The average probability rate of RLF is obtained from the overall simulation time for each UE speed. The results imply that the proposed algorithm achieves a significant reduction in RLF compared with other algorithms under all mobile speed scenarios. The algorithm proposed by Nie et al. achieves a higher RLF rate than other algorithms because only the HOM is adjusted according to UE speed. Therefore, both HCPs should be considered to manage the HO decision according to UE experience. In summary, the RLF rate of all compared algorithms gradually increases when the UE speed increases due to Doppler Effect and weak radio link connections. The proposed algorithm outperforms the other algorithms by more than 70%.

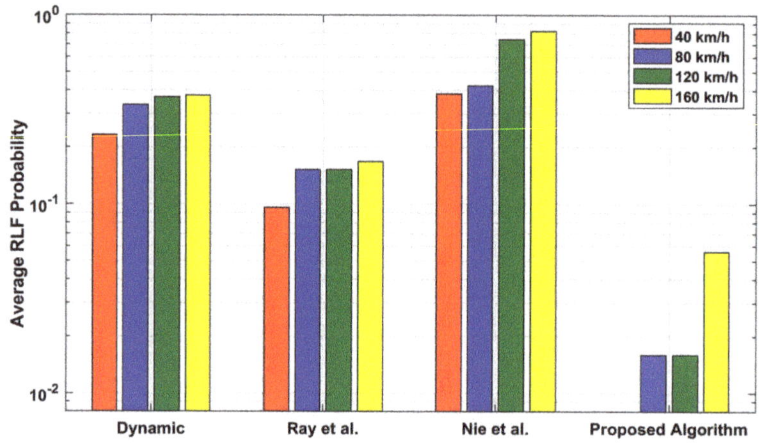

Figure 8. Average probability of RLF for different algorithms.

5. Conclusions

In this paper, a velocity-based self-optimization algorithm is proposed to adjust the HCP values in accordance with the UE speed and RSRP. The performances of HO-fixed HCP values and the proposed algorithm are investigated and evaluated. The advantages and disadvantages of the fixed HCPs are highlighted and discussed in terms of the total probability rate of HOP, HOPP, and RLF. The performance of the fixed HCPs results in the introduction of adaptive algorithms, which provides a satisfactory estimation of the HCP values. The performance of the proposed algorithm is analyzed under different mobile speeds and compared with other existing algorithms. The simulation results demonstrate that the proposed algorithm improves the overall system performance under all mobile speed scenarios and outperforms the existing algorithms by more than 70%. In future research, the proposed algorithm will be further improved by considering the traffic load of the serving BS and the target BS to increase the accuracy of the HO decision.

Author Contributions: Conceptualization, A.A. (Abdulraqeb Alhammadi), M.R and I.S.; methodology, A.A. (Abdulraqeb Alhammadi), M.R and M.Y.A; software, A.A. (Abdulraqeb Alhammadi), I.S.; validation, A.A. (Abdulraqeb Alhammadi), M.Y.A., and I.S.; formal analysis, A.A. (Abdulraqeb Alhammadi) and I.S.; investigation, A.A. (Abdulraqeb Alhammadi); writing—original draft preparation, A.A. (Abdulraqeb Alhammadi), M.R., I.S., and M.Y.A; writing—review and editing, I.S. and A.A (Abdullah Alquhali); supervision, M.R. and M.Y.A.; project administration, M.R.; funding acquisition, M.R. All authors have read and agreed to the published version of the manuscript.

Funding: This research was funded by Fundamental Research Grant Scheme (FRGS), Grant No. MMUE/170017, Under Ministry of Education, Malaysia.

Conflicts of Interest: The authors declare no conflict of interest.

Abbreviations

The following abbreviations are used in this manuscript:

3GPP	3rd generation partnership project
4G	Fourth generation
5G	Fifth generation
AWGN	Additive white Gaussian noise
BS	Base station
HCP	Handover control parameter
HetNets	Heterogeneous networks
HO	Handover
HOF	Handover failure
HOM	Handover margin
HOP	Handover probability
HOPP	Handover ping-pong
Mmwave	Millimeter wave
MR	Measurement report
MRO	Mobility robustness optimization
NSA	Non-standalone
QoS	Quality of service
RLF	Radio link failure
RSRP	Reference signal received power
SA	Standalone
SINR	Signal-to-noise-ratio
TTT	Time to trigger
UDSC	Ultra-dense small cell
UE	User equipment
UHO	Unnecessary handover

References

1. *Universal Mobile Telecommunications System (UMTS), Mobility Enhancementsin Heterogeneous Networks;* 3GPP TR 36.839, Tech. Report; 3GPP: Valbonne, France, 2012.
2. Ding, M.; Lopez-Perez, D.; Claussen, H.; Kaafar, M.A. On the fundamental characteristics of ultra-dense small cell networks. *IEEE Netw.* **2018**, *32*, 92–100. [CrossRef]
3. *Technical Specification Group Services and System Aspects*; Release 15 Description; 3GPP TR 21.915, Tech. Report; 3GPP: Valbonne, France, 2019.
4. *User Equipment (UE) Radio Transmission and Reception; Part 1: Range 1 Standalone*; 3GPP TR 38.101, Tech. Report; 3GPP: Valbonne, France, 2018.
5. Ericsson, H.E. *Modified RRH Arrangement for HST SFN*; 3GPP: Valbonne, France, 2019.
6. Castro-Hernandez, D.; Paranjape, R. Optimization of Handover Parameters for LTE/LTE-A In-building Systems. *IEEE Trans. Veh. Technol.* **2017**, *67*, 5260–5273. [CrossRef]
7. Zhang, H.; Huang, W.; Liu, Y. Handover Probability Analysis of Anchor-Based Multi-Connectivity in 5G User-Centric Network. *IEEE Wirel. Commun. Lett.* **2018**, *8*, 396–399. [CrossRef]
8. *Evolved Universal Terrestrial Radio Access (e-utra); User Equipment (ue) Procedures in Idle Mode*; release 14 (Tech. Rep. No. TS36.304); 3GPP: Valbonne, France, 2018.
9. Nie, S.; Wu, D.; Zhao, M.; Gu, X.; Zhang, L.; Lu, L. An Enhanced Mobility State Estimation Based Handover Optimization Algorithm in LTE-A Self-organizing Network. *Procedia Comput. Sci.* **2016**, *52*, 270–277. [CrossRef]
10. Tiwari, R.; Deshmukh, S. Analysis and Design of an Efficient Handoff Management Strategy via Velocity Estimation in HetNets. *Trans. Emerg. Telecommun. Technol.* **2019**, e3642. [CrossRef]
11. Ray, R.P.; Tang, L. Hysteresis Margin and Load Balancing for Handover in Heterogeneous Network. *Int. J. Future Comput. Commun.* **2016**, *4*, 231. [CrossRef]

12. Shayea, I.; Ismail, M.; Nordin, R.; Ergen, M.; Ahmad, N.; Abdullah, N.F.; Alhammadi, A.; Mohamad, H. New weight function for adapting handover margin level over contiguous carrier aggregation deployment scenarios in LTE-advanced system. *Wirel. Pers. Commun.* **2019**, *108*, 1179–1199. [CrossRef]
13. Shayea, I.; Ismail, M.; Nordin, R.; Mohamad, H.; Abd Rahman, T.; Abdullah, N.F. Novel Handover Optimization with a Coordinated Contiguous Carrier Aggregation Deployment Scenario in LTE-Advanced Systems. *Mob. Infor. Syst.* **2016**, *2016*, 4939872. [CrossRef]
14. Shayea, I.; Ismail, M.; Nordin, R.; Mohamad, H. Adaptive handover decision algorithm based on multi- influence factors through carrier aggregation implementation in LTE-advanced system. *J. Comput. Netw. Commun.* **2014**, *2014*, 739504. [CrossRef]
15. Su, D.; Wen, X.; Zhang, H.; Zheng, W. A Self-optimizing Mobility Management Scheme Based on Cell ID Information in High Velocity Environment. In Proceedings of the IEEE Second International Conference on Computer and Network Technology, Bangkok, Thailand, 23–25 April 2010; pp. 285–288.
16. Saeed, M.; Kamal, H.; El-Ghoneimy, M. Novel Type-2 Fuzzy Logic Technique for Handover Problems in a Heterogeneous Network. *Eng. Optim.* **2018**, *50*, 1533–1543. [CrossRef]
17. Chaudhuri, S.; Baig, I.; Das, D. Self organizing method for handover performance optimization in LTE-advanced network. *Comput. Commun.* **2017**, *110*, 151–163. [CrossRef]
18. Alhammadi, A.; Roslee, M.; Alias, M.Y.; Shayea, I.; Alriah, S.; Abas, A.B. Advanced Handover Self-optimization Approach for 4G/5G HetNets Using Weighted Fuzzy Logic Control. In Proceedings of the 2019 15th International Conference on Telecommunications (ConTEL), Graz, Austria, 3–5 July 2019; pp. 1–6.
19. Alhammadi, A.; Roslee, M.; Alias, M.Y.; Shayea, I.; Alraih, S. Dynamic Handover Control Parameters for LTE-A/5G Mobile Communications. In Proceedings of the 2018 Advances in Wireless and Optical Communications (RTUWO), Riga, Latvia, 15–16 November 2018; pp. 39–44.
20. Abdulraqeb, A.; Mardeni, R.; Yusoff, A.M.; Ibraheem, S.; Saddam, A. Self-optimization of Handover Control Parameters for Mobility Management in 4G/5G Heterogeneous Networks. *Autom. Control Comput. Sci.* **2019**, *53*, 441–451. [CrossRef]
21. *5G Study on Channel Model for Frequencies from 0.5 to 100 GHz*; 3GPP TR 38.901, Tech. Report; 3GPP: Valbonne, France, 2018.
22. Alhammadi, A.; Roslee, M.; Alias, M.Y.; Shayea, I.; Alraih, S.; Mohamed, K.S. Auto Tuning Self-Optimization Algorithm for Mobility Management in LTE-A and 5G HetNets. *IEEE Access* **2019**, *8*, 294–304. [CrossRef]
23. Lee, Y.L.; Loo, J.; Chuah, T.C.; Wang, L.C. Dynamic network slicing for multitenant heterogeneous cloud radio access networks. *IEEE Trans. Wirel. Commun.* **2018**, *7*, 2146–2161. [CrossRef]
24. *Universal Mobile Telecommunications System (UMTS), Radio Resource Control (RRC) Protocol Specification*; 3GPP TS 36.331, Tech. Report; 3GPP: Valbonne, France, 2018.
25. Hsieh, P.J.; Lin, W.S.; Lin, K.H.; Wei, H.Y. Dual-connectivity prevenient handover scheme in control/user-plane split networks. *IEEE Trans. Veh. Technol.* **2018**, *67*, 3545–3560. [CrossRef]
26. Polese, M.; Giordani, M.; Mezzavilla, M.; Rangan S.; Zorzi, M. Improved Handover Through Dual Connectivity in 5G MmWave Mobile Networks. *IEEE J. Sel. Areas Commun.* **2017**, *35*, 2069–2084. [CrossRef]

applied sciences

MDPI

Article

D-RoF and A-RoF Interfaces in an All-Optical Fronthaul of 5G Mobile Systems

Zbigniew Zakrzewski

Institute of Telecommunications and Computer Science, UTP University of Science and Technology, Kaliskiego 7, 85-796 Bydgoszcz, Poland; zbizak@utp.edu.pl; Tel.: +48-523-408-318

Received: 25 November 2019; Accepted: 7 February 2020; Published: 11 February 2020

Abstract: This paper presents a solution for enabling the coexistence of digitized radio-over-fiber (D-RoF) and analog radio-over-fiber (A-RoF) interfaces operating in the optical fronthaul of 5G mobile systems. In the first section, we formulate the need to introduce new technologies to the cloud/centralized radio access network (C-RAN) (Next Generation RAN (NG-RAN) in 5G systems). A proposition of construction of the optical remote radio head (O-RRH)/gNodeB—distributed unit (gNB-DU), which will enable the operation of digital Splits/Options and new proposed analog Splits/Options, is presented. The methods performing calculations of bit rate and optical bandwidth demand in the fronthaul/midhaul, with reference to the parameters of the new-radio-release-15 (NR-Rel-15) wireless interface and subsequent releases, towards the next generations, are presented. The bandwidth demands were calculated for selected Splits/Options, and the results are shown in diagrams. A special section is devoted to description of the results achieved and presenting potential applications of the proposed construction of a radio-photonic device as well as new Splits/Options of the next generation fronthaul/midhaul.

Keywords: fronthaul; midhaul; radio-over-fiber; optical gNB-DU; RRH; A-RoF; D-RoF; eCPRI; BBU; all-optical network

---

1. Introduction

In previous generations of mobile systems and networks (2G/3G), a special emphasis has been placed in the radio domain on the development of distributed radio access network (D-RAN) architecture. It consisted in the fact that devices processing signals in the baseband (BB), intermediate-band (IF), and radio-frequency-band (RF) were located near to the mast with the antennas. The significant increase in the demand for signal processing, especially in the baseband, meant that we had to look for different solutions. Existing terminal devices, called base stations, can no longer be developed in the traditional way, because they would have to be supercomputers with high computing power in a moment (with high network traffic), and in other cases idling equipment (e.g., with a negligibly small nighttime traffic) that would have to be turned off to save energy without the use of their computational potential. The solution to this problem is to transfer computing functions into the network so that the computers performing these operations may be used for other purposes in the absence of mobile network traffic. This requires central control architecture of remote radio modules whose functions are as limited as their physical equipment allows.

In the future Next Generation Radio Access Networks (NG-RANs) domain of the 5G systems, exactly centralized/cloud control technology (C-RAN) will be widely used with a large set of gNodeB (gNB) base stations that will also be working in distributed structures/architectures (not to be confused with D-RAN).

1.1. State-of-the-Art 5G C-RAN Solutions

The basic concept in the NG-RAN domain, which is promoted by 5GPPP, is the multi-layered architecture of XHaul [1,2]. This solution is based on an optical network that is designed to support traffic from various Split/Option interfaces. This architecture also includes D-RAN solutions, which means joint support for management of traffic from the fronthaul and backhaul. Based on this solution, the authors of publications [3–5] proposed network architectures based on optical and microwave (radio-line) transport. The presented solutions introduce traffic optimization, which, however, does not take into account the possibility of transmitting signals occurring in the A-RoF format in the network. The architecture of an Optical Transport Network (OTN) system is prepared for an efficient transport of digital traffic, which is provided at client access points. The situation is similar in the case of Time Shared Optical Network (TSON), which is a very good solution supporting the transport of information from radio-over-Ethernet (RoE) (enhanced common public radio interface (eCPRI) as well as next generation fronthaul interface (NGFI)) interfaces. A lot of research and development have been devoted to various optimization solutions for transportation systems of streams from D-RoF interfaces. The author reviewed the available studies [6–13], which contained the results of studies showing the undoubted legitimacy of using the D-RoF technique. In these studies, particular emphasis was put on showing that a particular type of Split/Option or method of digitizing a radio signal to a bit form gives the opportunity to increase a link efficiency. There is no difference in the selection of the Split/Option method in the context of the load on the fronthaul network or the RRH unit. In special solutions, attempts were made to use compression methods when processing from analog to digital [14]; which, however—with a very large number of digital streams delivered to RRH working in massive-MMIO format—will not bring much efficiency. There remains the A-RoF solution with the least research, especially for 5G C-RAN applications. The papers [15–19] show a focus on specific solutions, which consist in conducting experiments documenting that radio signals in the BB, IF, and RF bands can be transmitted in optical fiber paths. These solutions were already tested many years ago. Several books have also been written on this subject [20,21]. In each of the presented experiments, however, no attention was paid to what generation of optical fibers will be used and to what extent future xWDM networks can be used to transport de facto analog signals. Particular attention of the author was caught by the study [22], which indicates the next field of activity, where the use or disposition of an optical fiber of the appropriate generation, located in the optical path, can be decisive in the selection of the radio signal transport system in its original form.

1.2. Author's Contribution

The solution of the combined approach to the analysis of the needs of the A-RoF and D-RoF interfaces was not undertaken in the above-mentioned studies. The author proposed the coexistence of these interfaces in an optical network. This task is possible when the fiber optic RAN network is all-optical. Of course, we can use fiber-to-the-antenna (FFTA) or passive optical network (PON) architecture, but passive solutions do not provide such a wide scale of optical resource management. Implementation of the network in the XHaul architecture enables the transport of digital data (backhaul) and digitized signals (fronthaul—D-RoF). The combination of A-RoF and D-RoF traffic forces proper preparation of all nodes in the network from the NG-RAN domain. Therefore, the author proposed a special O-RRH construction that can be directly or through-connected to a all-optical network. This construction was first presented by the author at an optical conference in Prague [23]. This also applies to BBU and DU nodes, which by definition have adequate computing power and will perform most of the tasks on a software basis. The method of managing nodes that carry also A-RoF traffic was presented by the author at the fiber-optic conference in Suprasl [24]. In addition, in order to formalize the baseband-over-fiber (BBoF), intermediate-frequency-over-fiber (IFoF) and radio-frequency-over-fiber (RFoF) solutions, it was proposed to extend the function of the Option 8 and to introduce new Options 9 and 10 in relation to the 3GPP model. The most important component of the author's study are unique calculations that indicate the need to reserve optical resources when transporting digital streams from the D-RoF (CPRI, eCPRI) and A-RoF (BBoF, IFoF and RFoF) interfaces. Calculations were made for a variable width of radio channel according to the 5G-NR-Rel-15 baseband (CP-OFDM waveform). The results of the calculations can be of major application when scaling the resources of an all-optical network, whose task will be to transport any type of signal, including A-RoF. The calculation formulas presented below have been adapted to 5G modulation and code solutions or created from scratch as a result of appropriate transformations (applies to A-RoF interfaces). It should be noted that the effectiveness of the presented approach to handling A-RoF and D-RoF traffic will be leading when introducing O-RRH working in massive-MIMO format, where we will have to deal with handling large traffic from a large number of EUs (need to introduce IoT, which will be supported in the wireless part by 5G systems). The results of the calculations carried out will allow efficient decision-making about switching the connection to the selected type of options in the extended range by the author, i.e., Options 1–10.

1.3. NG-RAN Concept Description

The distributed architecture of the gNB base station makes it possible to surround the user equipment (UE) and thus more efficient management of spatial resources. This task is accomplished through the use of wideband and flexible all-optical networks based on the Dense Wavelength Division Multiplexing (DWDM) system with a flexible grid of optical channels [25] or new more flexible Elastic Optical Networks (EON) [26] working with the Ultra-Dense Wavelength Division Multiplexing (UDWDM) format and extra flexible reconfigurable transponders. The optical network is so versatile that it can be used to transmit information/signals both in the area of the backhaul (BH) as well as the fronthaul (FH)/midhaul (MH) of the future next generation mobile systems.

The purpose of the backhaul (inside the cloud in Figure 1) is to connect the next generation core (NGC/5GC) with the NG-RAN control units through the gNB-central units (gNB-CUs). The fronthaul is used in creation of fast and often synchronous links between the gNB-CU and components of the distributed 5G base stations, i.e., gNB-distributed unit (gNB-DU)—Figure 1 [27]. The architecture of the distributed NG-RAN occurs in two concepts, i.e., based on the fronhaul network (Figure 1a) and with the division of the distribution network on the fronthaul and midhaul—Figure 1b. The second concept indicates the possibility of using a distribution point/unit (DU), whose task will be to perform some activities related to local signal processing and their distribution to functionally limited network termination units such as active antenna unit (AAU)/remote radio unit (RRU)/remote radio head (RRH).

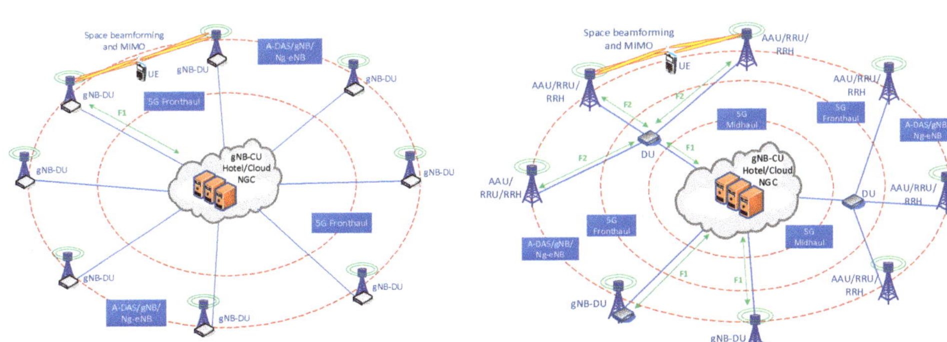

(a) (b)

Figure 1. NG-RAN architecture with distributed gNB: (**a**) concept of C-RAN controlled by gNB-CU connected to NGC (backhaul side) and gNB-DU (fronthaul side)—forced by 3GPP; (**b**) concept of C-RAN controlled by gNB-CU/BBU connected to NGC (backhaul side) and gNB-DU/AAU/RRH (fronthaul or/and midhaul side)—forced by 5GPPP.

Inside the gray cloud in Figure 1, there are a 5G-NGC network and a set of gNB-CU/ baseband unit (BBU) cooperating with each other. The method of cooperation between central units processing the radio signals symbolizes additional terms such as 'cloud' and 'hotel'. The term 'cloud' means that individual units can be located at a greater distance in separate buildings, which forces them to use quick connections with small delays. The term 'hotel' (alternatively often used as 'pool') symbolizes the placement of units in the same building, which is equivalent to a physical signal processing center, while BBUs/CUs are usually separated in distributed logical structures, which are managed by the so-called virtual machines.

From now, in order to simplify further description of the functioning of the network components, we will call the control units generally gNB-CU/BBU, and devices controlled by abbreviation gNB-DU/RRH. The name of the DU nodes that define the boundary between the F1 and F2 interfaces (Figure 1) will remain unchanged.

The fiber-optic FH can be a passive component (FTTA or PON) of the C-RAN, which is applicable in mobile networks of current generations (2G/3G/4G). The demand for more and more bandwidths in the backhaul networks forces the optimization of fiber network resources, which is why it is necessary to pay attention to the previously mentioned active flexible optical networks [25,26] using WDM technique.

The WDM technique enables the simultaneous transmission of several optical signals (different frequency channels) in one fiber-optic link or one optical path. Fiber-optic link is understood as a connection between nodes of the network, and the optical path is the path on which the optical signal travels passing through the optical nodes. Fiber-optic links are implemented using single-mode silica fibers, where single-modality is maintained for wavelengths in the range of 1260–1675 nm. An optical node determines whether a fiber-optic network can be called all-optical. If a node does not go to the electronic level (on the user's layer), to regenerate the signal or to reorganize the digital data, then this node can be called all-optical. Such nodes include optical add-drop multiplexer (OADM) or reconfigurable OADM (ROADM), as well as photonic cross-connects (PXCs). Of course, the process of signal regeneration can take place in all-optical nodes, but it must be implemented on the optical layer, which is currently extremely expensive (the need for precise recognition of the signal modulation format). ROADM and PXC are usually adapted to the so-called optical grids according to the standards defining transport systems. Optical channel grids, on the other hand, are adapted to the optimal parameters of optical fibers connecting nodes. In this case, it concerns the optical ranges of the S, C, L, and U bands [1]. If we release from these nodal devices the permanent set of optical filters and internal constraints to xWDM grids, then we will get full flexibility. Unfortunately, this is done at the expense of a high increase in the demand for effective photonic resources management in each of these nodes. If an optical node is able to switch any channel (in a technologically limited range—any optical carrier wavelength and any optical channel width), then it is sensible to use different modulation formats, in addition to the classic on-off keying (OOK). As the optical channels are spectrally independent with appropriately selected optical path or link parameters, they can carry completely different signals, i.e., modulated in various formats. Therefore, the flexibility of the optical network using the WDM technique enables an effective combination of traffic, in one fiber-optic link, coming from functionally different networks. Separation of different streams in an optical link can be accomplished by assigning separate optical carriers. An active network, in which there are various signals, must also be equipped with flexible nodes and transponders [26,28], which will be able to recognize signals and place them in the appropriate space of the grid of a given xWDM system (elastic or static). This assumption forces, on the side of gNB-DU/RRH, the use of all-optical nodes and optical terminations of the C-RAN. Of course, this does not exclude the possibility of mutual communication of C-RAN radio devices via radio interfaces operating in the "sub-6" (FR1) and "mmWave" (FR2) bands [29].

The following description presents the concept of integrated optical gNB-DU/RRH, that was presented by the author for the first time at the photonic conference in Prague [23] as the optical RRH (O-RRH). The material concerning this concept is also included in this work for the sake of completeness as well as due to the introduction of minor changes regarding the adaptation of the radio-photonic unit to flexible optical networks.

Optical gNB-DU/RRH enables communication with gNB-CU/BBU by using non-standardized interfaces and modulations. Integration of an optical multiplexer with a radio module enables the ingress to the optical fronthaul, connecting gNB-CU/BBU and gNB-DU/RRH, signals originating not only from a digitized radio-over-fiber (D-RoF) interfaces [30], i.e., common public radio interface (CPRI) [31] or evolved CPRI (eCPRI) [32], but also an analog radio-over-fiber (A-RoF) [30].

On the side of the signal processing center gNB-CU, the situation looks a little bit different, because here we have a lot of power in terms of signal processing. The task of the central unit will therefore be to collect all physical resource blocks (PRBs) (in terms of time and frequency) directed to specific UE on the wireless side, combining into one or several channels from the baseband and inserting these channels into the IF or RF band using the direct digital synthesis (DDS). In case of the D-RoF format, the process will be terminated on the digitization of the channel created in the baseband. In order to create an A-RoF signal, the other signal processing steps mentioned above must be implemented.

The purpose of performed analyses and calculations, the results of which are presented in the following paper, is to show that the signals of D-RoF and A-RoF formats can be transmitted in the same all-optical network, using the proposed construction of gNB-DU/RRH with optical termination, as well as the indication of scenarios in which it makes sense to use interfaces working in the A-RoF formats as those that provide a high degree of wavelength band savings compared to the D-RoF interfaces, in the fronthaul/midhaul optical path setup between gNB-CU and gNB-DU/RRH.

The motivation of the conducted research follows from the fact that the expansion of the C-RAN architecture is inevitable; therefore, it is necessary to search for solutions that will simplify the functions performed by a set of antenna modules and then move many more functions to signal processing centers that will lease computing power.

2. Radio and Photonic Components in Optical Massive-MIMO gNB-DU/RRH

The NG-RAN built on the basis of all-optical solutions enforces the introduction of the network termination in the form of an integrated optical gNB-DU/RRH. In this way, the created active-distributed-antenna-system (A-DAS) network (Figure 1) will enable solutions based on the spatial surrounding of the terminal by the so-called distributed gNB base station. The distribution of radio signals to/from gNB-DU/RRH via optical links will guarantee very high delay constancy, which in the situation of using digital beamforming (DBF) or hybrid beamforming (HBF) of radio beams directed to/from the UE is superior. An exception will occur in the case when the eCPRI interface is used, in which the synchronous ethernet (SyncE) technology is applied [33–35], creating a RoE link [36,37]. Of course, the stream of Ethernet frames can be transported through optical transport network (OTN) links [38], similarly to CPRI [39], but in case of the need to enter on the layer 2 (L2), in order to switch frame streams, the optical path will be terminated. Here, the guarantee of low variance of delay may not be possible, unless the Ethernet switch is equipped with optical ports that are connected through the linear clock on the layer 0 (L0) (the optical layer determines the transferred clock step).

The Integrated Optical gNB-DU/RRH can exist in several configurations, what depends on the construction of the radio massive-multiple-input–multiple-output (massive-MIMO) head and the photonic module. Figure 2 presents three types of optical gNB-DU/RRH, which are characterized by a four-sectors (a), one-sector (b) and six-sectors (c) radio head. Construction concepts (a) and (c) are very similar to each other, because they require the use of stepwise transfer of connections between sectors. In case (b), only one sector was used, which is equipped with, in proportion to other solutions, a large number of radio and aerial modules. The single-unit gNB-DU/RRH device will allow for smooth tracking through the UE radio beam around the mast. In this construction, it is required to smoothly switch off individual radio modules when the angle at which the UE terminal lies is exceeded. In each case shown (Figure 2), gNB-DU/RRH is directly connected to a fiber-optic link (at least one pair of optical fibers) in which the xWDM technique is applied (signals in accordance with the DWDM system or OTN/EON type development). In solutions (a) and (c) in the photonic part, the microROADM was used, which indicates the possibility of using the device in a double optical ring, where we will have a reserve route and the possibility of dynamic management of the add/drop band. The single-sector solution gNB-DU/RRH is based on the equipping of the photonic side in microOM/microDM [40,41], which indicates that this type of device will be able to be placed at the physical terminal of the optical path/link. The optical mux (OM)/optical demux (OD) terminating the optical path/link may be also applied in the cases (a) and (c)—Figure 2. This will depend only on the place where the gNB-DU/RRH device is connected to the optical network.

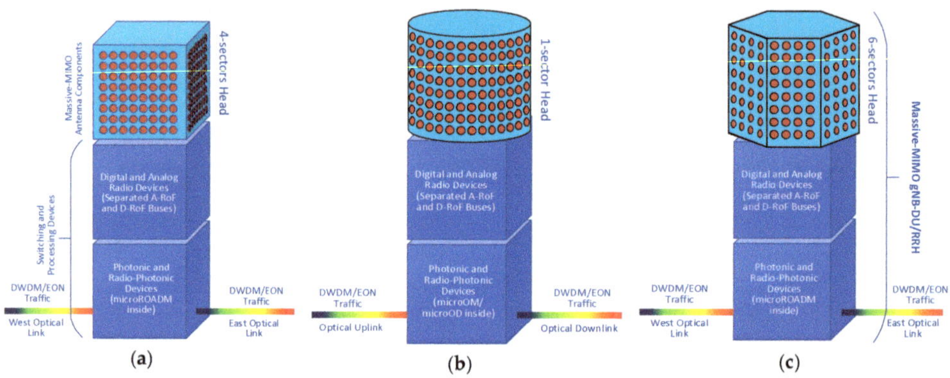

Figure 2. Block configurations of integrated massive-MIMO optical gNB-DUs/RRHs with the same photonic and radio-photonic modules (devices): (**a**) 4-sectors gNB-DU/RRH containing microROADM; (**b**) 1-sector gNB-DU/RRH containing microOM/microOD; (**c**) 6-sectors gNB-DU/RRH containing microROADM (concept first time presented in proceedings [6]).

The two downstream devices (modules) in optical a gNB-DU/RRH are described in more detail in Figure 3. The photonic switch (Figure 3a) allows a local optical channel redirection, depending on the need to direct concentrated radio traffic to a specific part of the antenna module assembly (Figure 4). A very important component is the array of media-converters (transceivers), whose task will be to shift from the photonic to the radio domain and vice versa. These systems will have to support signals modulated in D-RoF and A-RoF [30] formats, so they will have to work in a flexible mode. The lasers built into these converters should be longitudinally single-mode and have the possibility of dynamic tuning. In addition to the optical and radio bus, along the radio-photonic module, a bus runs with a control channel (Figure 3a, red—optical, green—electric), which was delivered as a dedicated optical channel from the controller working at the gNB-CU/BBU. Through this digital channel, we can control any component in the photonic and radio signal processing chain. Its function will primarily be to determine and establish the path on which radio frequencies from a specific radio band or baseband channels (in analog or digitized form) will travel.

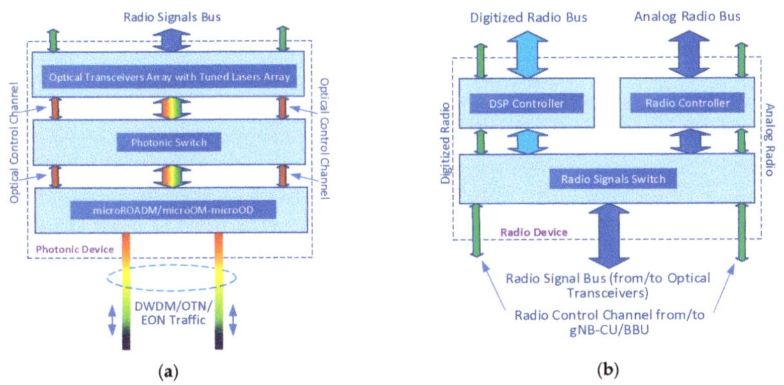

Figure 3. Switching and processing devices of integrated optical gNB-DU/RRH: (**a**) block construction of the photonic part with microROADM/microOM–microOD inside, (**b**) block construction of a common radio part with the digitized and analog radio signals switch (presented in proceedings [6]).

Next block (Figure 3b) is connected to an electric bus that has a large number of transmission lines. The number of transmission lines depends on the number of media converters included in the radio-photonic block. The transmission bus provides signals to the ports of the signal switch, whose task is to direct the appropriate signal for initial preparation or separation, before entering them to the specific radio antenna subunit. The digital signals go to the DSP controller, which demultiplexes/ multiplexes streams in the time domain and organizes them so that they reach the appropriate transmission line connecting with a specific radio-antenna module. The analog signals go to the radio controller, where they are pre-prepared (e.g., adaptation to the transmission line, correction of time and frequency parameters, or pre-amplification in a small range). The analog radio signal (A-RoF format) can come from the following ranges: BB, IF, or RF.

The control channel (Figures 3 and 4—green lines) enables the radio block subassemblies to operate and continues along the radio signal transmission bus. The functionality of the control channel does not terminate there. It is still appropriate to control the final radio systems closest to the antenna array assembly—Figure 4.

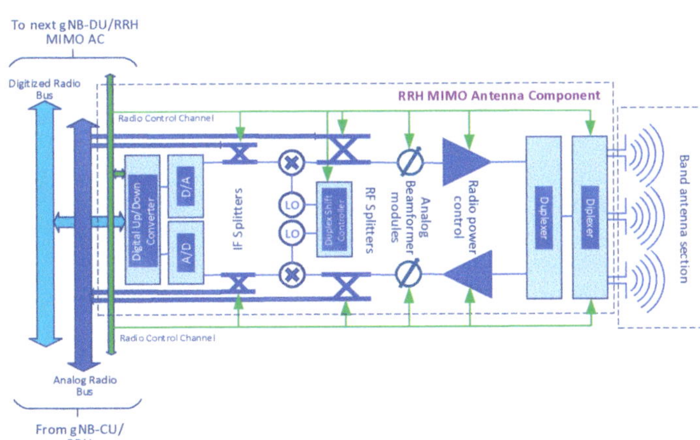

Figure 4. Massive-MIMO antenna component of optical gNB-DU/RRH—example of single radio-antenna module. Proposition of gNB-DU/RRH construction with components controlled in SDR mode by the radio control channel (green line) (presented in proceedings [23]).

The digital and analog buses are routed from the radio module to the radio-antenna head. Along the head with antennas, attachments should be placed to individual units converting radio signals prepared for emission in places of antennas.

Digitized radio signals (Figure 4) go to the digital block, where their duplex directions are subjected to separation and they are converted (processed) from digital to analog and vice versa. Here, there is an encoding/decoding operation for the partial streams that have reached the gNB-DU/RRH optical interface in CPRI/eCPRI format [31,32]. The signal from BB (uplink direction) gains its original form and is then transferred to the RF band. At this point, it is possible to use a greater number of the radio signal conversion degrees. It depends on the method of group transmission of baseband channels and the need for cooperation with analog signal transformation techniques. In the radio signal adaptation chain, there is also a section for forming the radiated beam. Each radio-antenna module is equipped with systems for two directions that allow analog beamforming and spatial beam control in cooperation with other radio-antenna modules. In the case when signals in a digitized form (prepared in gNB-CU/BBU) are delivered to the radio-antenna module, then DBF is additionally applied. The combination of these two techniques makes it possible to use hybrid solutions [42,43], which balances the load on the components of the fronthaul network and increase the precision of beam control on the free space side.

A single radio-antenna module must be equipped with a duplexer whose task is to separate and combine signals from different directions. The diplexers allow us to direct a specific radio band to the appropriate antenna dipole. Currently, diplexers are used as passive devices, but in the case of multi-band work (radio interfaces of next generations) of antenna array, we may need an active system, which will also be susceptible to control from the radio module or even gNB-CU/BBU.

The coupling of components shown in Figures 3 and 4, through fast and multi-track buses, gives the possibility to build a through or path terminating the optical gNB-DU/RRH. The optical channels will mostly be scaled at the gNB-CU/BBU level, but these functions may be partially moved to the gNB-DU/RRH. The control method of optical massive-MIMO gNB-DU/RRH depends on the method of controlling the entire domain of devices operating in the FH or MH network. The presented proposal assumes C-RAN control with the possibility of distributing the signal preparation centers gNB-CU/BBU Hotel/Cloud (Figure 1). From the radio termination point of the optical gNB-DU/RRH, despite the partially distributed network devices processing BB, IF, of RF signals (gNB-CU/BBU Cloud), gNB-CU/BBU nodes are seen as centrally located devices and communicate in parallel regarding the so-called gNB-CU/BBU clouds.

Figure 5 shows an example of a network where a cloud is located in the middle representing an evolved packet core (EPC)/next generation core (NGC), a network of software defined radio (SDR) devices (gNB-CU/BBU) and a central Open Flow (OF) controller [44,45] as a resource manager in the NG-RAN domain. The example network is based on the optical layer in the ring architecture, however, the logical structure of connections between NGC and gNB as well as gNB-CU/BBU and gNB-DU/RRH has tree architecture. Access to the optical network takes place through the so-called optical concentrators, i.e., optical multiplexers. The proposed gNB-DU/RRH have the built-in microROADM or microOM/microOD, which gives the possibility of their direct connection to the optical ring structure (microROADM case). If there is a need to connect the outside gNB-DU/RRH in the access mode (the so-called south direction), then it should have at least an optical multiplexer and demultiplexer (microOM/microOD case). The assumption that the network is all-optical requires the use of PXC or ROADM in nodes, which guarantees that there will be no conversion of the photonic to electric signal in these nodes and vice versa. Of course, this only applies to the transport of signals carrying user data (user plane).

The control layer using OF [44,45] must be organized so that all optical nodes together with gNB-DU/RRH and O-gNB remain under full control. As seen in the upper part of Figure 5, the optical network can connect gNB-DU/RRH and O-gNB to the core of the packet network and the surrounding cloud of the gNB-CU/BBU modules. All that remains is to design a mechanism for flexible resource management, which will decide how the optical resources will look like when simultaneously (hybrid) serving all streams from NG-RAN. PXC nodes equipped with DU functions will deal with the distribution of optical streams that will transport radio signals with high time requirements (physical layer) to the appropriate gNB-DU/RRH, whose functions can be dynamically reduced to AAU (depending on the load of links in network and signal processing centers in the cloud). In the situation that the midhaul network (serving traffic with lower time requirements) does not exist (Figure 1a), all the DU functions are taken over by gNB-DU. In this case, high time requirements are taken over by the network at the level of interface F1. The type of transport technology used (Figure 6) in particular interfaces and links will depend heavily on the type of Split/Option and its requirements. It should be noted that each stream of L1, L2, and L3 layers can be transported through an active optical network, which in the all-optical version represents the so-called L0 layer (beside the 7-layer ISO/OSI model).

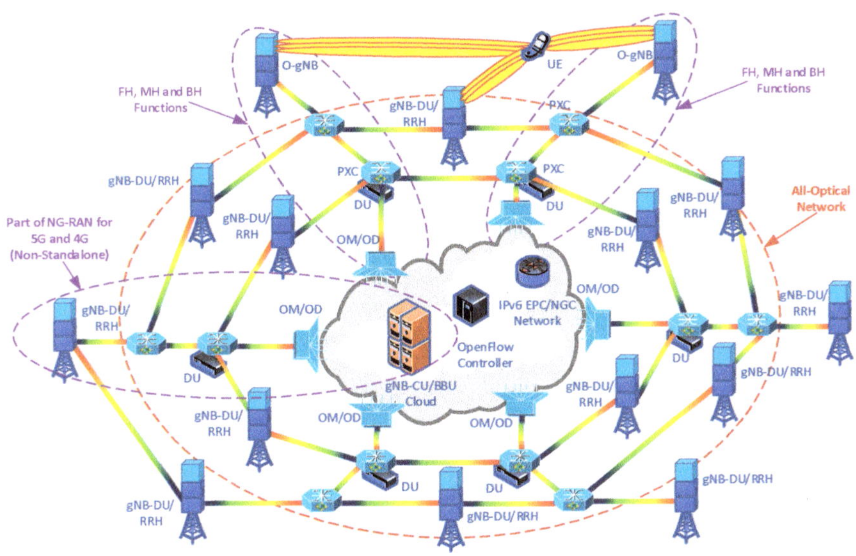

Figure 5. Example of flexible all-optical network connecting the integrated optical gNB-DU/RRH (with massive-MIMO functions) and O-gNB devices to gNB-CU/BBU/EPC/NGC cloud (presented in proceedings [23]).

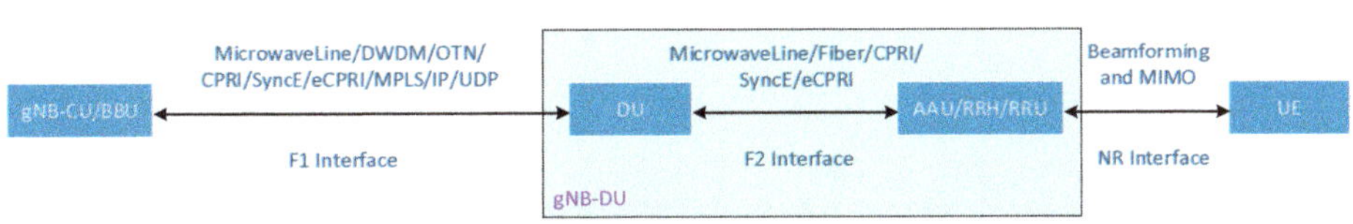

Figure 6. Transport technologies used in the F1 and F2 interfaces of fronthaul/midhaul as a components of the stationary part of NG-RAN (MicrowaveLine is the wireless transport option between gNB-CU/BBU and gNB-DUs/RRHs—wireless communication between gNB-DUs/RRHs is also possible—out of the scope of analysis).

3. Simple Model of Bandwidth Consumption Calculation in 5G Fronthaul Interfaces

The efficiency of the FH and MH links is the basis for a properly functioning C-RAN (in 5G-NG-RAN) structure. Currently, the CPRI format [31] is widely used in the links of fronthaul, the simplicity of which does not require using too large processing power from the signal processing systems, but the resulting streams are characterized by very high bit-rates. The demand for high CPRI bit-rate is primarily due to the need to connect to the gNB-CU/BBU cloud, independently, each antenna module (Figure 4) located in the radio-antenna head. In the Spits/Options model (Figure 7), the CPRI interface is at the top, which means that this digital stream will be directed to gNB-DU/RRH with low processing power. The functionality of this network termination will be limited to AAU. Currently, we have defined the fastest optical interface CPRI v.7.0 [31] with 24.3 Gbps. In many situations, especially when connecting the gNB-DU/RRH devices working in massive-MIMO format to fronthaul, the need for a larger number of the fastest CPRI streams is necessary. The introduction of faster bitstreams is pointless due to the emerging dispersal constraints of single-mode telecommunication optical fiber. In a DWDM network, we can concentrate more CPRI streams into one optical fiber, but with the dense location of the gNB-DU/RRH terminals, it is necessary to introduce new solutions in the transmission of radio signals, in conjunction with the introduction of optical nodes that can work more flexibly.

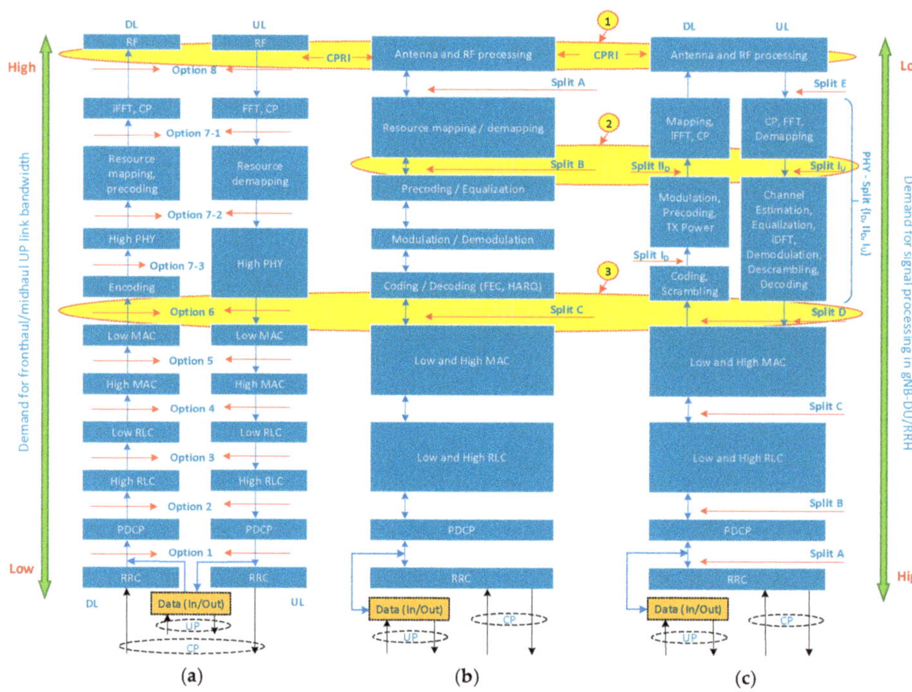

Figure 7. Models of Splits/Options and functions performed by 5G-FH/MH components in the scope of F1/F2 interfaces in the NG-RAN domain: (**a**) functional division according to 3GPP [46], (**b**) functional division according to 5GPPP [1], (**c**) functional division according to CPRI Industry Forum—interface eCPRI v.2.0 [32].

3.1. Bandwidth Consumption in D-RoF Interfaces

In the classical FH variant, usually one gNB-CU/BBU center will support interworking with several gNB-DU/RRH or DU radio terminations. In this sense, the total bit rate needed to service the massive-MIMO radio equipment can significantly exceed 10 Tbps. A single CPRI interface (Figure 7: (a) Option 8, (b) Split A, (c) Split E—yellow number 1) represents one antenna module in one sector of the radio-antenna head.

The digitized baseband signal must be delivered here independently to each antenna module so that MIMO techniques and spatial beamforming on the side of the wireless interface can function effectively. The total bit rate needed to deliver the appropriate number of fast CPRI streams to the gNB-DU/RRH mast, via the F1/F2 interface (Figure 1), can be determined based on the formula [47]

$$BR_8 = S \cdot A \cdot f_s \cdot b_s \cdot IQ \cdot HF \cdot LC \tag{1}$$

where S—the number of sectors per gNB-DU/RRH/AAU, A—the number of antenna modules in array per one sector, f_s—speed of sampling (in CPRI for 20 MHz radio baseband channel is equal to 30.72 MS/s $(3.84 \cdot (20/5) \cdot 2)$, in wider BB radio channel is proportionally higher [8,9]), b_s—number of bits per sample (depending on the format of the sampled signal: is equal 15 per one I/Q subcarrier for 4G/5G-Rel-15 (cyclic prefix orthogonal frequency division multiplexing (CP-OFDM))), IQ—a factor indicating a separate sub-sampling I as in-phase and Q as quadrature (is equal 2), HF (Headers Factor)—a factor indicating the redundancy of CPRI headers (redundancy is 1/15, therefore, amounts to 16/15), LC—alphabet nB/mB line code (8B/10B—ratio of 10/8—used in CPRI Options 1-7, 64B/66B—ratio of 66/64—used in CPRI Options 7A-10).

The relocation of the need of signal processing power to specific parts of the radio network chain (CU or DU) depends on the network architecture and, in SDR mode, enables balancing the power consumption and dynamic selection of signal processing centers in the gNB-CU/BBU cloud or gNB-DU/RRH terminal/node. In the situation when we transfer more functions related to the processing of baseband signals to the signal processing cloud, the stream that will only transmit data regarding the actual transmission can be significantly reduced. In the context of this case, we will consider Option 6 according to the 3GPP (Split C according to 5GPPP, Split D according to CPRI Forum—Figure 7—yellow number 3), which applies to both 'duplex' directions (uplink (UL) and downlink (DL)) and is located at the border between PHY and MAC layers. The division presented in Figure 7 indicates the places where the separation of the network elements functions can be made, which is also an important determinant of the directions of evolution and the emergence of NG-RAN Split/Option standards at the F1 and F2 interfaces [1,32,46]. The rate consumption calculations that will occur at the Split D during the maximum load, according to the CPRI Forum [9], can be carried out on the basis of a simplified and adapted formula [2,48]

$$BR_D = \frac{S \cdot N_L \cdot N_{SC} \cdot N_{SY} \cdot R_C \cdot K \cdot HF \cdot LC}{T_F} \tag{2}$$

where S—the number of sectors per gNB-DU/RRH, N_L—the number of layers (related to the number of layers needed to create and form space beams directed to mobile UE), N_{SC}—the number of active CP-OFDM subcarriers in BB channel (the number of subcarriers for the new waveform from the 5G-NR interface should be used—in the channel with a specified frequency bandwidth [MHz]), N_{SY}—the number of CP-OFDM symbols or newest waveform per standard time-frame (in the non-standalone 5G-Rel-15 interface a coherent value was assumed in relation to FDD-LTE), R_C—the factor of FEC code efficiency, $K = \log_2 M$—bits per modulation symbol, where M—modulation order (usually for M-QAM format), HF (Headers Factor)—CPRI frame redundancy factor (redundancy at 1/15 for CPRI, so the ratio is 16/15—much smaller and variable for the eCPRI, depending on the size of the charge in a frame [32] matched to the Ethernet frame and/or OTN [36,39]), LC—a line code also used as a scrambling (for faster streams it is 64B/66B, so the code rate is 66/64) and a physical Ethernet link control (also applicable to the RoE technology [32,36]). The line code in the optical Ethernet link applies only to the LAN format. Ethernet WAN interface is devoid of this code, because physical layer functions are taken over by the transport system, e.g., OTN. When the radio samples are transported in the fronthaul/midhaul paths using Ethernet (RoE) frames only, the LC value is included in the HF redundancy.

When the level of the Split/Option increases, the total data rate related to the user's data plane will increase, which results from the need to send additional data defining the radio signal. At Split I_U and II_D (Figure 7c—eCPRI 1.2—yellow number 2) an additional parameter appears which determines the number of quanta in the process of converting the frequency sub-carriers. Thus, the coding and modulation rules will not be taken into account, as the frequency components will be quantized. In order to estimate the bit rate that will occur in the fronthaul/midhaul path type Split I_U/II_D, an approximate formula [2] can be used

$$BR_{I_U/II_D} = \frac{2 \cdot S \cdot N_P \cdot N_{SC} \cdot N_{SY} \cdot N_{QF} \cdot HF}{T_F} \quad (3)$$

where S—the number of sectors per gNB-DU/RRH/AAU, N_P—the number of ADC/DAC chains (used in digital beamforming (DBF)—special application in massive-MIMO mode), N_{SC}—the number of active CP-OFDM subcarriers in BB channel (the number of subcarriers for the 5G-NR waveform interface should be used), N_{SY}—the number of CP-OFDM or newest waveform symbols per standard 4G/5G time-frame, N_{QF}—the quantizer resolution in the frequency domain, HF (Headers Factor)—eCPRI frame header redundancy factor [32] and higher IP/Ethernet network layers, T_F—frame duration (4G/5G system parameter).

3.2. Bandwidth Consumption in A-RoF Interfaces

Digital fronthaul/midhaul interfaces are very demanding in terms of throughput, which is why in this part of the work we will consider the implementation of A-RoF solutions. In the analog mode of the photonic carrier modulation, we can use a modulating signal located in the baseband, the intermediate band, or the radio frequency band [49]—Figure 8.

In the case of photonic carrier modulation with a radio signal coming from the baseband, we have the largest saving of the photonic band. This technique requires the use of separate optical resources for each frequency BB channel. This causes complications in the need to use a separate optical channel or a separate optical fiber for each BB channel. As in the D-RoF interface, a modulator must be used on the gNB-DU/RRH side to allow the BB signal to be applied to the RF band. The second solution indicates the use of an IF intermediate frequency. In this technique, all frequency channels can be sent simultaneously to the selected RRH. In a remotely managed radio module, we convert the signal that transfers data from the intermediate band to the target RF band. This solution is quite economical, due to the possibility of optical transmission of many radio channels focused around one unified, in a given network, intermediate frequency, which can be selected depending on the needs. The solution that most occupies the optical band remains, that is, simultaneous transmission of all components of the radio signal, including the RF carrier. This solution is the least effective from the point of view of the FH network, but allows the use of maximally simplified AAU, whose tasks are reduced to optical modulation and demodulation, amplification, pre-amplification, and shaping the beam emitted by the antenna array.

In comparison to the D-RoF solutions, the signals in the A-RoF format do not occupy too much bandwidth in the optical channel, however, they are characterized by much higher requirements for distortions of modulating signals. The modulation signal is specific, because it resembles an analog format in spite of the fact that it transfers digital data. For many years, scientists have been working on the optical OFDM (O-OFDM) technique, which is very similar to the A-RoF such as baseband-over-fiber (BBoF) solution, but the On/Off Keying (OOK) method with a direct detection is still popular and the cheapest one. The modulation of the photonic carrier with a signal from the BB is therefore similar to the O-OFDM, however, there is only one source of carrier surrounded by two bands (double-side band (DSB) mode), thus the orthogonality of subcarriers remains on the side of the modulating signal. Care for the orthogonality of radio subcarriers, during transport in the optical path, is not a problem due to their very low frequency distance and the occurrence of multi-path effects in a small range (chromatic dispersion and power penalty influence—DSB mode), which occur on a much larger scale on the side of the wireless interface. Thus, the only limitations here are signal attenuation, its delay and dispersion, which must be taken into account in particular when using a very broadband modulation signal.

The use of high radio frequency allows the creation of more wider frequency channels (e.g., the need to use mmWave ranges in 5G); however, when using DSB modulation this results in a doubled increase in the width of the optical channel. This can be limited by using the SSB modulation technique by introducing Mach–Zehnder Modulation (MZM) in the Hilbert configuration or by filtering out one sideband. Such a treatment, however, weakens the modulating signal, at the expense of improving the bandwidth efficiency and reducing the impact of the chromatic dispersion of the fiber-optic paths. The large number of systems (2G–5G) supported by the base station causes its high degree of complexity and hinders its quick reconfiguration. In this case, there is a need to move all system functions to the CU cloud, where radio resource allocation will take place to individual antenna matrices mounted on the mast. The question arises, whether it is possible to coordinate all spatial streams using only one optical channel? Of course, this is not physically feasible, therefore there is a need to use the optical wavelength division multiplexing (WDM) technique or space division multiplexing (SDM) on the optical side (a larger number of optical fibers in the cable, or the use of multi-core optical fibers [50]).

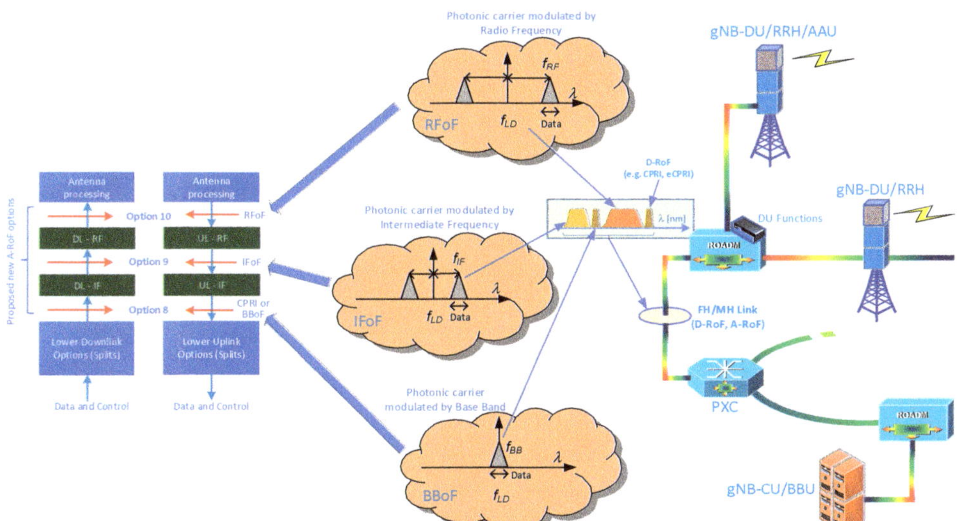

Figure 8. Proposition of new Options (Splits) for realization of analog interfaces working in A-RoF techniques. Splits mainly used for fronthaul/midhaul to connect DU and AAU (as a gNB-DU)—simple A-DAS.

Optical resources are presented in the wavelength domain, because the optical fiber influences the change of the wavelength as a function of the optical frequency of the carrier. In order to evaluate the frequency bandwidth of the modulating signal, a basic relationship can be used

$$\Delta f = 2 \cdot f_b + 2 \cdot f_k \tag{4}$$

where f_b—radio frequency (RF) carrier (identifying the RF channel number), f_k—the extreme right frequency of the CP-OFDM sub-carrier of the baseband channel. The transition to the wavelength domain can be accomplished using the following approximate dependence

$$\Delta\lambda = \lambda_2 - \lambda_1 = \frac{c \cdot \Delta f}{(f_2 - \Delta f) \cdot f_2} = \frac{c \cdot \Delta f}{(f_1 + \Delta f) \cdot f_1} \cong \frac{c \cdot \Delta f}{f_{LD}^2} \tag{5}$$

where $f_{LD} = \lambda_{LD}/c$—the frequency of the optical carrier emitted by the laser radiation source. By substituting relation (4) into (5), we get

$$\Delta\lambda = \frac{\Delta f \cdot \lambda_{LD}^2}{c} = \frac{2(f_b + f_k) \cdot \lambda_{LD}^2}{c} \tag{6}$$

On this basis, it can be seen that the width of the optical channel is primarily dependent on the component that constitutes the radio frequency (RF). Regarding the relation (1), it can be seen that the efficiency of the analog solution is higher if the radio frequency used is higher, too. The simple design of the RRH/AAU connected via the A-RoF interface to the fronthaul/midhaul, enforces the use of optical multiplexing techniques, which will ensure full service of the MIMO radio technology. In case of the D-RoF interface, the use of CPRI/eCPRI enables the introduction of multiplexing of temporary TDM, which limits the need for multiple laser sources, but creates a very broadband optical channel. For 5G mobile systems, even at the level of a few-kilometer fiber-optic link, the D-RoF interface will need WDM/SDM multiplexing. Another issue is related to the multi-sectority of the RRH, which makes it necessary to use RF channel separating systems and direct them to the appropriate array of antennas in the sector. We can solve this by using a flexible grid of DWDM channels [25], where the minimum frequency bandwidth of the optical channel is 12.5 GHz. The grid for the optical carrier is 6.25 GHz, but for the simplification resulting from the analog DSB modulation in the A-RoF interface, we assume that the carrier will change the frequency at least every 12.5 GHz. On this basis, the total occupancy of the optical band in the wavelength domain, using the formulas (4) and (5), can be written as the optical wavelength spectrum (OWS) [23]

$$\text{OWS} = \sum_{i=1}^{S \cdot A} \frac{\left\lceil \Delta f_i / \Delta f_{B\min} \right\rceil \cdot \Delta f_{B\min} \cdot c}{\left[f_0 \pm \sum_{j=1}^{i} \left(\left\lceil \Delta f_j / \Delta f_{B\min} \right\rceil \cdot \Delta f_{B\min} \right) \right]^2} \tag{7}$$

where S—a number of sectors per RRH, A—antenna modules array per sector; Δf_i, Δf_j—radio frequency bandwidth of *i,j*-th modulating signals; $\Delta f_{B\min}$—minimal frequency bandwidth of optical channel, which according to the DWDM grid is 12,5 GHz [25] or smaller in the future EONs; f_0—the reference frequency for the optical carrier equal to 193,1 THz—according to [25] or other band carrier; c—a speed of light in free space equal to $3 \cdot 10^8$ mps.

The Formula (7) shown above allows the frequency bands to be combined with the count-down (sign "+") or count-up (sign "−") of the optical channel number relative to the reference frequency f_0.

4. Calculation Results and Partial Discussion

In order to perform exemplary calculations indicating the bandwidth of fronthaul/midhaul links, using the previously presented interfaces, the initial parameters of radio signals that will occur in the 5G-NR Rel-15 wireless interface and subsequent releases should be set first.

In 5G mobile systems, a 4096-point FFT and a minimum 15 kHz spacing between the CP-OFDM subcarriers was introduced [51]. Thanks to such arrangements, 5G-NR Rel-15 can cooperate with 4G (LTE Rel-8 and newer). The subcarrier spacing can be increased in a coherent proportion [51], allowing the creation of broader combined frequency baseband channels—Table 1.

Table 1. Main parameters of NR-PHY Rel-15 interface and radio-antenna modules needed to calculate FH/MH link bandwidth consumption [2,29,51].

BB Channel Size	Layers N_L	Subcarrier Spacing	Active Subcarriers N_{SC}	Symbols/ Frame N_{SY}	Code Rate R_C	Modulation Order M	Antennas/ Sector A
(MHz)	-	(kHz)	-	-	-	-	-
20	16	15	1272/3300	140	0.85	256	96
50	24	15	3240/3300	140	0.85	256	128
80	24	30	2604/3300	280	0.85	256	128
100	16	30	3276/3300	280	0.85	1024 [1]	256
200	12	60	3168/3300	560	0.85	256	256
400	10	120	3168/3300	1120	0.85	256	256
800 [1]	10	240	3168/3300	2240	0.85	256	256

[1] The option is expected to be introduced in subsequent 5G releases.

The 3GPP Rel-15 of mobile systems has defined a physical layer and higher layers, but there are no restrictions that would indicate the number of antenna modules per sector. The extensive arrays of antennas make it possible to create systems for mass communication of the base station with many UE devices. Thanks to the usage of a higher number of antenna modules, the more accurately spatial beams can be determined. However, this entails the need to create a large number of data transmission channels and control channels. The demand for bit-rate is strictly dependent on the type of fronthaul/midhaul interface (type of Split/Option) that is between the mast with radio modules and the device preparing the radio signal for emission. The study assumes that the number of antenna modules per sector in the near future (5G Rel-15/16) should not exceed 256—Table 1.

If a well-known CPRI interface is used in the fronthaul/midhaul, then the required split speed (Option 8/Split E) can be calculated using (1). Here, each antenna module has its own dedicated stream with a digitized signal from the BB range. This gives the possibility of convenient scaling with spatial resources (MIMO and beamforming) but forces the use of an optical network with very high bandwidth links.

Figure 9a shows the results of calculations that visualize the total bit rate that will be generated by the heavily loaded, on the radio side, 4-sector gNB-DU/RRH, and Figure 9b shows the number of optical DWDM-OOK-50GHz-grid channels (for CPRI stream/Option 10 [31]—24.33 Gbps), which will have to be created in the FH/MH link to allow communication with the signal processing center. The calculations were made for 7 cases of BB channels (Table 1), where only the last one is not compatible with the 5G-NR Rel-15.

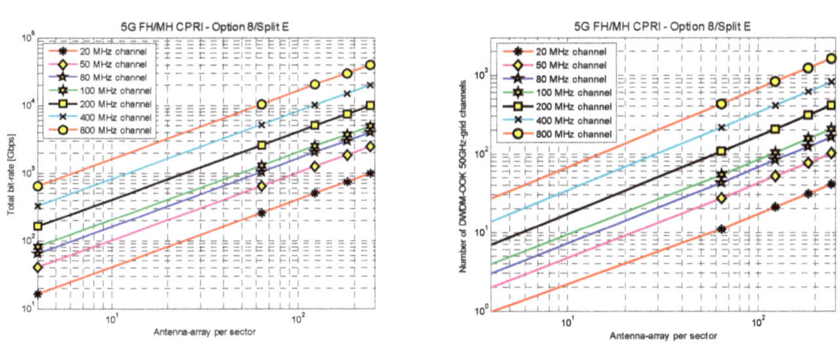

(a) (b)

Figure 9. Calculation results of optical resources needed to connect 4-sector gNB-DU/RRH to optical FH/MH as a function of antenna elements number of massive-MIMO array: (**a**) total bit-rate for radio BB channels digitized to CPRI format (LC = 66/64)—3GPP—Option 8/CPRI Forum—Split E (Figure 7a,c); (**b**) number of DWDM-OOK 50GHz-grid channels for radio BB channels digitized to CPRI format (LC = 66/64) – 3GPP—Option 8/CPRI Forum—Split E (Figure 7a,c).

Based on Formula (1) and Figure 9, it can be seen that the increase in the bandwidth required for the CPRI link increases very strongly with the increase of the basic bandwidth (proportional increase in the sampling frequency). It should be noted that when the BB width increases, the number of CP-OFDM subcarriers is not significantly increased, therefore, when expanding the BB channel, the efficiency of using the CPRI interface becomes smaller and smaller.

It is possible to reduce the bandwidth demand when we lower the number of Split/Option level in the fronthaul/midhaul interface. This will, of course, entail an increase in the functions related to signal processing at the gNB-DU/RRH. In order to compare the degree of reduction in the bandwidth demand, calculations were made, the results of which in the form of diagrams are shown in Figures 10 and 11.

At the discretion of the radio device communicating with the UE in the massive-MIMO mode, which will be connected to the gNB-CU cloud via the eCPRI-Split D/3GPP-Option 6, it cause to process the PHY signal in the UL and DL directions. This contact lies at the border between PHY and MAC layers, which is why it is clearly defined functional boundary (Figure 7). The dependency (2) was used to calculate the bit rate generated at the level of this Split/Option and some assumptions were made, which are included in Tables 1 and 2. The eCPRI interface is prepared for transporting of radio signals via Ethernet (RoE) paths, therefore the total redundancy of the frame headers go to 1.33 (Table 2). At the Split D level, the number of active subcarriers in the specified BB channel is significant (in Table 1 this is the first N_{SC} value, the second is the permissible number of subcarriers in NR-Rel-15, i.e., 3300). Therefore, the frequency bandwidth of the BB channel is not the reason for the bit rate increase in the Split/Option, but the number of active subcarriers and the number of layers created in order to be able to use MIMO and beamforming techniques at the level of radio-antenna modules. In this case, it was assumed that the number of layers is equal to the number of chains converting the signal into a digital form and vice versa (ADC/DAC)—the values were assumed approximately. The increase in the demand for spatial multiplexing in the wireless link will entail the need to increase these values. The larger bandwidth of the BB channel is applicable in higher millimeter RF bands (FR2) [29], where the devices from the internet-of-things (IoT) group will be communicated, therefore the number of layers is much lower. The TDM technique parameters in the 5G-NR frames were kept the same as in LTE (4G), with the difference that when the frequency distance between CP-OFDM subcarriers increased, the duration of the symbols shortens proportionally, and their number in the time-frame increases. As a result, broader BB channels will be used in the massive machine-type communications (mMTC) and ultra-reliable and low-latency communications (URLLC), where OFDM symbols do not have to be too resistant to multipath effects occurring in the wireless channel. The results of the calculations presented in Figure 10 show only the demand for the bandwidth of fronthaul/midhaul links at the level of the user plane (UP). The final bitrate must be increased by the control stream of the components on the gNB-DU/RRH board (control plane (CP)), but the increase in demand at the level of this Split/Option should not exceed 10% of the calculated value—Figure 10.

Table 2. Additional parameters of FH/MH radio streams for calculation of bandwidth consumption in the lower Split/Option interfaces [29,32,51].

BB Channel Size	ADC/DAC Chains N_p	Quantizer Resolution N_{QF}	Frame Duration	Headers Factor *HF*
(MHz)	-	-	(ms)	-
20	16	9	10	1.33
50	24	9	10	1.33
80	24	8	10	1.33
100	16	8	10	1.33
200	12	7	10	1.33
400	10	7	10	1.33
800 [1]	10	7	10	1.33

[1] The option is expected to be introduced in subsequent 5G releases.

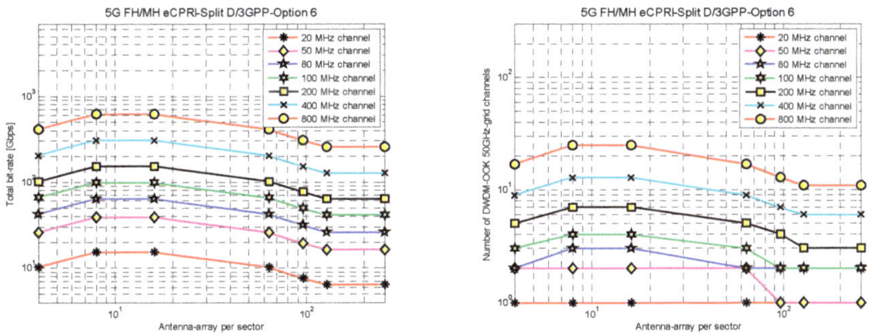

(a) (b)

Figure 10. Calculation results of optical resources needed to connect 4-sector gNB-DU/RRH to optical FH/MH as a function of antenna elements number of massive-MIMO array (calculation parameters changing according to Tables 1 and 2): (**a**) total bit-rate for radio BB channels digitized to eCPRI format—3GPP—Option 6/CPRI Forum—Split D (Figure 7a,c); (**b**) number of DWDM-OOK 50GHz-grid channels for radio BB channels digitized to eCPRI format—3GPP—Option 6/ CPRI Forum—Split D (Figure 7a,c).

Intermediate Splits/Options were introduced in all reference models (Figure 7), however, they function differently. In this analysis, additional calculations were made for the eCPRI Split I_U/II_D (directions UL and DL), which is functionally similar to 5GPPP Split B (Figure 7b,c). The results of calculations, carried out on the basis of (3), were placed in the form of diagrams in Figure 11. It can be noticed that in this Split/Option the demand for fronthaul/midhaul the bandwidth is increasing slightly. This increase is mainly due to the fact that each mapped OFDM subcarrier must be quantized. The number of quantization bits per IQ subcarrier can be found in Table 2. The increase of the speed of this Split/Option on the control plane (CP) will be significant, as it can reach even 30% of the UP flow.

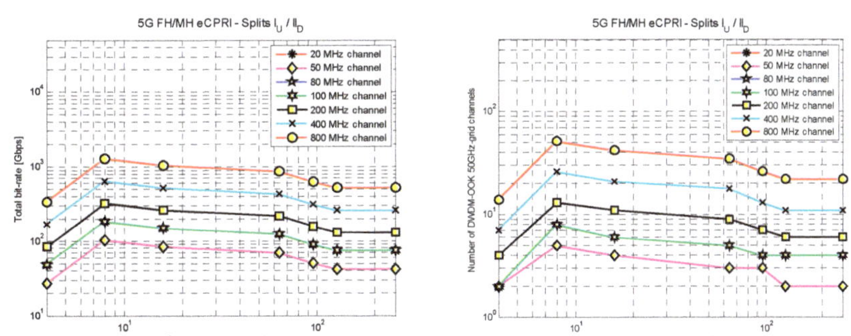

(a) (b)

Figure 11. Calculation results of optical resources needed to connect 4-sector gNB-DU/RRH to optical fronthaul/midhaul as a function of antenna elements number of massive-MIMO array (calculation parameters changing according to Tables 1 and 2)—there are overlapping lines in graphs for 20 MHz and 50 MHz, as well as 80 MHz and 100 MHz: (**a**) total bit-rate for radio BB channels digitized to eCPRI format—Splits I_U/II_D (Figure 7c); (**b**) number of DWDM-OOK 50GHz-grid channels for radio BB channels digitized to eCPRI format—Splits I_U/II_D (Figure 7c).

Two new Option 9/10, as well as the extension of the functionality of Option 8 split were proposed in the paper—Figure 8. To this end, a special design of an optical gNB-DU/RRH as a network node or network termination is proposed, which is equipped with appropriate radio components and photonics—Figures 2–4. Proposals are based on A-RoF solutions and are achievable if the fronthaul/midhaul links are based on an all-optical network using WDM technique. The calculations were carried out on the basis of a derived relation (7), and their results are presented in the form of diagrams shown in Figure 12. In the Option/Split 10, the radio-frequency-over-fiber (RFoF) technique is used, which enables transporting radio signals through optical links exactly in the form in which they have be radiated in place of the radio-antenna device.

However, this solution has significant limitations, as with the increase of RF frequencies, photonic resources in fronthaul/midhaul links are very quickly occupied. This is due there being more and more optical channels from the DWDM standard grid—Figure 12a. In the case of the RFoF, the use of photonic resources is effective when the radio carrier is located in the sub-6 (FR1) band—Figure 12a—area marked with the number 1.

This technique fits very well in the simple and low energy analog beamforming (ABF) implementation on the AAU side, however, for each space beam (layer) a separate RFoF channel will have to be used. With Option 10, the DBF technique can only be implemented on the gNB-CU cloud side.

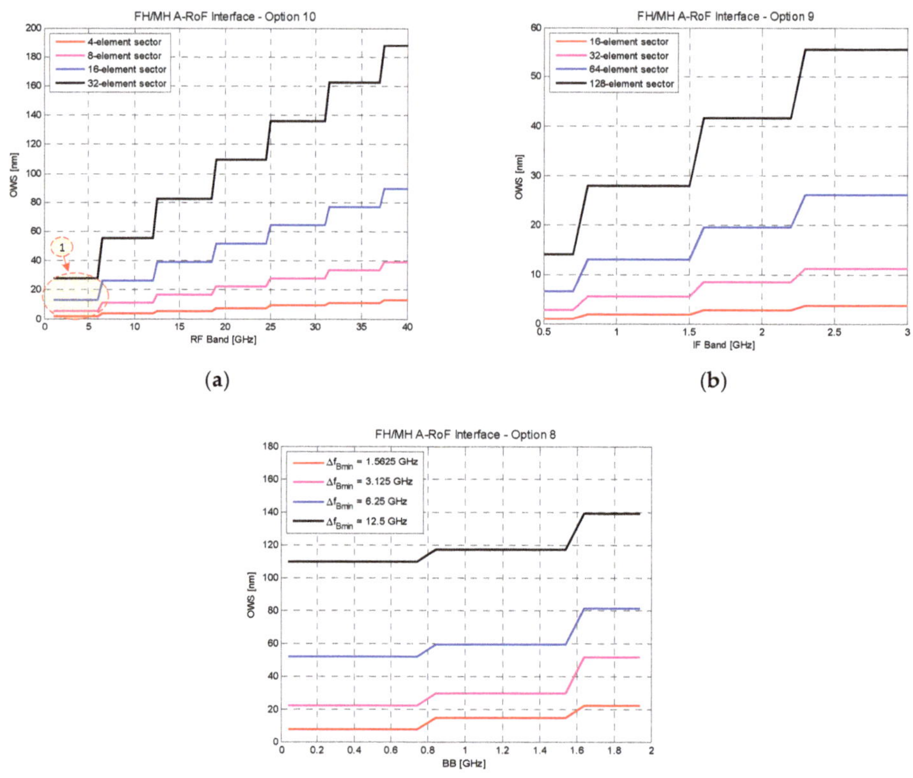

Figure 12. Calculation results of needs for optical resources in optical A-RoF fronthaul/midhaul path from gNB-CU/BBU/DU to 4-sector gNB-DU/RRH/AAU presented as Optical Wavelength Spectrum (OWS)—count-down from 1675 nm optical carrier (band U): (**a**) Option 10 (proposition—Figure 8)—analog transport of RF over optical DWDM link (min. Δf = 12.5 GHz); (**b**) Option 9 (proposition—Figure 8)—analog transport of IF over optical DWDM/UDWDM link (min. Δf = 1.5625 GHz); (**c**) analog transport of baseband over optical DWDM link (128 antennas per sector)—extension of Option 8 (proposition—Figure 8).

In the case of the need to use a higher radio carrier (>10 GHz), the RFoF method becomes ineffective. Here, it is worth using the intermediate frequency (intermediate-frequency-over-fiber (IFoF) technique) around 3 GHz (Figure 12b)—we assume that the DWDM system scope will be exceeded due to the lack of the need for optical line amplifiers—short optical paths. The introduction of a much denser DWDM grid (evolution towards EON/UDWDM) will further increase the efficiency of optical path utilization. However, only if we use a lower intermediate frequency (IF) or use only basebands (BBoF technique) directed to individual antenna modules. In Figure 12c, we see the cumulative OWS, but for zero RF carrier. In this case, the A-RoF mode was only used for BB signals, where a single BB signal occupies one optical channel with a width resulting from DSB modulation. The calculations were performed for the optical bandwidths of channels lying outside the standard DWDM grid [25]—Figure 12c. In a situation where one classic fiber is insufficient, it is recommended to introduce more fibers connected to one gNB-DU/RRH or use multi-core fiber in proportion to the number of sectors [50].

It can be noticed that the analysis of the D-RoF and A-RoF interfaces is slightly different, because in the case of digitized solutions, the bit rate is the starting point, while in analog solutions we operate on the frequency band occupancy. Optical links based on the DWDM/UWDM network can be used to transport signals of different formats. It is only important that signals at the nodes of this network are correctly recognized due to the photonic band they occupy in the core of the optical fiber.

5. Applications of the Proposed Solutions

All the solutions proposed in the work above can be applied in developing mobile communication and wireless access systems. The main destination is the RAN domain, whose architecture evolves strongly towards C-RAN. The result is a search for solutions that will allow for an efficient transmission of processed radio signals to maximally simplified radio modules. The presented concept of the network (Figure 5) is an unavoidable step that will allow virtualization of the computing power needed for the processed signals. The proposal for construction of the gNB-DU/RRH and coexistence of A-RoF and D-RoF links/paths in the same network is a combined approach to the solution that will enable dynamic management of optical resources of the fronthaul/midhaul, available cheap CU computing power, and radio resources.

Basic application scenarios for fronthaul/midhaul C-RAN networks (we assume that optical network (ON) and signal processing center (SPC) provide many other services to other recipients) are the following:

1. ON is heavily loaded, and SPC has a lot of free computing power resources. In this case, we can afford cheap signal processing even to the A-RoF. The A-RoF signals in most cases represent a high degree of spectral compression, so it will not be heavily loaded the ON links.
2. ON has a lot of free resources, and SPC is heavily loaded with other services. In this case, we can go to D-RoF modes, that is, lower the level of the split to the ceiling enabling the operation of the gNB-DU/RRH (even to Option 6 (eCPRI)—Figure 7).
3. Both ON and SPC are not overloaded. A case in which the D-RoF with Option 8 (CPRI) support may be enabled—Figure 7.
4. Both ON and SPC are heavily loaded. This is a very uncomfortable situation, because RAN has to go into the D-RAN mode, where the gNB-DU/RRH turns into the function of a classic gNB. A base station that does not have too much computing power and have limited access to the optical network must limit the supported wireless traffic.

Scenarios of detailed applications enabling switching the format in the fronthaul/midhaul link of the C-RAN (comparison with reference to the bandwidth in FH/MH links/paths of optical networks, without analyzing the demand for computing power during signal processing to the target form) are as follows:

1. The A-RoF and D-RoF interfaces on the same level of Option 8 (Figure 7), i.e., CPRI and BBoF. The comparative calculations are shown in Table 3.

2. The A-RoF and D-RoF interfaces located at different levels of split (Figure 7), i.e., eCPRI and BBoF. Comparative calculations are shown in Table 4.
3. The A-RoF and D-RoF interfaces located at different levels of split (Figure 7), i.e., CPRI and RFoF. Comparative calculations are shown in Table 5.
4. The A-RoF and D-RoF interfaces located at different levels of split (Figure 7), i.e., eCPRI and RFoF. Comparative calculations are shown in Table 6.
5. The A-RoF and D-RoF interfaces located at different levels of split (Figure 7), i.e., CPRI and IFoF. Comparative calculations are shown in Table 7.
6. The A-RoF and D-RoF interfaces located at different levels of split (Figure 7), i.e., eCPRI and IFoF. Comparative calculations are shown in Table 8.

Table 3. Calculated comparative parameters for detailed scenario No. 1 to switch from D-RoF to A-RoF format (the optical channel count-down is starting from wavelength equal to 1675 nm)—Figures 9 and 12 (128-element antenna-array and 4-sector unit).

Interface	BB	Optical Channel BW	OWS	Number of Optical Channels	Format/Split
-	(MHz)	(GHz)	(nm)	-	-
BBoF	400	1.5625	7.4535	512	A-RoF/Option 8
CPRI	400	50	178.4642	427	D-RoF/Option 8
BBoF	400	12.5	57.8231	512	A-RoF/Option 8

Table 4. Calculated comparative parameters for detailed scenario No. 2 to switch from D-RoF to A-RoF format (the optical channel count-down is starting from wavelength equal to 1675 nm)—Figures 10 and 12 (128-element antenna-array and 4-sector unit).

Interface	BB	Optical Channel BW	OWS	Number of Optical Channels	Layers	Format/Split
-	(MHz)	(GHz)	(nm)	-	-	-
BBoF	400	1.5625	0.1462	10	10	A-RoF/Option 8
eCPRI	400	50	2.8021	6	10	D-RoF/Option 6
BBoF	400	12.5	1.1689	10	10	A-RoF/Option 8

Table 5. Calculated comparative parameters for detailed scenario No. 3 to switch from D-RoF to A-RoF format (the optical channel count-down is starting from wavelength equal to 1675 nm)—Figures 9 and 12 (32-element antenna-array and 4-sector unit).

Interface	BB	Optical Channel BW	OWS	Number of Optical Channels	RF Carrier	Format/Split
-	(MHz)	(GHz)	(nm)	-	(GHz)	-
RFoF	100	12.5	14.84	128	3.6	A-RoF/Option 10
CPRI	100	50	12.5360	27	-	D-RoF/Option 8
CPRI	400	50	48.6018	107	-	D-RoF/Option 8
RFoF	400	50	57.8112	128	22	A-RoF/Option 10

Table 6. Calculated comparative parameters for detailed scenario No. 4 to switch from D-RoF to A-RoF format (the optical channel count-down is starting from wavelength equal to 1675 nm)—Figures 10 and 12 (32-element antenna-array and 4-sector unit).

Interface	BB	Optical Channel BW	OWS	Number of Optical Channels	RF Carrier	Layers	Format/Split
-	(MHz)	(GHz)	(nm)	-	(GHz)	-	-
RFoF	100	12.5	0.9353	8	3.6	8	A-RoF/Option 10
eCPRI	100	50	0.9351	2	-	8	D-RoF/Option 6
eCPRI	400	50	1.4022	2	-	4	D-RoF/Option 6
RFoF	400	50	1.8691	4	22	4	A-RoF/Option 10

Table 7. Calculated comparative parameters for detailed scenario No. 5 to switch from D-RoF to A-RoF format (the optical channel count-down is starting from wavelength equal to 1675 nm)—Figures 9 and 12 (128-element antenna-array and 4-sector unit).

Interface	BB	Optical Channel BW	OWS	Number of Optical Channels	IF Carrier	Format/Split
-	(MHz)	(GHz)	(nm)	-	(GHz)	-
IFoF	100	3.125	14.8408	512	1	A-RoF/Option 9
CPRI	100	50	48.6018	107	-	D-RoF/Option 8
CPRI	400	50	178.4642	427	-	D-RoF/Option 8
IFoF	400	6.25	29.4204	512	2	A-RoF/Option 9

Table 8. Calculated comparative parameters for detailed scenario No. 6 to switch from D-RoF to A-RoF format (the optical channel count-down is starting from wavelength equal to 1675 nm)—Figures 10 and 12 (32-element antenna-array and 4-sector unit).

Interface	BB	Optical Channel BW	OWS	Number of Optical Channels	IF Carrier	Layers	Format/Split
-	(MHz)	(GHz)	(nm)	-	(GHz)	-	-
IFoF	100	3.125	0.4678	16	1	16	A-RoF/Option 9
eCPRI	100	50	1.4022	3	-	16	D-RoF/Option 6
eCPRI	400	50	2.8021	6	-	10	D-RoF/Option 6
IFoF	400	6.25	0.9353	10	2	10	A-RoF/Option 9

For the calculations related to the conversion from the number of channels realizing communication in the optical fronthaul/midhaul link/path to the OWB, a simplified version of the Formula (7) was applied.

The proposed entirely optical network in this study may also have other applications. If the gNB-DU terminating devices are modified accordingly, then we can use them in other work areas. Light fidelity (Li-Fi) networks can be such an example. If the optical stream from the photonic switch (Figure 3a) is guided directly to the light emitter, then it will be possible to implement the Li-Fi network. Before radiating, however, the signal should be properly amplified and scattered, because in the optical path of the xWDM-based network we cannot use too much optical power. An additional limitation is that the operating ranges of radiators that are in general use are not adapted to the optical bands used in fiber optic communication. If the Li-Fi network needs to be moved to the visible light communication (VLC) systems, basically all photonic elements of the proposed network will be subject to changes. Only the all-optical network architecture will remain unchanged. In the case of small-scale VLC networks, it is possible to use polymer optical fibers (POFs) and PXC/ROADM/OM/OD operating in the visible light range [52].

6. Conclusions and Final Discussion

The proposed construction of the optical gNB-DU/RRH as O-RRH in [23] should meet the set requirements, which mainly consist in handling optical signals occurring in the D-RoF and A-RoF formats. It is unlikely that such devices could be installed as part of the first release of 5G mobile systems, i.e., Rel-15/16, but in the future mass production of technologically advanced devices will certainly significantly reduce their unit price. The proposed radio-photonic device can be connected to the optical network as a node or termination of the path, which means that it is simply a radio access node whose operation can be implemented in the majority of software (SDN/SDR). Available transmission optical network resources and the available signal and data processing power of the gNB-CU/BBU cloud will be a decisive factor in the introduction of the type of Split/Option at a given moment.

The results of calculations presented in this paper show what capacity will be needed in fronthaul/midhaul networks, so that communication with such extended radio terminations will be effective. The first versions of the CPRI interface were not characterized by excessive bit rate, but they were to apply to 3G (UMTS) and 2G (GSM) mobile systems. Subsequent versions had to be adapted to 4G (LTE), where MIMO techniques and spatial beamforming were used more and more. Massive-MIMO technology, which is to be widely introduced in 5G, in order to start IoT communication, forces the use of very fast optical networks already in the local domain and access. This is caused by the need for surrounding the UE devices by antenna arrays that must be connected using very fast, synchronous, and flexible fronthaul/midhaul networks. Therefore, the CPRI splits will only apply where new techniques do not have too high requirements.

Figure 9a shows that with a number of more than 100 antenna modules, a 4-sector device supporting 400 MHz BB channels will need about 10 Tbps, which is equivalent to about 400 optical OOK-DSB-50GHz-grid channels—Figure 9b. Of course, we can use coherent interfaces with high-order modulation and polarization multiplexing, which will reduce the demand for frequency bandwidth in the optical link several times, but the numbers will still be very large. Therefore, it is very important to introduce new solutions, among which one should reduce the throughput by introducing new Splits/Options.

Figures 10 and 11 show the calculation results for two selected eCPRI interface splits, where we see a very large drop in the demand for bit-rate. This is done at the expense of the need to perform more operations in the antenna-radio terminal, but the profit is significant, because the demand for bit-rate at the maximum values of the radio signal balances around 1 Tbps. This is associated with the occupancy of an acceptable number of optical DWDM/UDWDM channels in the fronthaul/midhaul optical path.

In order to increase the transparency of the potential application of the presented solutions, some exemplary scenarios (Section 5) are given in which the transition from the D-RoF format to the A-RoF format is suggested. The basic scenarios constitute the merits of undertaking the research topic, as they show the direction of the RAN network evolution along with the increase in the use of fronthaul/midhaul links for this purpose, as well as the signal processing centers. Detailed scenarios are examples in which a specific D-RoF interface can be replaced with the A-RoF interface due to a smaller occupancy of the optical band in the fronthaul/midhaul link/path.

Scenario No. 1 concerns the comparison of BBoF and CPRI interfaces that work at the same split level—Table 3. When using the BBoF optical channel with the minimum width matched to the modulation signal band, the gain is almost 24-fold.

Scenario No. 2 shows a comparative overview of BBoF interfaces (Option 8) and eCPRI (Option 6)—Table 4. The eCPRI interface uses the fact that the so-called layers are most often associated with the number of radiated beams by use of DBF or HBF method. A large part of operations related to processing a radio signal is transferred to the gNB-DU/RRH. In this case, the gain for the optical UDWDM channel (1.5625 GHz) is over 19, while for the DWDM channel (12.5 GHz) only 2.4. In this example, it is clearly visible what effect the channel width has on the optical bandwidth saving, which is a serious technological limitation in the production of tunable lasers, photonics switches, and filters used in PXC/ROADM.

Scenario No. 3 concerns the comparison of RFoF (Option 10) and CPRI (Option 8) interfaces—Table 5. In this comparison it is difficult to find a profit which for example BB 100 and 400 MHz channels shows losses. It should be added, however, that the RRH/AAU in the case of RFoF is almost completely unloaded from the signal processing. Here, even the RF carrier is built on the SPC/CU side. The calculations were carried out assuming a separate transfer of the BB channel with the RF carrier. If we send more BB channels on one RF carrier, then the profit will be significant, but we will limit the possibilities of independent beamforming in particular BB ranges. We eliminate this problem in the next scenario.

Scenario No. 4 includes the technique of creating layers that manage beams. Here we compare the RFoF (Option 10) and eCPRI (Option 6) interfaces—Table 6. The situation is similar, because there is no profit here or there is a loss. Therefore, the situation is similar to scenario 3, and the significant difference in the occupation of the optical band between the scenarios results mainly from the compression consisting in the BB encoding in the layer-related stream.

Scenario No. 5 concerns the comparison between IFoF (Option 9) and CPRI (Option 8)—Table 7. The introduction of the IFoF format for the 100 MHz BB channel gives about 3.3-fold gain on the optical FH/MH band. At the 400 MHz BB channel, the profit is already 6 times. In the case of the introduction of the BB channel multiplexing technique around IF subcarriers, the gain can be improved [53].

Scenario No. 6 shows the comparison of IFoF (Option 9) and eCPRI (Option 6) interfaces, also for 2 sample BB channel widths—Table 8. For both cases, we have a similar gain of around 3. The amount of profit will depend heavily on the selection of the IF subcarrier in relation to the selected optical channel width from the grid of the next generation of EON/UDWDM networks.

In the scenario examples we did not use the D-RoF digital indirect interface type eCPRI Split I_U/II_D—Figure 7. In this interface there is a strict dependence of the final FH/MH link bandwidth on the BB channel width and the number of active CP-OFDM sub-carriers—Figure 11. In the eCPRI Split D (Option 6) interface, the fronthaul/midhaul link rate depends only on the number of active CP-OFDM sub-carriers. Considering that in the 5G-NR wireless interface the frequency distance between the CP-OFDM subcarriers increases faster than the number of subcarriers themselves, therefore with the increasing BB channel width the eCPRI Split I_U/II_D interface will become increasingly less efficient with A-RoF (profit results will be between scenarios 1–2, 3–4, and 5–6 respectively).

Some of the issues related to the influence of the optical path or fiber-optic link on the behavior of the wireless radio channel has already been addressed in sub-Section 3.2. However, to a large extent these issues are not covered by the topic of this paper. It can be assumed that an all-optical path/link guarantees the stability of the delay. The amount of delay depends on the optical path length or link. This parameter, however, seriously affects the work of the DBF or HBF (regardless of the method used [54]), which requires conducting many calculations and practical tests in any implementation of subsequent C-RAN solutions. This applies to both the D-RoF format and the A-RoF format. With the narrowing the spatial beam on the side of the wireless link (massive MIMO), a faster feedback in the DBF control channel with the update of channel state information (CSI) data is needed—especially in the fast-moving UE. The use of HBF is helpful here, since the instant compensation resulting from the delay of the wired RAN part can be implemented with the ABF support.

Interferences that can occur between adjacent cells or sectors can be reduced by allocating other physical resource blocks (PRBs) (time-frequency isolation). In the case of sectors belonging to the same gNB-DU/RRH or gNB-CU and spaced apart, interference is even indicated due to the constructive spatial surrounding of the UE. Time-space synchronization in the area of the distributed base station ensures the implementation of joint transmission (JT) and high spatial gain. By means of an antenna-array consisting of a large number of modules, a very narrow beam can be radiated, thanks to which the channel will be resistant to multi-path. In the optical path or fiber-optic link, the previously mentioned phenomenon of chromatic dispersion has a static character. During the DSB transmission, a relative delay of the side bands occurs, in which there is a signal coming from the wireless channel. CP-OFDM symbols are protected against this phenomenon by inserting a guard interval with a cyclic prefix. The optical path enhances the multi-path phenomenon, but it is a constant parameter that can be quickly determined and compensated by the exchange of CSI. The compensation of optical fiber chromatic dispersion can also be implemented permanently at the end of the optical path/link, which will ensure a constant minimization of the impact on the transported radio channel signal in the target format.

The radio signal transported in the optical path in analog format is also influenced by noise. The main source of noise are the radio-photonic converters. The noise factor (NF) of the converter therefore decides on the extent of reduction of the SNR and indirectly influences the BER in the user channel. Optical power density should also be controlled due to the potential effect of nonlinear distortions. This is a price that in many cases is worth paying for a significant profit resulting from the compression of the used optical spectrum in the optical path/link. Studies of A-RoF interfaces in this area have been conducted for many years, and their results can be found in many papers and books [34,35].

The proposals for analog Splits/Options with high ID numbers (Option 8/9/10) may be useful in many solutions, but they cannot be universal, as in the case of CPRI. With the currently used grid of optical DWDM channels, the IFoF technique seems to be the most universal, since the intermediate frequency can be selected systematically, depending on the demand and available photonic resources. In this case, it is also possible to use a multistage BB signal transfer to the intermediate band. According to the calculations, the results of which are shown in Figure 12 and Tables 3 and 4, the use of the BBoF technique will become particularly important when a much denser UDWDM grid is available.

The possibility of introducing the O-OFDM format to optical links has been working for many years, and the results of research are more and more promising. This type of format does not differ significantly from the BBoF technique, because we have been hosting the OFDM format for over 25 years in wireless interfaces.

In order to indicate the degree of use of the optical channel by interfaces from the D-RoF family and others, used in classic NG-RAN solutions, the author carried out calculations, the results of which are presented in [55]. The bit rates that will occur in the 5G network will undoubtedly be high. When approaching the next generation of mobile systems, i.e., 6G, we need to look for new solutions for optical networks. The issue of future capacity crunch of optical networks has been raised for several years [56–58]. However, previous analyses concerned solutions for broadband backbone networks. The emergence of heavy IT traffic in the access domain is another new challenge. A serious generator of this traffic in the RAN domain is the widespread use of transporting digitized radio signals to an antenna MIMO modules. The introduction of radio signal transport in its original form will be a must. The flexibility of future all-optical EON networks will enable the introduction of another degree of freedom in optical resources management. As indicated above, depending on the type of Option engaged at a given moment, the occupation of the optical channel will vary within wide limits. In addition to WDM multiplexing, TDM technique can be introduced in digital channels. This will provide us with OTN or its newer generations with EON interfaces, in which optical OFDM may be used [55]. In the case of A-RoF signals we can only use WDM technique. The flexibility of the optical network consisting in the optimal selection of the optical channel width for the A-RoF signal, as well as packing the channels independently of the DWDM grid, significantly increases the use of resources lying in the scope of optical nodes and the range of optical fiber single-modality. The latter is particularly important in network solutions with FTTA or PON architectures. An additional advantage of the optical link with the irregular grid of optical channels (EON feature) is a significant increase in resistance to non-linear distortions resulting from the Kerr effect. Interleaving of A-RoF and D-RoF optical channels will increase this resistance. With a large number of optical channels, the total power can reach a high density in the fiber core, but in fronthaul there is no need for high optical power. It is anticipated that on 5G fronthaul connections (high number Options) the transport of signals will take place over a distance of not more than 20 km. Therefore, problems in access networks will differ from those we see in wide area networks.

Another way to avoid capacity crunch of the fiber-optic networks in the NG-RAN access domain is to introduce new fiber technologies, i.e., multicore fibers (MCFs). The author proposed this type of solution for the first time in [50]. Connecting O-RRH/gNB-DU to fiber optic cables based on multi-core structures will enable the implementation of the next SDM stage. If we combine the above dynamic interface selection solutions with new fiber-optic cable standards, there will be no problem with network congestion in the future. The next challenges to overcome are the construction of all-optical node devices that will be able to effectively switch the movement on the optical layer with flexible changing the optical channel grid. Switching will have to take place not only in the optical channel range, but also optical frames. For A-RoF signals from the NG-RAN domain, the 5G-NR interface frame synchronization with the optical layer switching system will be required.

7. Future Work

The paper presents proposed solutions that can be used in the design of 5G system components intended for the use in radio domain networks. The calculations and simulations performed show the potential applications of the proposed interfaces. In the next stage of work on the proposed network solutions, extended simulations should be carried out, which will demonstrate the possibility of a correct physical operation of the A-RoF and D-RoF interfaces. It should be noted here that the coexistence of these interfaces in the DWDM/EON link is preferred. Simulations will be conducted in the photonic domain and partly in the radio domain using the VPIphotonics or OptiWave platforms. Having in mind that the operation of digital and analog interfaces is completely different, this will require appropriate selection of simulation parameters due to the way the signal samples are created. In the optical path, linear and nonlinear phenomena should be taken into account. The main components of the network that will bring these distortions will be single-mode telecommunications optical fibers and, to a lesser extent, lasers, external optical modulators, and optical multiplexers.

Funding: This research received no external funding.

Conflicts of Interest: The author declares no conflict of interest.

References

1. 5GPPP, Architecture Working Group. View on 5G Architecture. 2017. Available online: https://5g-ppp.eu/wp-content/uploads/2017/07/ (accessed on 25 September 2019).
2. 5G-XHaul, D2.3. Architecture of Optical/Wireless Backhaul and Fronthaul and Evaluation. 2017. Available online: https://www.5g-xhaul-project.eu/download/ (accessed on 9 August 2019).
3. Camps-Mur, D.; Gutierrez, J.; Grass, E.; Tzanakaki, A.; Flegkas, P.; Choumas, K.; Giatsios, D.; Beldachi, A.F.; Diallo, T.; Legg, P.; et al. 5G-XHaul: A novel wireless-optical SDN transport network to support joint 5G backhaul and fronthaul services. *IEEE Commun. Mag.* **2019**, *57*, 99–105. Available online: https://ieeexplore.ieee.org/document/ (accessed on 15 January 2020). [CrossRef]
4. Eramo, V.; Listanti, M.; Lavacca, F.G.; Iovanna, P. Dimensioning Models of Optical WDM Rings in Xhaul Access Architectures for the Transport of Ethernet/CPRI Traffic. *Appl. Sci.* **2018**, *8*, 612. Available online: https://www.mdpi.com/2076-3417/8/4/612 (accessed on 15 January 2020). [CrossRef]
5. De La Oliva, A.; Perez, X.C.; Azcorra, A.; Di Giglio, A.; Cavaliere, F.; Tiegelbekkers, D.; Lessmann, J.; Haustein, T.; Mourad, A.; Iovanna, P. Xhaul: Toward an integrated fronthaul/backhaul architecture in 5G networks. *IEEE Wirel. Commun.* **2015**, *22*. Available online: https://ieeexplore.ieee.org/abstract/document/ (accessed on 12 January 2020). [CrossRef]
6. Pérez, G.O.; López, D.L.; Hernández, J.A. 5G New Radio Fronthaul Network Design for eCPRI-IEEE 802.1CM and Extreme Latency Percentiles. *IEEE Access* **2019**, *7*, 82218–82230. Available online: https://ieeexplore.ieee.org/abstract/document/ (accessed on 11 January 2020). [CrossRef]
7. Civerchia, F.; Kondepue, L.; Giannone, F.; Doddikrinda, S.; Castoldi, P.; Valcarenghi, L. Encapsulation Techniques and Traffic Characterisation of an Ethernet-Based 5G Fronthaul. In Proceedings of the 2018 20th International Conference on Transparent Optical Networks (ICTON), Bucharest, Romania, 1–5 July 2018. Available online: https://ieeexplore.ieee.org/document/ (accessed on 10 January 2020).
8. Chih-Lin, I.; Li, H.; Korhonen, J.; Huang, J.; Han, L. RAN Revolution With NGFI (xhaul) for 5G. *J. Lightwave Technol.* **2018**, *36*. Available online: https://ieeexplore.ieee.org/document/ (accessed on 14 January 2020). [CrossRef]
9. Assimakopoulos, P.; Zou, J.; Habel, K.; Elbers, J.P.; Jungnickel, V.; Gomes, N.J. A Converged Evolved Ethernet Fronthaul for the 5G Era. *IEEE J. Sel. Areas Commun.* **2018**, *36*, 2528–2537. Available online: https://ieeexplore.ieee.org/document/8482245 (accessed on 16 January 2020). [CrossRef]
10. Sarmiento, S.; Altabas, J.A.; Spadaro, S.; Lazaro, J.A. Experimental Assessment of 10 Gbps 5G Multicarrier Waveforms for High-Layer Split U-DWDM-PON-Based Fronthaul. *J. Lightwave Technol.* **2019**, *37*, 2344–2351. [CrossRef]

11. Salama, A.I.; Elmesalawy, M.M. Experimental OAI-based Testbed for Evaluating the Impact of Different Functional Splits on C-RAN Performance. In Proceedings of the 2019 Novel Intelligent and Leading Emerging Sciences Conference (NILES), Giza, Egypt, 28–30 October 2019. Available online: https://ieeexplore.ieee.org/document/8909310 (accessed on 13 January 2020).
12. Musumeci, F.; Bellanzon, C.; Carapellese, N.; Tornatore, M.; Pattavina, A.; Gosselin, S. Optimal BBU Placement for 5G C-RAN Deployment Over WDM Aggregation Networks. *J. Lightwave Technol.* **2015**, *34*, 1963–1970. Available online: https://ieeexplore.ieee.org/document/7368094 (accessed on 11 January 2020). [CrossRef]
13. Roldan, M.J.; Leithead, P.; Mack, J. Experiments and results of a mmW transport platform to enable 5G cloud RAN lower layer splits. In Proceedings of the 2018 IEEE Long Island Systems, Applications and Technology Conference (LISAT), Farmingdale, NY, USA, 4 May 2018. Available online: https://ieeexplore.ieee.org/abstract/document/8378032 (accessed on 17 January 2020).
14. Wu, C.Y.; Li, H.; Van Kerrebrouck, J.; Breyne, L.; Bogaert, L.; Demeester, P.; Torfs, G. Real-Time 4 × 3.5 Gbps Sigma Delta Radio-over-Fiber for a Low-Cost 5G C-RAN Downlink. In Proceedings of the 2018 European Conference on Optical Communication (ECOC), Rome, Italy, 23–27 September 2018. Available online: https://www.researchgate.net/publication/ (accessed on 17 January 2020).
15. Mitsolidou, C.H.; Vagionas, C.; Mesodiakaki, A.; Maniotis, P.; Kalfas, G.; Roeloffzen, C.; van Dijk, P.W.L.; Oldenbeuving, R.M.; Miliou, A.; Pleros, N. A 5G C-RAN Optical Fronthaul Architecture for Hotspot Areas Using OFDM-Based Analog IFoF Waveforms. *Appl. Sci.* **2019**, *9*, 4059. Available online: https://www.mdpi.com/2076-3417/9/19/4059 (accessed on 10 January 2020). [CrossRef]
16. Zeb, Z.; Zhang, X.; Lu, Z. High Capacity Mode Division Multiplexing Based MIMO Enabled All-Optical Analog Millimeter-Wave Over Fiber Fronthaul Architecture for 5G and Beyond. *IEEE Access* **2019**, *7*, 89522–89533. Available online: https://ieeexplore.ieee.org/document/ (accessed on 12 January 2020). [CrossRef]
17. Giannoulis, G.; Argyris, N.; Iliadis, N.; Poulopoulos, G.; Kanta, K.; Apostolopoulos, D.; Avramopoulos, H. Analog Radio-over-Fiber Solutions for 5G Communications in the Beyond-CPRI Era. In Proceedings of the 2018 20th International Conference on Transparent Optical Networks (ICTON), Bucharest, Romania, 1–5 July 2018. Available online: https://ieeexplore.ieee.org/document/ (accessed on 14 January 2020).
18. Proietti, R.; Lu, H.; Liu, G.; Castro, A.; Shamsabardeh, M.; Yoo, S.J.B. Experimental Demonstration of Elastic RF-Optical Networking (ERON) for 5G mm-wave Systems. In Proceedings of the 2017 European Conference on Optical Communication (ECOC), Gothenburg, Sweden, 17–21 September 2017. Available online: https://ieeexplore.ieee.org/document/8346014 (accessed on 11 January 2020).
19. Bekkali, A.; Ishimura, S.; Tanaka, K.; Nishimura, K.; Suzuki, M. Multi-IF-Over-Fiber System with Adaptive Frequency Transmit Diversity for High Capacity Mobile Fronthaul. *J. Lightwave Technol.* **2019**, *37*, 4957–4966. Available online: https://ieeexplore.ieee.org/abstract/document/8753579 (accessed on 14 January 2020). [CrossRef]
20. Fernando, X.N. *Radio over Fiber for Wireless Communications: From Fundamentals to Advanced Topics*; Wiley-IEEE Press: Chichester, UK, 2014; ISBN 978-1-118-79706-8.
21. Urick, V.J.; Mckinney, J.D.; Williams, K.J. *Fundamentals of Microwave Photonics*; Wiley: Hoboken, NJ, USA, 2015; ISBN 978-1-118-29320-1.
22. Ishimura, S.; Kim, B.G.; Tanaka, K.; Nishimura, K.; Kim, H.; Chung, Y.C.; Suzuki, M. Broadband IF-Over-Fiber Transmission With Parallel IM/PM Transmitter Overcoming Dispersion-Induced RF Power Fading for High-Capacity Mobile Fronthaul Links. *IEEE Photonics J.* **2018**, *10*, 1–9. Available online: https://ieeexplore.ieee.org/document/ (accessed on 10 January 2020). [CrossRef]
23. Zakrzewski, Z. Optical RRH working in an all-optical fronthaul network. In Proceedings of the 8th International Conference on Photonics, Devices and Systems, SPIE—The International Society for Optical Engineering, Prague, Czech Republic, 1 December 2017; Volume 10603, pp. 1–12. Available online: https://www.spiedigitallibrary.org/conference-proceedings-of-spie (accessed on 26 September 2019).
24. Zakrzewski, Z. Fronthaul optical networks working with use of the hybrid analog and digital radio-over-fiber techniques. In Proceedings of the International Society for Optical Engineering, 17th Conference on Optical Fibers and Their Applications, Supraśl, Poland, 10 February 2017; Volume 10325. Available online: https://www.spiedigitallibrary.org/conference-proceedings-of-spie/10325/103250X/ (accessed on 14 January 2020).

25. ITU-T, G.694.1. Spectral Grids for WDM Applications: DWDM Frequency Grid. 2012. Available online: https://www.itu.int/rec (accessed on 5 September 2019).
26. Napoli, A.; Bohn, M.; Rafique, D.; Stavdas, A.; Sambo, N.; Potì, L.; Nölle, M.; Fischer, J.K.; Riccardi, E.; Di Giglio, A.; et al. Next generation elastic optical networks: The vision of the European research project IDEALIST. *IEEE Commun. Mag.* **2015**, *53*, 152–162. Available online: https://ieeexplore.ieee.org/document/7045404 (accessed on 2 September 2019). [CrossRef]
27. 3GPP, TS 38.401 V15.6.0. Technical Specification Group Radio Access Network: NG-RAN Architecture Description. Rel-15. 2019. Available online: https://portal.3gpp.org/desktopmodules/Specifications (accessed on 26 September 2019).
28. Layec, P.; Morea, A.; Vacondio, F.; Rival, O.; Antona, J.C. Elastic Optical Networks: The Global Evolution to Software Configurable Optical Networks. *Bell Labs Tech. J.* **2013**, *18*, 133–151. Available online: https://ieeexplore.ieee.org/document/6772713 (accessed on 5 September 2019). [CrossRef]
29. 3GPP, TS 38.104 V16.1.0. Technical Specification Group Radio Access Network; NR; Base Station (BS) Radio Transmission and Reception. 2019. Available online: https://portal.3gpp.org/desktopmodules/Specifications (accessed on 15 October 2019).
30. ITU-T, Series G, Supplement 55. Study Group 15. Radio-over-Fibre (RoF) Technologies and Their Applications. 2015. Available online: https://www.itu.int/rec/T-REC-G.Sup55/en (accessed on 10 September 2019).
31. CPRI Industry Forum (Ericsson, Huawei, NEC, and Nokia). CPRI Specification 7.0. 2015. Available online: www.cpri.info (accessed on 10 September 2019).
32. CPRI Industry Forum (Ericsson, Huawei, NEC, and Nokia). eCPRI Specification 2.0. 2019. Available online: www.cpri.info (accessed on 10 September 2019).
33. IEEE, 1588v2. IEEE Standard for a Precision Clock Synchronization Protocol for Networked Measurement and Control Systems. PNCS—Precise Networked Clock Synchronization Working Group. 2008. Available online: https://standards.ieee.org/standard (accessed on 21 August 2019).
34. ITU-T, G.8261. Timing and Synchronization Aspects in Packet Networks. 2019. Available online: https://www.itu.int/rec/T-REC-G.8261 (accessed on 12 October 2019).
35. ITU-T, G.8262. Timing Characteristics of Synchronous Ethernet Equipment Slave Clock. 2018. Available online: https://www.itu.int/rec/T-REC-G.8262 (accessed on 12 September 2019).
36. IEEE. 1914 (1914.1 Standard for Packet-Based Fronthaul Transport Networks, 1914.3 Standard for Radio Over Ethernet Encapsulations and Mappings). Next Generation Fronthaul Interface. 2018. Available online: https://standards.ieee.org/standard/ (accessed on 11 September 2019).
37. IEEE, P802.1CM. Draft V2.2. Time-Sensitive Networking for Fronthaul. 2018. Available online: www.ieee802.org (accessed on 25 August 2019).
38. ITU-T, G.709 and Amendments 1–3. Interfaces for the Optical Transport Network. 2016. Available online: http://www.itu.int/rec/T-REC-G.709/ (accessed on 22 August 2019).
39. ITU-T, Series G, Supplement 56. Study Group 15. OTN Transport of CPRI Signals. 2016. Available online: https://www.itu.int/rec/T-REC-G.Sup56/en (accessed on 20 August 2019).
40. Zaken, B.B.B.; Zanzury, T.; Malka, D. An 8-Channel Wavelength MMI Demultiplexer in Slot Waveguide Structures. *Materials* **2016**, *9*. Available online: https://www.mdpi.com/1996-1944/9/11/881 (accessed on 26 September 2019). [CrossRef]
41. Malka, D.; Sintov, Y.; Zalevsky, Z. Design of a 1 × 4 silicon-Alumina wavelength demultiplexer based on multimode interference in slot waveguide structures. *J. Opt.* **2015**, *17*. Available online: http://iopscience.iop.org/article/10.1088/2040-8978/17/12/125702 (accessed on 26 September 2019). [CrossRef]
42. Shuangfeng, H.; Chih-Lin, I.; Zhikun, X.; Corbett, R. Large-Scale Antenna Systems with Hybrid Analog and Digital Beamforming for Millimeter Wave 5G. *IEEE Commun. Mag.* **2015**, *53*, 186–194. Available online: https://ieeexplore.ieee.org/document/7010533 (accessed on 17 September 2019). [CrossRef]
43. Sohrabi, F.; Yu, W. Hybrid Analog and Digital Beamforming for mmWave OFDM Large-Scale Antenna Arrays. *IEEE J. Sel. Areas Commun.* **2017**, *35*, 1432–1443. Available online: https://ieeexplore.ieee.org/document/7913599 (accessed on 25 August 2019). [CrossRef]
44. ONF. OpenFlow Switch Specification. 2015. Available online: www.opennetworking.org (accessed on 24 August 2019).
45. ONF. Optical Transport Protocol Extensions. 2015. Available online: www.opennetworking.org (accessed on 24 August 2019).

46. 3GPP, TR 38.801 V14.0.0. Study on New Radio Access Technology: Radio Access Architecture and Interfaces. 2017. Available online: https://portal.3gpp.org/desktopmodules/Specifications/ (accessed on 24 August 2019).
47. Pfeiffer, T. Next Generation Mobile Fronthaul and Midhaul Architectures. *J. Opt. Commun. Netw.* **2015**, *7*. Available online: https://ieeexplore.ieee.org/document/7331128 (accessed on 10 August 2019). [CrossRef]
48. Miyamoto, K.; Kuwano, S.; Terada, J.; Otaka, A. *Performance Evaluation of Mobile Fronthaul Optical Bandwidth Reduction and Wireless Transmission in Split-PHY Processing Architecture*; IEEE: Anaheim, CA, USA, 2016. Available online: https://ieeexplore.ieee.org/document/7537663 (accessed on 17 August 2019).
49. Lee, C.H. *Microwave Photonics*; CRC Press: Boca Raton, FL, USA, 2007; ISBN 978-1466502864.
50. Zakrzewski, Z. Microwave-photonic networks based on single-mode multi-core optical fibers. *Photonics Lett. Pol.* **2013**, *5*, 161–163. Available online: http://photonics.pl/PLP/index.php/letters/article/view/5-57 (accessed on 26 September 2019). [CrossRef]
51. 3GPP, TS 38.211 V15.7.0. Technical Specification Group Radio Access Network; NR; Physical Channels and Modulation. 2019. Available online: https://portal.3gpp.org/desktopmodules/Specifications (accessed on 24 October 2019).
52. Shoresh, T.; Katanov, N.; Malka, D. 1 × 4 MMI visible light wavelength demultiplexer based on GaN slot waveguide structures. *Photonics Nanostructures Fundam. Appl.* **2018**, *30*, 45–49. Available online: https://www.journals.elsevier.com/photonics-and-nanostructures-fundamentals-and-applications (accessed on 27 September 2019). [CrossRef]
53. Sung, M.; Cho, S.H.; Kim, J.; Lee, J.K.; Lee, J.H.; Chung, H.S. Demonstration of IFoF-Based Mobile Fronthaul in 5G Prototype With 28-GHz Millimeter wave. *J. Lightwave Technol.* **2018**, *36*, 601–609. Available online: https://ieeexplore.ieee.org/abstract/document/8068208 (accessed on 17 September 2019). [CrossRef]
54. Molisch, A.F.; Ratnam, V.V.; Han, S.; Li, Z.; Nguyen, S.L.H.; Li, L.; Haneda, K. Hybrid Beamforming for Massive MIMO: A Survey. *IEEE Commun. Mag.* **2017**, *55*, 134–141. Available online: https://ieeexplore.ieee.org/document/8030501 (accessed on 18 September 2019). [CrossRef]
55. Zakrzewski, Z. Effectiveness of optical fiber networks in fifth and next generations of mobile systems. In Proceedings of the International Society for Optical Engineering, 18th Conference on Optical Fibers and Their Applications 2018, Nałeczow, Poland, 15 March 2019. Available online: https://www.spiedigitallibrary.org/conference-proceedings-of-spie/11045/110450M/ (accessed on 14 January 2020).
56. Chralyvy, A. Plenary Paper: The Coming Capacity Crunch. In Proceedings of the 2009 35th European Conference on Optical Communication, Vienna, Austria, 20–24 September 2009. Available online: https://ieeexplore.ieee.org (accessed on 12 January 2020).
57. Ellis, A.D.; Mac Suibhne, N.; Saad, D.; Payne, D.N. Communication networks beyond the capacity crunch. *Philos. Trans. R. Soc.* **2016**, *374*, 20150191. Available online: https://royalsocietypublishing.org/doi/full/10.1098/rsta.2015.0191 (accessed on 17 January 2020). [CrossRef] [PubMed]
58. Waldman, H. The Impending Optical Network Capacity Crunch. In Proceedings of the 2018 SBFoton International Optics and Photonics Conference (SBFoton IOPC), Campinas, Brazil, 8–10 October 2018. Available online: https://ieeexplore.ieee.org/document/8610949 (accessed on 17 January 2020).

applied sciences

Article

Towards Efficiently Provisioning 5G Core Network Slice Based on Resource and Topology Attributes

Xin Li [1], Chengcheng Guo [1,*], Jun Xu [1], Lav Gupta [2] and Raj Jain [3]

1 School of Electronic Information, Wuhan University, Wuhan 430072, China; xinli1105@whu.edu.cn (X.L.); eisxujun@whu.edu.cn (J.X.)
2 Department of Mathematics and Computer Science, University of Missouri-St. Louis, St. Louis, MO 63130, USA; lgyn6@umsl.edu
3 Department of Computer Science and Engineering, Washington University in St. Louis, St. Louis, MO 63130, USA; jain@wustl.edu
* Correspondence: netccg@whu.edu.cn

Received: 6 September 2019; Accepted: 11 October 2019; Published: 16 October 2019

Abstract: Efficient provisioning of 5G network slices is a major challenge for 5G network slicing technology. Previous slice provisioning methods have only considered network resource attributes and ignored network topology attributes. These methods may result in a decrease in the slice acceptance ratio and the slice provisioning revenue. To address these issues, we propose a two-stage heuristic slice provisioning algorithm, called RT-CSP, for the 5G core network by jointly considering network resource attributes and topology attributes in this paper. The first stage of our method is called the slice node provisioning stage, in which we propose an approach to scoring and ranking nodes using network resource attributes (i.e., CPU capacity and bandwidth) and topology attributes (i.e., degree centrality and closeness centrality). Slice nodes are then provisioned according to the node ranking results. In the second stage, called the slice link provisioning stage, the k-shortest path algorithm is implemented to provision slice links. To further improve the performance of RT-CSP, we propose RT-CSP+, which uses our designed strategy, called *minMaxBWUtilHops*, to select the best physical path to host the slice link. The strategy minimizes the product of the maximum link bandwidth utilization of the candidate physical path and the number of hops in it to avoid creating bottlenecks in the physical path and reduce the bandwidth cost. Using extensive simulations, we compared our results with those of the state-of-the-art algorithms. The experimental results show that our algorithms increase slice acceptance ratio and improve the provisioning revenue-to-cost ratio.

Keywords: 5G core network slice; network slicing; resource attributes; slice provisioning; topology attributes

---

1. Introduction

Information and communication technologies (ICTs) are infiltrating many fields, including governance, economics, defense, media, social media, health care, industry, education, etc. [1–4]. These fields are undergoing continuous digitalization and pervasive interconnection, making communication networks an indispensable infrastructure. The coming 5G networks will promote the further upgrade of human interaction. More importantly, 5G will support a variety of vertical services, such as self-driving cars, augmented reality, live video, telemedicine, and financial transactions [5]. While 5G will improve productivity and optimize business processes, it will inevitably bring new legal and ethical issues that cannot be ignored [6,7].

The 5th generation (5G) mobile networks are expected to handle the tremendous growth of data from diverse and heterogeneous services. Softwarization, virtualization, and cloud-based 5G architecture design [8,9] are considered to be promising technologies to address the challenges introduced by the diversified service demands. Network slicing is one of the key concepts that can be realized by these techniques to support the specific needs of vertical industries. End-to-end network slicing enables multiple network services to share a single physical network infrastructure (also called the substrate network) including radio access networks (RAN) and core networks [10,11]. The big idea behind network slicing is to allow the shared 5G physical network infrastructure to be *sliced* into multiple logical networks, each of which is a collection of virtual computing and networking resources capable of supporting a specific type of service. It is, therefore, believed that network slicing will be an indispensable enabler of 5G network architecture to meet the diverse requirements of vertical applications.

We can broadly divide network slicing into two categories: radio access network slicing and core network slicing. In this paper, we focus on 5G core network slicing. A three-layer 5G core network slicing system model proposed by us has been elaborated in [12] and illustrated in Figure 1. There are three administrative roles in this model: 5G core infrastructure provider, 5G core slice provider, and slice tenants. The Infrastructure Provider (InP) owns the 5G core infrastructure and can lease physical resources such as computing and networking resources to the slice providers. A Slice Provider (SP) can be regarded as a virtual telecommunications service provider (TSP). The SP controls the virtualization of the resources to form network slices and provides services for users. Slice tenant is the consumer of an application specific network slice. It informs the slice provider of the characteristics of the service it needs. The slice provider requests physical resources from the infrastructure provider to create a network slice to provide the service according to the tenant's demands. The slice provisioning system interacts with the three roles to orchestrate and manage physical resources.

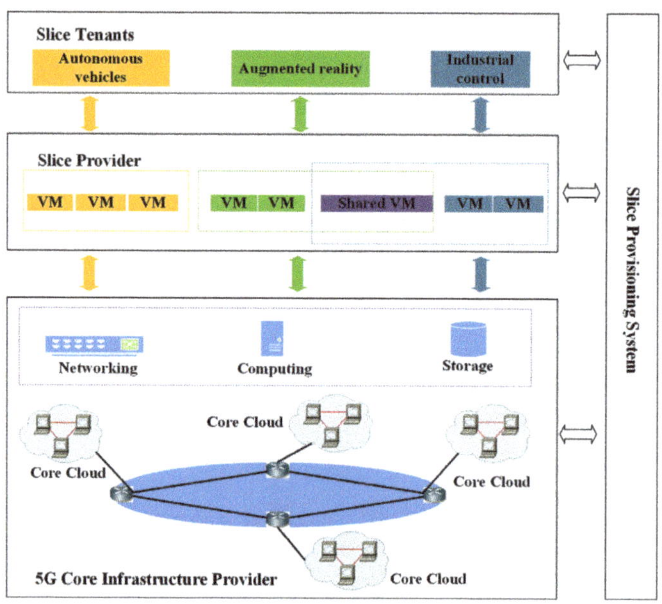

Figure 1. Illustration of three-layer 5G core network slicing system model.

Although network slicing has attracted increasing attention from both academia and industry [13], slice provisioning is a key issue to be addressed [14]. Slice provisioning is an approach to creating separate virtual networks based on service requirements using common physical computing and networking resources. Two sub-tasks in slice provisioning are slice node provisioning and slice link provisioning. From the perspective of InP, since the computing and networking capacities of the physical network are limited, increasing physical resource utilization to provision more slices is crucial to raising its revenue. Therefore, in this article, we study how to efficiently provision 5G core network slices to optimize resource utilization of the 5G physical network infrastructure, thus, increasing the revenue of InP.

The slice provisioning problem in 5G network slicing is essentially the same as the traditional virtual network embedding (VNE) problem [15] in network virtualization (NV) (we rename the virtual network embedding (VNE) problem to slice provisioning problem in network slicing). Most previous VNE methods have only considered the resource attributes of the network and ignored its topology attributes to allocate physical resources to virtual network requests. Notwithstanding that several approaches consider the resource and topology attributes, the local and global resource attributes as well as the local and global topology attributes are not reasonably defined, which causes these methods to be not effective.

Based on the above considerations, we have designed a heuristic 5G core network slice provisioning strategy based on the local and global network resource attributes and topology attributes including the product of the CPU of the node and all its adjacent links, i.e., local resource attribute, the minimum bandwidth of the links in the shortest path of the node to all other nodes and the minimum CPU of the nodes along the shortest path, i.e., global resource attribute, node degree centrality, and node closeness centrality. When a 5G core slice request arrives at the slice provisioning system, the system uses resource attributes and topology attributes to perform comprehensive node evaluation and ranking, and then slice nodes are provisioned according to the ranking results. Next, slice links are provisioned using the k-shortest path algorithm. Our contributions are summarized as follows:

- We propose a network node scoring and ranking method by jointly considering local and global network resource and topology attributes. Specifically, we introduce a *cooperative provisioning coefficient* for the physical node scoring to enhance the efficiency of provisioning slice links.
- We design a two-stage 5G core slice provisioning algorithm, called RT-CSP, which includes a heuristic slice node provisioning algorithm and a k-shortest path based slice link provisioning algorithm. In the first stage, slice nodes are provisioned in a heuristic manner in accordance with the network node ranking results. In the second stage, the k-shortest path algorithm is used to provision slice links.
- To further improve the performance of RT-CSP, we propose RT-CSP+ slice provisioning algorithm based on our designed *minMaxBWUtilHops* strategy in the slice link provisioning stage. The strategy selects the physical path which has the minimum product of the maximum link bandwidth utilization and its hop count from the candidate physical paths obtained by the k-shortest path algorithm to host the slice link.
- We verify the performance of our proposed algorithm through extensive simulations and prove that our algorithm can increase the slice request acceptance ratio and, hence, the revenue of physical network provider.

The remainder of the paper is organized as follows. Section 2 discusses the related work. In Section 3, we describe the 5G core slice provisioning problem and present the system model. The heuristic 5G core slice provisioning algorithms based on network resource attributes and topology attributes are presented in Section 4. In Section 5, we present simulation experiments and the experimental results. Finally, the conclusions and future work are laid out in Section 6.

2. Related Work

In this section, we first summarize several classic methods to solve the VNE problem with special regard to those methods considering network resource attributes or topology attributes. Then, we review the latest work related to the resource allocation in network slicing.

2.1. VNE Methods

Since VNE problem has been proven to be NP-hard by being reduced to multiway separator problem [16], its solutions can be grouped into three categories: the exact solutions, meta-heuristic solutions and heuristic solutions. Houidi et al. [17] proposed and evaluated an exact algorithm for the VNE problem using integer linear programming (ILP). They used the branch and bound method to embed virtual network requests to multiple physical networks. The results showed that the proposed exact algorithm was effective for solving small-scale problem instances. The VNE problem considered in [18] was formalized as an ILP model. Due to the complexity of ILP, the authors proposed a discrete particle swarm optimization (DPSO) method to solve the VNE problem. Simulation results showed that the DPSO method could get a better convergence performance than existing PSO methods.

Some heuristic algorithms used network resource attributes or topology attributes to rank nodes in the node mapping stage in VNE. Yu et al. [19] employed the product of the node CPU capacity and its adjacent link bandwidth for node ranking. This method was used in many other research works. However, it only considered the local resource attributes of the node. Cheng et al. [20] introduced topology attributes for the first time to embed virtual networks. Referring to the Google PageRank algorithm, they used Markov Random Walk method to rank nodes and performed node mapping in a greedy manner according to the ranking results. Wang et al. [21] introduced the network centrality in complex network theory into the VNE problem and ranked the nodes by calculating their closeness centrality. However, the closeness centrality is only one of the topology properties of the network, which measures the distance between a node and other nodes. A network topology attribute and network resource-considered algorithm was proposed to embed virtual networks in [22]. However, they only defined the local resource attributes. None of these algorithms considered the global network resource attributes as well as reasonably combined the resource attributes and topology attributes to comprehensively evaluate the importance of nodes.

2.2. Resource Allocation in Network Slicing

The authors of [23] specifically studied the problem of provisioning slice links with splittable flows. Since this problem is NP-hard, using the idea of the multipartite graph, they proposed a polynomial heuristic algorithm based on linear relaxation and randomized rounding. It was verified by simulation that the algorithm could achieve good performance. However, this study did not solve the slice node provisioning. The work in [24] used a mixed integer linear programming (MILP) model for the dynamic slicing problem and proposed several heuristic algorithms for it considering temporal variations of the virtual resource requirements. The simulation results showed that the dynamic slice provisioning could increase the slice acceptance ratio and, thus, enabled the physical network provider to increase their revenue. However, dynamic slicing comes at a cost of service quality degradation. Danish Sattar and Ashraf Matrawy [25] proposed an optimal slice allocation strategy for the 5G core network concerning the intra-slice isolation and delay requirement of slices. They formulated the problem as a MILP model and solved it with CPLEX. Their results showed that the resource utilization of the physical network would improve if the slice isolation was not considered, and stricter delay requirement also affected the slice acceptance ratio as well as resource utilization. In the very latest study [26], the authors presented a latency-optimal resource allocation method for 5G transport network slices to support URLLC services. They introduced the network resource attributes and topology attributes to resource allocation in network slicing, but they did not delve into the impact of network resource attributes and topology attributes on slice provisioning.

3. Problem Description and System Model

In this section, we first describe the 5G core slice provisioning problem, and then present the system model. A summary of used notations is presented in Table 1.

Table 1. System model notations.

Notation	Description
G^I	5G core infrastructure topological graph.
V^I	Set of physical nodes.
E^I	Set of physical links.
$c_0(v^I)$	Initial total CPU capacity of physical node v^I.
$c_a(v^I)$	Available CPU capacity of physical node v^I.
$c_u(v^I)$	Total CPU capacity of physical node v^I allocated to slice nodes.
$loc(v^I)$	Location of physical node v^I.
$\phi(v_i^I, v_j^I)$	Euclidean distance between physical nodes v_i^I and v_j^I.
$b_0(e^I)$	Initial total bandwidth of physical link e^I.
$b_a(e^I)$	Available bandwidth of physical link e^I.
$b_u(e^I)$	Total bandwidth of physical link e^I allocated to slice links.
$P^I(v_i^I, v_j^I)$	Set of loop-free physical paths between v_i^I and v_j^I.
$L(p^I(v_i^I, v_j^I))$	Set of links in $p^I(v_i^I, v_j^I)$.
G^S	5G core network slice request topological graph.
V^S	Set of slice nodes.
E^S	Set of slice links.
$c(v^S)$	CPU capability required by slice node v^S.
$loc(v^S)$	Expected deployed location of slice node v^S.
$r(v^S)$	Maximum deployed deviation allowed by slice node v^S.
$b(e^S)$	Bandwidth required by slice link e^S.

3.1. 5G Core Slice Provisioning Problem Description

The InP supplies 5G core physical infrastructure. It consists of physical nodes deployed in different locations and physical links connecting physical nodes. Physical nodes have computing, storage, and network forwarding capabilities. Virtual machine (VM) or container [27] technologies can be used to enable a physical node to host logically isolated virtual routers or VNFs (e.g., firewall, proxy, etc.). High-speed fiber optic cables are deployed in 5G core networks as physical links which have attributes such as bandwidth. Slice links are hosted on physical links or paths.

Slice tenants request 5G core network slices from the SP. A core network slice instance consists of virtual network functions and virtual links. In this study, we assume that slice nodes offer the same type of virtual network function, which is virtual computing function. Slice nodes and links request computing resources, storage resources, bandwidth resources, etc. from the slice provisioning system. Without loss of generality, we only consider computing resources and bandwidth resources here. In addition, a slice node may have a location constraint instead of being arbitrarily deployed.

The slice provisioning process includes mapping slice nodes to physical network nodes that satisfy the resource and deployment location requirements and mapping slice links onto physical paths that meet the bandwidth requests. The slice request is only accepted if the requirements of all nodes and links of the slice are satisfied; otherwise, it is rejected. We make the following assumptions for the slice provisioning problem:

- The topology of the slice remains unchanged during the life cycle of the slice, which means slice reconfiguration is not considered here.
- Slice nodes from the same 5G core network slice request can only be mapped to different physical nodes, that is, co-hosting is not allowed [28].
- Slice links cannot be split. They can only be hosted by one physical path [28].

3.2. System Model

3.2.1. 5G Core Infrastructure

The 5G core infrastructure topology is represented by a weighted undirected graph $G^I = (V^I, E^I)$, where V^I is the set of physical nodes and E^I is the set of physical links. For each physical node $v^I \in V^I$, its initial total and available computing capacities are represented as $c_0(v^I)$ and $c_a(v^I)$, respectively. Its location is denoted by $loc(v^I)$, which is represented as Cartesian coordinates, i.e., $loc(v^I) = (x(v^I), y(v^I))$. The distance of the physical link with physical nodes v_i^I and v_j^I as endpoints is the Euclidean distance between them, expressed as $\phi(v_i^I, v_j^I)$. For each physical link $e^I \in E^I$, its initial total and available bandwidth are represented as $b_0(e^I)$ and $b_a(e^I)$, respectively. The set of all loop-free paths in the infrastructure is denoted as P^I. $P^I(v_i^I, v_j^I)$ represents the set of loop-free physical paths between v_i^I and v_j^I. For each path $p^I(v_i^I, v_j^I) \in P^I(v_i^I, v_j^I)$, $L(p^I(v_i^I, v_j^I))$ is the set of links in $p^I(v_i^I, v_j^I)$. Then, the bandwidth of $p^I(v_i^I, v_j^I)$ is defined as $b(p^I(v_i^I, v_j^I)) = \min_{e^I \in L(p^I(v_i^I, v_j^I))} b(e^I)$.

3.2.2. 5G Core Slice Request

5G core slice requests arrive dynamically at the resource provisioning system. The ith slice request is represented by a triplet $SR_i = (G_i^S, t_i^a, t_i^l)$, where G_i^S, t_i^a, and t_i^l represent the topology of the ith slice, its arrival time and its lifetime, respectively. The slice topology is represented by a weighted undirected graph $G^S = (V^S, E^S)$. Here, V^S is the set of slice nodes and E^S is the set of slice links. For each slice node $v^S \in V^S$, the CPU capability required by it is $c(v^S)$. Its expected deployed location is $loc(v^S) = (x(v^S), y(v^S))$ and the maximum deployed deviation allowed is $r(v^S)$, that is, the slice node can be deployed at the location within a circle whose center is the expected location $loc(v^S)$ and the radius is $r(v^S)$. Each slice link $e^S \in E^S$ is characterized by the amount of required bandwidth $b(e^S)$.

3.2.3. Slice Provisioning Process

The resource provisioning for 5G core slice task contains two sub-tasks: mapping slice nodes to physical nodes and mapping slice links to physical paths. Slice nodes from one slice request cannot be mapped to the same physical node. A slice link is mapped to a physical path with endpoints that host two slice nodes connected by the slice link. Once the resource provisioning for a slice request is successful, the allocated resource will be dedicated for the slice during its lifetime. When the lifetime ends, the allocated resource is released.

The slice node mapping function is defined as follows:

$$\boldsymbol{M}(V): V^S \to V', \qquad V' \subseteq V^I.$$

Slice nodes are mapped to V' that is a subset of the physical node set V^I. Since $\boldsymbol{M}(V)$ is an injective function, $\forall v_k^S, v_l^S \in V^S$, we have:

$$\boldsymbol{M}(v_k^S) = \boldsymbol{M}(v_l^S), \qquad \text{if and only if } v_k^S = v_l^S.$$

The slice link mapping function is defined as follows:

$$\boldsymbol{M}(E): E^S \to P', \qquad P' \subseteq P^I.$$

Slice links are mapped to P' that is a subset of all loop-free paths set P^I in the infrastructure. Thus, we define 5G core network slice mapping function as:

$$\boldsymbol{M}(S): (V^S, E^S) \to (V', P').$$

The calculations of resource updating during mapping process are as follows. We use $c_u(v^I, t)$ to indicate the total computing resources that the slice provisioning system assigns to all the mapped slice nodes from different slice requests at time t. It is defined below:

$$c_u(v^I, t) = \sum_{v^S} c(v^S).$$

Then, the available computing resources the physical node v^I has at time t can be calculated as:

$$c_a(v^I, t) = c_0(v^I) - c_u(v^I, t).$$

Similarly, we use $b_u(e^I, t)$ to represent the bandwidth resources the slice provisioning system has allocated to all the slice links at time t.

$$b_u(e^I, t) = \sum_{e^S} c(e^S).$$

Then, the available bandwidth the physical link e^I has at time t can be calculated as:

$$b_a(e^I, t) = b_0(e^I) - b_u(e^I, t).$$

To ensure a successful 5G core slice provisioning, all node and link constraints need to be met. The corresponding constraints are defined as follows.

Slice node mapping: Each slice node should be mapped to one physical node.

$$\sum_{v_i^I} x_i^k = 1, \qquad \forall v_k^S \in V^S, \tag{1}$$

where x_i^k indicates whether the slice node v_k^S is mapped to the physical node v_i^I or not. If v_k^S is mapped to v_i^I, x_i^k is 1. Otherwise, it is 0.

One-to-one node mapping: Each physical node can only host one slice node from the same slice request.

$$\sum_{v_k^S} x_i^k \leq 1, \qquad \forall v_i^I \in V^I. \tag{2}$$

CPU capacity: The allocated CPU capacity for slice nodes at a physical node cannot exceed the available CPU capacity of that physical node.

$$\sum_{v_k^S} x_i^k \cdot c(v_k^S) \leq c_a(v_i^I), \qquad \forall v_i^I \in V^I. \tag{3}$$

Location constraint: The distance between the mapped location of a slice node and its expected deployment location cannot exceed the maximum allowed deviation.

$$x_i^k \cdot dis(v_k^S, v_i^I) \leq r(v^S), \tag{4}$$

where $dis(v_k^S, v_i^I)$ is calculated as:

$$dis(v_k^S, v_i^I) = \sqrt{(x(v_k^S) - x(v_i^I))^2 + (y(v_k^S) - y(v_i^I))^2}. \tag{5}$$

Bandwidth: The sum of bandwidth allocated to all the slice links that are mapped to one physical link cannot exceed its available bandwidth.

$$\sum_{e^S_{kl}} y^{kl}_{ij} \cdot b(e^S_{kl}) \le b_a(e^I_{ij}), \qquad \forall e^I_{ij} \in E^I. \tag{6}$$

If the physical link e^I_{ij} hosts the slice link e^S_{kl}, y^{kl}_{ij} is 1. Otherwise, it is 0.

3.2.4. Performance Metrics

The 5G core network infrastructure provider, while providing physical resources to the tenants, attempts to maximize its operating profit. As such, the main goal of the resource provisioning for slices is to maximize resource provisioning revenue by provisioning as many slice requests as possible. In this paper, we use slice acceptance ratio, long-term average provisioning revenue and provisioning revenue-to-cost ratio as metrics to evaluate the performance of the provisioning algorithms. They are defined below.

Slice acceptance ratio (λ): It is the ratio of the number of slices successfully provisioned to the total number of slice requests that arrive over a period of time. Then,

$$\lambda = \lim_{T \to +\infty} \frac{\sum_{t=0}^{T} S_m(t)}{\sum_{t=0}^{T} S(t)}, \tag{7}$$

where $S(t)$ is the total number of slice requests at time t and $S_m(t)$ is the number of slice requests provisioned successfully at time t.

Long-term average provisioning revenue (μ): Here, we assume the unit price of CPU capacity and bandwidth is 1. Then, the provisioning revenue of slice request G^S at time t is defined as:

$$REV(G^S, t) = \sum_{v^S \in V^S} c(v^S) + \sum_{e^S \in E^S} b(e^S). \tag{8}$$

The long-term average provisioning revenue is represented as:

$$\mu = \lim_{T \to +\infty} \frac{\sum_{t=0}^{T} \sum_{G^S \in S_m(t)} REV(G^S, t)}{T}. \tag{9}$$

Provisioning revenue-to-cost ratio (η): The provisioning cost of slice request G^S at time t is

$$COST(G^S, t) = \sum_{v^S \in V^S} c(v^S) + \sum_{e^S \in E^S} |L(p^I(e^S))| b(e^S), \tag{10}$$

where $p^I(e^S)$ is the physical path hosting the slice link e^S and $L(p^I(e^S))$ denotes the set of physical links in $p^I(e^S)$. Hence, we define provisioning revenue-to-cost ratio η as:

$$\eta = \frac{REV}{COST} = \lim_{T \to +\infty} \frac{\sum_{t=0}^{T} \sum_{G^S \in S_m(t)} REV(G^S, t)}{\sum_{t=0}^{T} \sum_{G^S \in S_m(t)} COST(G^S, t)}. \tag{11}$$

4. Heuristic 5G Core Network Slice Provisioning Algorithm Design

In this section, we describe our heuristic 5G Core Slice Provisioning algorithm based on network Resource and Topology attributes called RT-CSP in detail. Accordingly, we first present a method for ranking network nodes by using these attributes. Then, we elaborate our two-stage algorithm consisting of algorithms for heuristic slice node provisioning and k-shortest path based slice link provisioning. Finally, we analyze the time complexity of the RT-CSP algorithm and prove that it can run in polynomial time.

4.1. Node Ranking Based on Network Resource Attributes and Topology Attributes

In the slice node provisioning phase, a physical node for hosting a slice node needs to be carefully selected in order to meet its required CPU capacity and provisioned location requirement. Many studies in the VNE research area only consider the local resource attributes of a network node such as its CPU capacity and its adjacent link bandwidth. These works map virtual network nodes according to node ranking results based on the local resource. However, these studies do not consider global resource attributes. Moreover, the topology properties of nodes also affect the evaluation of the importance of nodes, such as degree centrality, betweenness centrality, and closeness centrality [29]. Only considering local resource attributes cannot accurately rank nodes. Therefore, we take into consideration both the local and the global resource attributes and topology attributes to comprehensively evaluate the importance of nodes.

4.1.1. Local Resource Attributes

The local resource metric of a node is obtained by multiplying the CPU capacity of the node by the sum of bandwidths of all its adjacent links.

$$LR(v_i) = c(v_i) \sum_{e \in E(v_i)} b(e), \tag{12}$$

where $E(v_i)$ is the set of all the adjacent links of v_i. The reason we define this metric is that the larger $LR(v_i)$ is, the more slice nodes can be hosted by the physical node.

4.1.2. Global Resource Attributes

Considering only the local resources of a node can cause load imbalance and resource fragmentation in the physical network. To address this, we take the minimum bandwidth of the links in the shortest path of the node to all other nodes and the minimum computing capacity of the nodes along the shortest path as the global resource metric. The following formula is its normalized definition.

$$GR(v_i) = \frac{\sum_{i \neq j} [b(p(v_i, v_j)) + c(p(v_i, v_j))]}{|V| - 1}, \tag{13}$$

where $b(p(v_i, v_j))$ is the minimum bandwidth of the links and $c(p(v_i, v_j))$ is the minimum CPU of the nodes in the shortest path between v_i and v_j.

4.1.3. Degree Centrality

In an undirected graph, the degree centrality of a node indicates the ratio of the number of its adjacent links to the total number of links in the graph, i.e., normalized degree centrality.

$$DC(v_i) = \frac{\sum_{v_j} a_{ij}}{|V| - 1}, \tag{14}$$

where a_{ij} is 1 if node v_i and node v_j are connected by a link; otherwise, it is 0. The degree centrality measures the local topological importance of the node in the network. The greater it is, the more connected the node is and the more likely it is to be selected.

4.1.4. Closeness Centrality

Closeness centrality is a method of measuring the importance of a node from a global topological perspective.

The closeness centrality of a node is obtained by first calculating the sum of the shortest paths from the node to all other nodes in the graph and then taking the reciprocal of the sum. The normalized closeness centrality is:

$$CC(v_i) = \frac{|V| - 1}{\sum\limits_{i \neq j} d(v_i, v_j)}, \tag{15}$$

where $d(v_i, v_j)$ is the length of the shortest path between node v_i and node v_j.

Thus, the nodes that are near the geometric center of the graph have higher closeness centrality.

4.1.5. Node Ranking Strategy

Our node ranking strategy combines all of the above four attributes. We rank the nodes as follows:

$$S(v_i) = \alpha LR(v_i) \cdot DC(v_i) + \beta GR(v_i) \cdot CC(v_i). \tag{16}$$

In this strategy, we have integrated the local resources, global resources, local topology attributes, and global topology attributes. It can systematically evaluate nodes in the physical network and slice requests. α and β are used to weigh the relative importance of local attributes and global attributes of the network.

4.2. Heuristic Slice Provisioning

The proposed two-stage slice provisioning algorithm is described in detail below.

4.2.1. Slice Node Provisioning

When a slice request arrives at the slice provisioning system, each slice node in the slice request is scored according to Equation (17), and then the slice nodes are ranked according to the score from high to low. The higher the score of the slice node is, the more preferentially it is provisioned. Here, $\alpha = \beta = 0.5$.

$$S(v_i) = \alpha LR(v_i^S) \cdot DC(v_i^S) + \beta GR(v_i^S) \cdot CC(v_i^S). \tag{17}$$

If physical network nodes are also scored according to Equation (17), the selected physical nodes hosting slice nodes in the final provisioning result may be far apart. Long physical paths would have to be provisioned under this circumstance, resulting in low utilization of physical network resources. To overcome this issue, we use the following cooperative provisioning method. When the current slice node is to be provisioned, the candidate physical node-set that can host the slice node is attained. The sum of the hop counts of the shortest path between the physical nodes hosting all the neighbor slice nodes of current slice node and the candidate physical node is calculated as the cooperative provisioning coefficient. The candidate physical node having the smaller coefficient may be a good hosting node. This cooperative way is beneficial to obtain a shorter physical path to host the slice link in the slice link provisioning stage, thereby improving the utilization of network bandwidth. Thus, we introduce cooperative provisioning coefficient for scoring the physical nodes.

$$H(v_i^I) = \sum_{v_j^I \in M(Adj(v^S))} h(v_i^I, v_j^I). \tag{18}$$

$$S(v_i^I) = \frac{\alpha LR(v_i^I) \cdot DC(v_i^I) + \beta GR(v_i^I) \cdot CC(v_i^I)}{H(v_i^I) + \epsilon}. \quad (19)$$

$H(v_i^I)$ is the cooperative provisioning coefficient. v_i^I represents the candidate physical node that satisfies the CPU and location requirements of slice node v^S. $M(Adj(v^S))$ represents a physical node set hosting all the neighbor slice nodes of the slice node v^S. $h(v_i^I, v_j^I)$ is the hop counts of the shortest path between the physical nodes v_i^I and v_j^I. ϵ is set to be 10^{-5} to prevent divisor from being 0. Here, $\alpha = \beta = 0.5$. The slice node provisioning algorithm is described in Algorithm 1.

Algorithm 1 Slice node provisioning based on network resource and topology attributes.

Input: Infrastructure network G^I and slice request G^S
Output: Slice node provisioning solution

for each slice node $v^S \in V^S$ **do**
  $S(v^S)$ is calculated based on Equation (17).
end for
Rank all the slice nodes in descending order of $S(v^S)$ value.
Put the ranking results into *sliceNodeRankList*.
for each physical node $v^I \in V^I$ **do**
  RTScore based on resource and topology attributes is calculated based on Equation (16).
end for
for each slice node $v^S \in sliceNodeRankList$ **do**
  Obtain the candidate physical nodes $candidate(v^S)$ for v^S meeting its CPU capacity and provisioned location demands.
  if $candidate(v^S)$ is not empty **then**
    Obtain the physical nodes $M(Adj(v^S))$ hosting the neighbor slice nodes of v^S.
    for each physical node $v^I \in candidate(v^S)$ **do**
      Cooperative provisioning coefficient $H(v^I)$ is calculated based on Equation (18).
      $S(v^I)$ is calculated based on Equation (19).
    end for
    Provision v^S onto the candidate physical node which has the largest S.
    Put the provisioning result of v^S into *sliceNodeProvisioningList*.
  else
    return *sliceNodeProvisioningFailed*
  end if
end for
return *sliceNodeProvisioningList*

4.2.2. Slice Link Provisioning

In the slice link provisioning stage, since the slice link with a larger amount of bandwidth resource demand is more difficult to be provisioned, the slice links are first ranked according to the bandwidth requirements from large to small. The k-shortest path algorithm [30] is then implemented to provision slice links. Thus, our RT-CSP algorithm includes the heuristic slice node provisioning algorithm and the basic k-shortest path-based slice link provisioning algorithm.

To further improve the performance of RT-CSP, we propose a novel strategy, called *minMaxBWUtilHops*, in the slice link provisioning stage. After the k-shortest path algorithm obtains k candidate physical paths for each slice link satisfying its bandwidth demand, *minMaxBWUtilHops* evaluates each candidate physical path as follows:

$$\Gamma_{p^I} = (1 - \frac{b_a(e^I)}{b_0(e^I)})_{max} \cdot |L(p^I)|. \quad (20)$$

Γ_{p^I} is the product of the maximum link bandwidth utilization of the candidate physical path p^I and its hop counts. The candidate physical path with the smallest Γ_{p^I} is selected to host the slice link. The reason we propose this strategy is that the physical link with large bandwidth utilization in the physical path becomes the bottleneck of the path, which causes the provisioning of the other slice links to fail easily, resulting in a decrease in the slice acceptance ratio. In addition, selecting the physical path with fewer hops can reduce the provisioning cost. The slice link provisioning algorithm based on the *minMaxBWUtilHops* is described in Algorithm 2. The slice provisioning algorithm with heuristic slice node provisioning algorithm and the *minMaxBWUtilHops* based slice link provisioning algorithm is named as RT-CSP+.

Algorithm 2 Slice link provisioning based on *minMaxBWUtilHops*.

Input: Infrastructure network G^I, slice request G^S, and slice node provisioning results $sliceNodeProvisioningList$
Output: Slice link provisioning solution
Rank all the slice links in E^S based on bandwidth requirements from large to small.
Put the ranking results into $sliceLinkRankList$.
for each slice link $e^S \in sliceLinkRankList$ **do**
  k shortest path algorithm is implemented to obtain the candidate substrate paths $subPathList$ for e^S meeting its bandwidth demand.
  if $subPathList$ is not empty **then**
    for each substrate path $subPath \in subPathList$ **do**
      Calculate Γ_{p^I} based on Equation (20).
    end for
    Provision e^S onto the candidate substrate path with the minimum Γ_{p^I}.
  else
    return $sliceLinkProvisioningFailed$
  end if
end for
return $sliceLinkProvisioningList$

4.2.3. Slice Provisioning

When the ith slice request $SR_i = (G_i^S, t_i^a, t_i^l)$ arrives at the slice provisioning system, the system first checks the already provisioned slices whose lifetime ends at time t_i^a and releases the physical resources they occupied. Then, slice nodes and slice links of G_i^S are, respectively, provisioned according to the above heuristic node provisioning and the link provisioning algorithms. The slice is only accepted by the slice provisioning system if the nodes and the links are both provisioned successfully. The slice provisioning algorithm RT-CSP+ is described in Algorithm 3. The only difference between RT-CSP and RT-CSP+ is that RT-CSP uses the basic k-shortest path-based slice link provisioning while RT-CSP+ uses the *minMaxBWUtilHops* based slice link provisioning.

4.2.4. Time Complexity of RT-CSP+ Algorithm

In this section, we analyze the time complexity of the RT-CSP algorithm. Its time complexity is the sum of time complexities of slice node provisioning algorithm (i.e., Algorithm 1) and the slice link provisioning algorithm (i.e., Algorithm 2). The complexity of Algorithm 1 is dominated by the calculation of closeness centrality, whose complexity is $O(|V^I||E^I| + |V^I|^2)$. Then, the

k-shortest path algorithm [30] is implemented in the slice link provisioning stage. Its complexity is $O(k|V^I|(|E^I| + |V^I|log|V^I|))$. Therefore, the time complexity of RT-CSP algorithm is $O(|V^I||E^I| + |V^I|^2) + O(k|V^I|(|E^I| + |V^I|log|V^I|))$. It can run in polynomial time.

Algorithm 3 Slice provisioning algorithm RT-CSP.

Input: Infrastructure network G^I and the ith slice request $SR_i = (G_i^S, t_i^a, t_i^l)$
Output: Slice provisioning result

Check slice requests whose lifetime ends at t_i^a, release physical resources they occupied and update physical resources.
Provision slice nodes of G_i^S using Algorithm 1.
if Slice nodes of G_i^S provisioning failed **then**
    return $sliceProvisioningFailed$
else
    Provision links of slice G_i^S using Algorithm 2.
    if Links of slice G_i^S provisioning failed **then**
        return $sliceProvisioningFailed$
    else
        Provision slice request G_i^S, allocate physical resources and update physical resources.
        return $sliceProvisioningSucceeded$
    end if
end if

5. Performance Evaluation

In this section, we evaluate the performance of the proposed heuristic 5G core slice provisioning algorithms RT-CSP+ and RT-CSP. First, we describe the experimental settings for implementing our algorithms. Then, we present the results obtained from extensive evaluation experiments and analyze the results by comparing them with the state-of-the-art algorithms.

5.1. Evaluation Settings

We developed a discrete event simulator using Java to evaluate our algorithms and ran all the experiments on a Windows 10 laptop with Intel Core i7-6820HQ CPU and 24 GB RAM. The topology generation package "Brite" [31] was integrated with our simulator to generate the 5G core infrastructure topology and the 5G core slice requests based on the Waxman topology model [32].

To compare our results with those of existing research, the simulation parameters were set according to the parameter settings widely used in previous research [19,28,33]. They are described as follows and summarized in Table 2.

The physical network nodes are randomly deployed in a rectangular area of 500 by 500. The initial total available CPU capacities of the nodes are real numbers uniformly distributed between 50 and 100. Adjacent nodes are connected by a probability of 0.5 to form physical links, whose initial total available bandwidths are real numbers uniformly distributed between 50 and 100.

The 5G core slice requests arrive following a Poisson process. The number of nodes in the slice request is a uniformly distributed integer between 2 and 10. For each slice request, the slice nodes allow the provisioned position to have a deviation of less than 80. The CPU demands of the slice nodes are real numbers uniformly distributed between 1 and 20. Slice nodes are connected by a probability to form slice links. The bandwidth requirement of each slice link takes a uniformly distributed real number in the range [1, 20]. The lifetime of the slice request follows the exponential distribution with a mean of 500 time units. We have 2000 slice requests in total in the experiments.

Table 2. Notations of system model.

Parameter	Description
Number of substrate nodes	50/100/150
Probability of connecting substrate nodes	0.5
Substrate node CPU	$U[50, 100]$
Substrate link bandwidth	$U[50, 100]$
Lifetime of slice requests obeying Exponential distribution	500 time units in average
Number of slice requests	2000
Number of slice nodes in each slice	$U[2, 10]$
Probability of connecting slice nodes	0.2/0.5/0.8
Slice node CPU demand	$U[1, 20]$
Slice link bandwidth demand	$U[1, 20]$

5.2. Evaluation Results and Analysis

To evaluate the experimental results, we compared the state-of-the-art algorithms, as listed in Table 3. The RT-CSP+ and RT-CSP algorithms are our proposed algorithms. First, we evaluated the performances of these algorithms in the scenario where the slice request arrival rate is four requests per 100 time units. Next, we changed the slice link connected probability to study its effects on the performance of the algorithms. Then, to verify the scalability of our algorithms, we examined simulation scenarios with different slice arrival rates and different sizes of the substrate network. We ran each experiment for 10 times to analyze experimental results.

Table 3. Algorithms for comparison.

Notation	Description
RT-CSP+	The provisioning algorithm considering resource and topology attributes with the *minMaxBWUtilHops* based slice link provisioning
RT-CSP	The provisioning algorithm considering resource and topology attributes with the basic *k*-shortest path-based slice link provisioning
VNE-DCC	The algorithm considering local resource and topology attributes in [33]
NRM-VNE	The algorithm only considering local resource attributes in [28]
CC	The provisioning algorithm in [21] considering classic closeness centrality

5.2.1. Experiments in the Scenario where the Slice Request Arrival Rate Is Four Requests Per 100 Time Units

In this scenario, there are 100 substrate nodes in the substrate network and the slice nodes are connected by a probability of 0.5. The results of slice acceptance ratio, long-term average revenue and the revenue-to-cost ratio of the algorithms are shown in Figures 2 and 3a,b, respectively.

Figure 2 shows our algorithm RT-CSP+ has the best slice acceptance performance over the entire simulation time. The acceptance ratio of all algorithms is relatively high at the beginning of the simulation because the CPU and bandwidth of the physical network are sufficient. As the simulation progresses, the available resources of the physical network gradually reduce due to the occupation of the active slice requests in the provisioning system, resulting in a gradual decrease in the slice reception ratio. After 10,000 time units, the slice acceptance ratio tends to stabilize. The reason is that the arrival and departure of the slices reach a relatively balanced state, and thus the available resources of the physical network are relatively stable. When the simulation time reaches 40,000 time units, the slice acceptance ratio of RT-CSP+ is 91.52%, which is 15.06%, 17.93%, and 51.25% higher than those of VNE-DCC, NRM-VNE, and CC, respectively. Our algorithm can comprehensively evaluate nodes from the perspective of local and global resource and topology attributes, making node provisioning more optimized. Thus, our algorithm can increase the slice acceptance ratio. The slice acceptance ratio of RT-CSP+ is higher than RT-CSP, which shows that our *minMaxBWUtilHops* strategy in the link provisioning stage can further enhance the performance of RT-CSP.

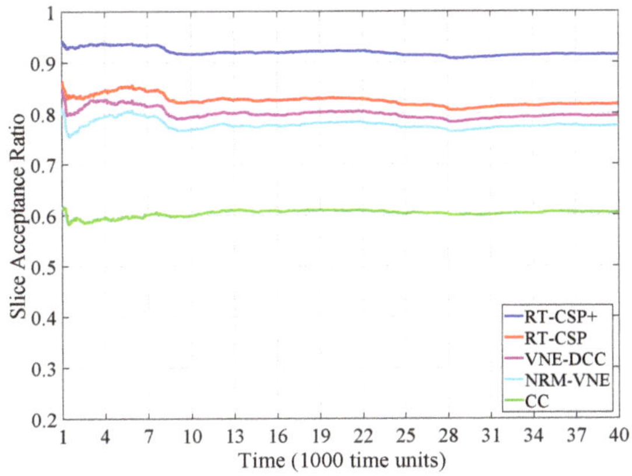

Figure 2. Slice acceptance ratio.

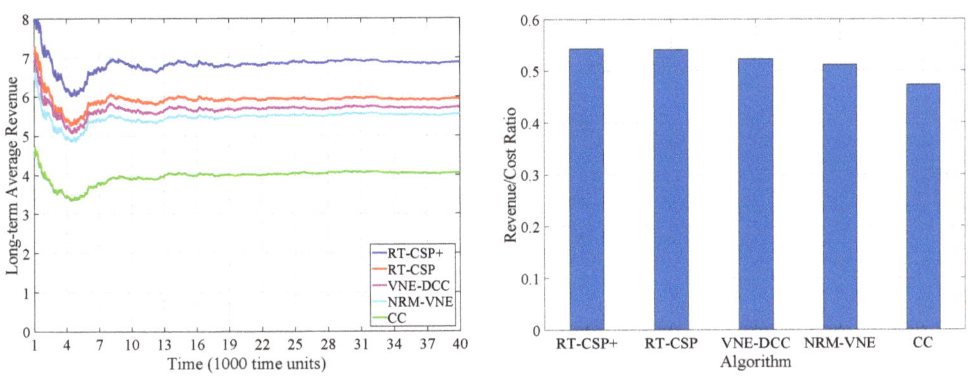

(a) (b)

Figure 3. (**a**) Long-term average slice provisioning revenue; and (**b**) slice provisioning revenue-to-cost ratio.

As shown in Figure 3a, the RT-CSP+ algorithm has the largest long-term average slice provisioning revenue. In the early stage of the simulation, the long-term average revenue decreases rapidly. The reason is that, as the slice arrives, the physical resources are consumed. The subsequent arriving slices are easy to be rejected, which decreases the provisioning revenue. When the simulation time reaches 10,000 time units, it tends to be stable because the arrival and departure of the slices reach a relatively balanced state. In the final steady state, the long-term average revenue of RT-CSP+ algorithm is 20.01%, 23.86% and 69.88% higher than those of VNE-DCC, NRM-VNE, and CC, respectively. Similar to the slice acceptance ratio and the long-term average revenue, the revenue-to-cost ratio also tends to be stable after 10,000 time units. Therefore, we show the average revenue-to-cost ratio histogram during the steady stage in Figure 3b. The RT-CSP+ and RT-CSP algorithms have better performance than others in terms of this metric. This is consistent with the long-term average revenue performance. Furthermore, since the revenue-to-cost ratio depends on the revenue and cost, the larger revenue-to-cost ratio is not only because our algorithms can achieve higher revenue, but also because it can reduce the provisioning cost.

5.2.2. Experiments in the Different Slice Link Connected Probability Scenario

We experimented on the different slice link connected probability scenario, in which the slice link connected probability is 0.2, 0.5, and 0.8, respectively, to investigate its impact on the performance of the algorithms. Figures 4 and 5 present the results of the slice acceptance ratio and slice provisioning revenue performance.

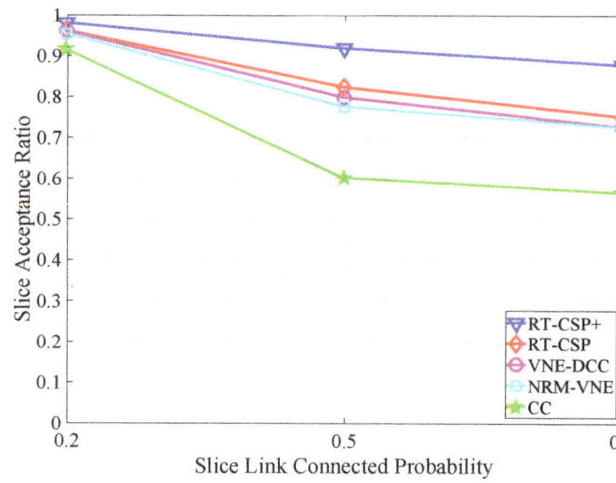

Figure 4. Slice acceptance ratio in the different slice link connected probability scenario.

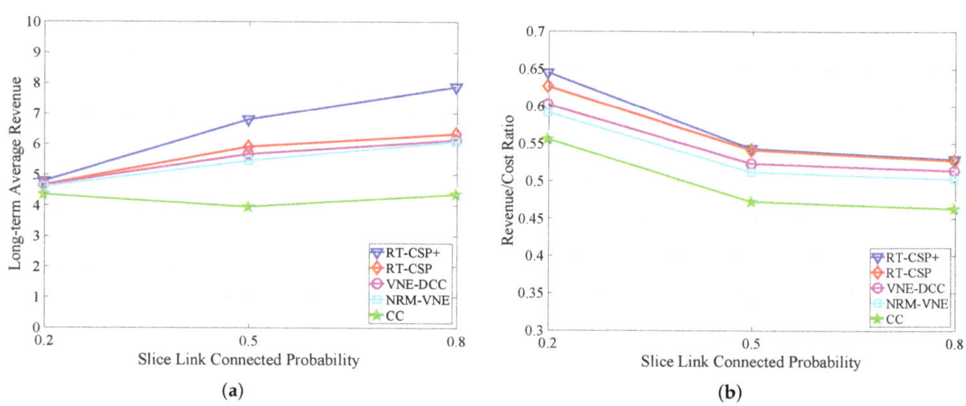

Figure 5. (**a**) Long-term average slice provisioning revenue in the different slice link connected probability scenario; and (**b**) slice provisioning revenue-to-cost ratio in the different slice link connected probability scenario.

Figure 4 shows that the slice acceptance ratio decreases as the slice link connection probability increases. This is because slice requests with more slice links demand more bandwidth resources, which makes the physical network difficult to satisfy bandwidth demands, resulting in more rejected slice requests. On the other hand, RT-CSP+ always has the highest slice acceptance ratio because of its efficiency. Figure 5a shows that, as the slice link connection probability increases, the long-term average revenue of all algorithms increases except for that of CC. For algorithms except CC, although the slice acceptance ratio is smaller at larger slice link connection probability, more slice links are provisioned in this case, which brings more provisioning revenue. For CC, when the slice link connection probability is 0.2, it can obtain much better slice acceptance ratio compared with 0.5 and 0.8, which contributes a lot to provisioning revenue. The long-term average revenue of CC has a similar trend as other algorithms when the slice link connection probability gets larger. With regard to the long-term average provisioning revenue, RT-CSP+ still outperforms others. Figure 5b shows that the revenue-to-cost ratio decreases as the slice link connection probability increases. The reason more revenue cannot result in larger revenue-to-cost ratio is that more slice links should be provisioned when the slice link connection probability is larger, in which case slice links are easier to be provisioned to a longer physical path, resulting in more provisioning bandwidth cost.

5.2.3. Experiments in the Different Slice Request Arrival Rates Scenario

We further validated the performance of our proposed algorithm by experimenting with different slice arrival rates. There are 100 substrate nodes in the substrate network and the slice nodes are connected by a probability of 0.5 in this scenario. Figures 6 and 7 show the results of the slice acceptance ratio and slice provisioning revenue performance with mean slice arrival rates of 0.02, 0.04, 0.06, 0.08, and 0.1.

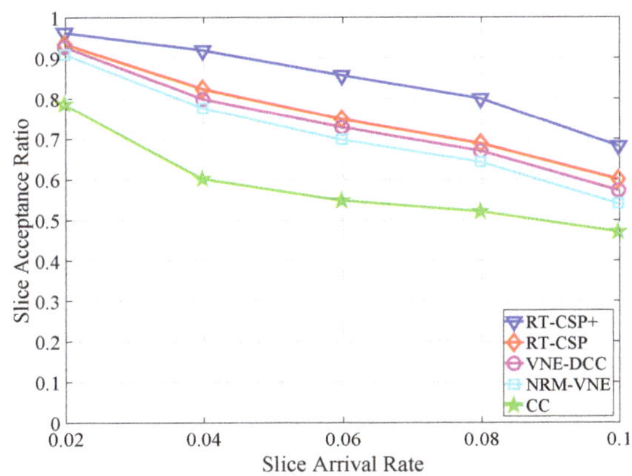

Figure 6. Slice acceptance ratio in the different arrival rates scenario.

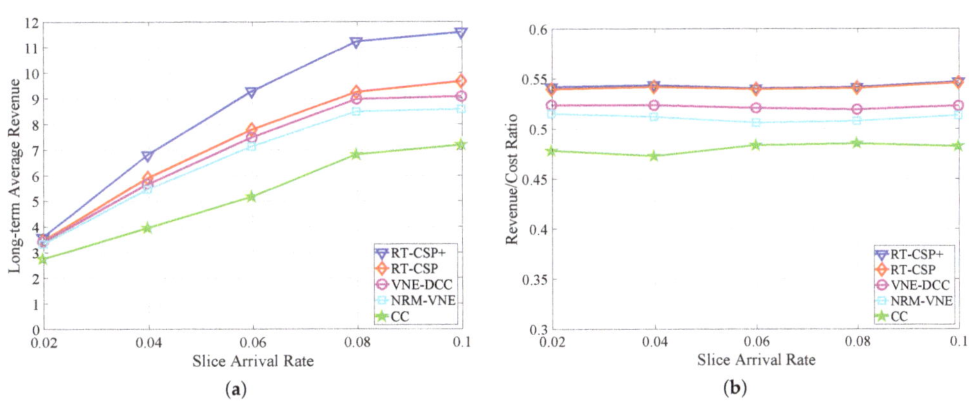

Figure 7. (**a**) Long-term average slice provisioning revenue in the different arrival rates scenario; and (**b**) slice provisioning revenue-to-cost ratio in the different arrival rates scenario.

As can be seen in Figure 6, RT-CSP+ algorithm always has the highest slice acceptance ratio when slices arrive at different rates. For example, when the slice request arrival rate is 0.06, the slice acceptance ratio of RT-CSP+ is 85.82%, which is 14.43%, 17.59%, 22.65%, and 56.15% higher than those of RT-CSP, VNE-DCC, NRM-VNE, and CC, respectively. This is because RT-CSP+ can comprehensively optimize node provisioning using the resource and topology attributes and the *minMaxBWUtilHops* strategy increases the probability of successfully provisioning slice links. In addition, slice acceptance ratio of all algorithms decreases as the slice arrival rate increases. The reason is that the larger the slice arrival rate, the more slices enter the slice provisioning system per unit time. Due to the limited physical resources, the probability of slice provisioning failure increases when more slices compete for limited physical resources, resulting in low slice acceptance ratio.

Figure 7a shows that the RT-CSP+ and RT-CSP algorithm always have better long-term average slice provisioning revenue with different slice arrival rates. This is because RT-CSP+ and RT-CSP can reasonably evaluate nodes in the node provisioning stage, resulting in more slices to be received. For each algorithm, the reason the long-term average revenue grows as the arrival rate increases is that more slice requests arrive per time unit under higher arrival rate scenario. Thus, more revenue can be obtained per time unit. Figure 7b presents that the average slice provisioning revenue-to-cost ratio during the steady stage is relatively stable with different slice arrival rates because the arrival and departure of the slices can reach a relatively balanced state. Our algorithms still have higher revenue-to-cost ratio.

5.2.4. Experiments in The Different Sizes of Substrate Network Scenario

The slice nodes are connected by a probability of 0.5 in this scenario. Figures 8 and 9 show the results of the slice acceptance ratio and slice provisioning revenue performance when the number of substrate network nodes is 50, 100, and 150, which represent small-, medium-, and large-sized physical network, respectively.

Figure 8 shows that, when the size of the physical network gets larger, all the algorithms have higher slice acceptance ratio. This is because the physical network with larger size has sufficient resources to host slice requests, which makes it easier to accept more slice requests. In the scenario with different sizes of substrate network, RT-CSP+ always has best slice acceptance ratio. For instance, when the substrate network has 150 nodes, the slice acceptance ratio of RT-CSP+ is 98.30%, which is 6.02%, 6.58%, 7.92%, and 17.44% higher than those of RT-CSP, VNE-DCC, NRM-VNE, and CC, respectively. The reason is that RT-CSP+ can efficiently provision slice requests based on the resource and topology attributes. In accordance with better slice acceptance ratio, our algorithms can produce better revenue performance as shown in Figure 9. From another aspect, the better performance of our algorithm in this scenario verifies its the scalability.

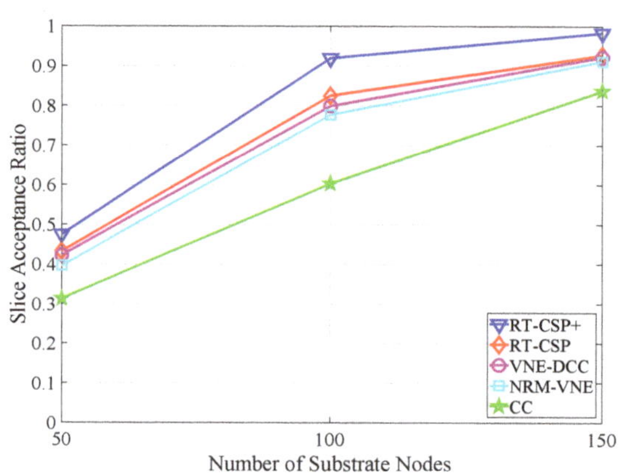

Figure 8. Slice acceptance ratio in the different sizes of substrate network scenario.

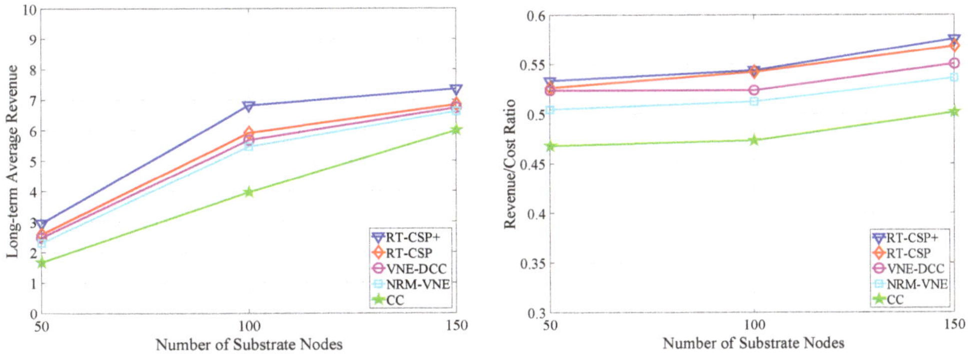

(**a**) (**b**)

Figure 9. (**a**) Long-term average slice provisioning revenue in the different sizes of substrate network scenario; and (**b**) slice provisioning revenue-to-cost ratio in the different sizes of substrate network scenario.

6. Conclusions

5G will be a disruptive technology in many ways. It has the potential to shakeup the telecommunications industry but would require significant investments. Consumers, both businesses and individuals, expect new opportunities from massive, ultra low latency and high density Internet of Things, as a run up to ambitious use cases such as smart cities and autonomous vehicles. The potential of 5G can only be truly realized if telecommunications service providers build in economies in the new deployments. Network slicing would be a key factor in achieving increased efficiencies and revenues through service specific offerings.

We have worked on the slice-provisioning problem by taking into account both the slice node provisioning and the slice link provisioning aspects. Accordingly, we have proposed a two-stage slice-provisioning algorithm called RT-CSP. As far as provisioning of slice nodes is concerned, our method takes into account the compute capacities, link bandwidths, degree centrality, and closeness centrality for comprehensive evaluation and ranking of nodes. These amount to jointly considering the local and global network resource attributes along with the topology attributes. Along with the heuristic slice node provisioning algorithm, RT-CSP uses the *k*-shortest path based slice link provisioning algorithm. An enhancement developed by us called RT-CSP+, based on *minMaxBWUtilHops* strategy designed by us, improves the performance further by selecting the physical path that has the minimum product of the maximum link bandwidth utilization and its hop count from the candidate physical paths obtained by the *k*-shortest path algorithm.

Extensive evaluations were carried out to compare both of our algorithms with other state-of-the-art algorithms and prove that the proposed algorithm does increase the slice request acceptance ratio and consequently the revenue of the network infrastructure provider. As far as acceptance ratio is concerned, both RT-CSP and RT-CSP+ perform better than other algorithms with the latter consistently giving the best performance. As the slice request arrival rate increases, the acceptance ratio of all the algorithms goes down but RT-CSP+ retains its supremacy. In terms of provisioning revenue, RT-CSP+ excels in long-term average slice provisioning revenue and revenue-to-cost ratio. Both RT-CSP and RT-CSP exhibit better revenue performance than other algorithms as the arrival rate increases. These results verify that our algorithms can comprehensively optimize node provisioning using the resource and topology attributes.

We are enthused with the good performance of our algorithms and, in the future, we plan to propose an efficient provisioning solution for latency-sensitive slices to satisfy low-latency 5G applications.

Author Contributions: Conceptualization, X.L. and R.J.; Methodology, X.L., R.J. and C.G.; Software, X.L.; Validation, X.L.; Writing—Original Draft Preparation, X.L. and L.G.; Writing—Review and Editing, all authors; and Supervision, C.G. and R.J.

Funding: This work was supported by the NPRP grant #NPRP 8-634-1-131 from the Qatar National Research Fund (a member of The Qatar Foundation), NSF grant #CNS-1718929, Huawei Technologies, and China Scholarship Council (No. 201506270075).

Conflicts of Interest: The authors declare no conflict of interest.

References

1. Amendola, C.; Calabrese, M.; Caputo, F.; Fabrizio, D. Fashion companies and customer satisfaction: A relation mediated by Information and Communication Technologies. *J. Retail. Consum. Serv.* **2018**, *43*, 251–257. [CrossRef]
2. Del Giudice, M.; Caputo, F.; Evangelista, F. How are decision systems changing? The contribution of social media to the management of decisional liquefaction. *J. Decis. Syst.* **2016**, *25*, 214–226. [CrossRef]
3. Stone, D.L.; Deadrick, D.L.; Lukaszewski, K.M.; Johnson, R. The influence of technology on the future of human resource management. *Hum. Resour. Manag. Rev.* **2015**, *25*, 216–231. [CrossRef]
4. Susskind, R.E.; Susskind, D. *The Future of the Professions: How Technology Will Transform the Work of Human Experts*; Oxford University Press: New York, NY, USA, 2015.
5. Papagiannis, H. *Augmented Human: How Technology Is Shaping The New Reality*; O'Reilly Media, Inc.: Farnham, UK, 2017.
6. Selin, C. The Ethics of Invention Technology and the Human Future. *Science* **2016**, *353*, 756. [CrossRef]
7. Johnson, D.G. Technology with no human responsibility? *J. Bus. Ethics* **2015**, *127*, 707–715. [CrossRef]
8. Rost, P.; Banchs, A.; Berberana, I.; Breitbach, M.; Doll, M.; Droste, H.; Mannweiler, C.; Puente, M.A.; Samdanis, K.; Sayadi, B. Mobile network architecture evolution toward 5G. *IEEE Commun. Mag.* **2016**, *54*, 84–91. [CrossRef]
9. Jain, R.; Paul, S. Network virtualization and software defined networking for cloud computing: A survey. *IEEE Commun. Mag.* **2013**, *51*, 24–31. [CrossRef]
10. NGMN. Description of Network Slicing Concept. Available online: https://www.ngmn.org/fileadmin/user_upload/160113_Network_Slicing_v1_0.pdf (accessed on 2 May 2019).
11. 3GPP. *Study on Management and Orchestration of Network Slicing for Next Generation Network (Release 15)*; Technical Specification (TS) 28.801, 3rd Generation Partnership Project (3GPP); 3GPP: Sophia Antipolis, France, 2018 .
12. Li, X.; Samaka, M.; Chan, H.A.; Bhamare, D.; Gupta, L.; Guo, C.; Jain, R. Network slicing for 5G: Challenges and opportunities. *IEEE Internet Comput.* **2017**, *21*, 20–27. [CrossRef]
13. Afolabi, I.; Taleb, T.; Samdanis, K.; Ksentini, A.; Flinck, H. Network slicing and softwarization: A survey on principles, enabling technologies, and solutions. *IEEE Commun. Surv. Tutor.* **2018**, *20*, 2429–2453. [CrossRef]
14. Kim, Y.; Kim, S.; Lim, H. Reinforcement Learning Based Resource Management for Network Slicing. *Appl. Sci.* **2019**, *9*, 2361. [CrossRef]
15. Fischer, A.; Botero, J.F.; Beck, M.T.; De Meer, H.; Hesselbach, X. Virtual network embedding: A survey. *IEEE Commun. Surv. Tutor.* **2013**, *15*, 1888–1906. [CrossRef]
16. Andersen, D.G. *Theoretical Approaches to Node Assignment*; Carnegie Mellon University: Pittsburgh, PA, USA, 2002. Available online: http://repository.cmu.edu/compsci/86/ (accessed on 16 October 2019).
17. Houidi, I.; Louati, W.; Ameur, W.B.; Zeghlache, D. Virtual network provisioning across multiple substrate networks. *Comput. Netw.* **2011**, *55*, 1011–1023. [CrossRef]
18. Wang, L.; Qu, H.; Zhao, J.; Guo, Y. Virtual network embedding with discrete particle swarm optimisation. *Electron. Lett.* **2014**, *50*, 285–286. [CrossRef]
19. Yu, M.; Yi, Y.; Rexford, J.; Chiang, M. Rethinking virtual network embedding: Substrate support for path splitting and migration. *ACM SIGCOMM Comput. Commun. Rev.* **2008**, *38*, 17–29. [CrossRef]
20. Cheng, X.; Su, S.; Zhang, Z.; Wang, H.; Yang, F.; Luo, Y.; Wang, J. Virtual network embedding through topology-aware node ranking. *ACM SIGCOMM Comput. Commun. Rev.* **2011**, *41*, 38–47. [CrossRef]

21. Wang, Z.; Han, Y.; Lin, T.; Tang, H.; Ci, S. Virtual network embedding by exploiting topological information. In Proceedings of the 2012 IEEE Global Communications Conference (GLOBECOM), Anaheim, CA, USA, 3–7 December 2012; pp. 2603–2608.
22. Cao, H.; Yang, L.; Zhu, H. Novel node-ranking approach and multiple topology attributes-based embedding algorithm for single-domain virtual network embedding. *IEEE Internet Things J.* **2017**, *5*, 108–120. [CrossRef]
23. Paschos, G.S.; Abdullah, M.A.; Vassilaras, S. Network Slicing with Splittable Flows is Hard. In Proceedings of the 2018 IEEE 29th Annual International Symposium on Personal, Indoor and Mobile Radio Communications (PIMRC), Bologna, Italy, 9–12 September 2018; pp. 1788–1793.
24. Raza, M.R.; Fiorani, M.; Rostami, A.; Öhlen, P.; Wosinska, L.; Monti, P. Dynamic slicing approach for multi-tenant 5G transport networks. *J. Opt. Commun. Netw.* **2018**, *10*, A77–A90. [CrossRef]
25. Sattar, D.; Matrawy, A. Optimal Slice Allocation in 5G Core Networks. *arXiv* **2018**, arXiv:1802.04655.
26. Li, W.; Zi, Y.; Feng, L.; Zhou, F.; Yu, P.; Qiu, X. Latency-Optimal Virtual Network Functions Resource Allocation for 5G Backhaul Transport Network Slicing. *Appl. Sci.* **2019**, *9*, 701. [CrossRef]
27. Bernstein, D. Containers and cloud: From lxc to docker to kubernetes. *IEEE Cloud Comput.* **2014**, *3*, 81–84. [CrossRef]
28. Zhang, P.; Yao, H.; Liu, Y. Virtual network embedding based on computing, network, and storage resource constraints. *IEEE Internet Things J.* **2017**, *5*, 3298–3304. [CrossRef]
29. Newman, M. *Networks*; Oxford University Press: New York, NY, USA, 2018.
30. Yen, J.Y. Finding the k shortest loopless paths in a network. *Manag. Sci.* **1971**, *17*, 712–716. [CrossRef]
31. Medina, A.; Lakhina, A.; Matta, I.; Byers, J. *BRITE: Universal Topology Generation from a User's Perspective*; Technical Report; Boston University Computer Science Department: Boston, MA, USA, 2001.
32. Waxman, B.M. Routing of multipoint connections. *IEEE J. Sel. Areas Commun.* **1988**, *6*, 1617–1622. [CrossRef]
33. Zhang, P.; Yao, H.; Liu, Y. Virtual network embedding based on the degree and clustering coefficient information. *IEEE Access* **2016**, *4*, 8572–8580. [CrossRef]

applied sciences

MDPI

Article

Performance Improvement of Ethernet-Based Fronthaul Bridged Networks in 5G Cloud Radio Access Networks

Muhammad Waqar and Ajung Kim *

Department of Optical Engineering, Sejong University, Seoul 143747, Korea
* Correspondence: akim@sejong.ac.kr; Tel.: +82-10-4597-7542

Received: 9 June 2019; Accepted: 12 July 2019; Published: 15 July 2019

Featured Application: The results of the research can be used in the practical deployments of the leveraging Ethernet fronthaul bridged networks in the 5G cloud radio access networks.

Abstract: Cloud radio access networks (C-RANs) are emerging architectural solutions to anticipate the increased capacity and quality demands of future 5G cellular networks at a reduced cost. In C-RANs, a transport segment referred to as fronthaul has been defined, which become a major constraint in practical implementations due to its high cost. A transport protocol referred to as eCPRI (enhanced common public radio interface), which was specifically designed for the fronthaul networks, imposes stringent end-to-end (E2E) latency and capacity requirements, which can be satisfied through the extortionate optical links. The high implementation cost of optical fronthaul networks significantly increased the system cost and made the fronthaul a hurdle to accomplish the cost–benefits of the C-RANs' architecture. The globally deployed Ethernet networks could be leveraging solutions, but are inadequate to comply with the eCPRI requirements in fronthaul bridged networks and result in intolerable latencies due to ineffectual traditional quality of service aware forwarding schemes. Therefore, to realize the cost–benefits of ubiquitously deployed Ethernet infrastructure, this paper proposes the E2E latency aware path computation and packet forwarding schemes, which ameliorate the performance of Ethernet-based fronthaul bridged networks to transport the eCPRI traffic at tolerable latencies. The simulation results verify the feasibility of low-cost Ethernet to carry the eCPRI traffic streams up to 100 Gbps with the proposed schemes in fronthaul bridged networks.

Keywords: 5G networks; BBU centralization; cloud radio access networks; eCPRI; fronthaul

1. Introduction

According to a recent prediction [1], the mobile data traffic will increase 10 times from 2017 to 2022, and more than 50 billion devices will be connected to the existing infrastructure by 2020. A transition from the static radio access networks (RANs) to flexible and reconfigurable networks through the virtualizations and cloudification was highly desirable to anticipate the traffic demands of future 5G cellular networks. Recently, cloud radio access networks (C-RANs) have been proposed, which are enhancements of classical RAN architectures through the cloudification and virtualizations techniques to comply with the requirements of envisioned 5G mobile networks. In C-RANs, the base station functionalities are split into distributed units known as the remote radio heads (RRHs) and central units known as the baseband units (BBUs), which are virtualized and pool at a central cloud [2]. The C-RANs approach brings several advantages for the network operators, including a reduction in the operational, management, and energy costs, as well as effective implementations of the coordinated multipoint (CoMP) transmission and reception schemes for the inter-cell interference mitigations [3]. With the virtualization of base station functionalities, the scalability of RAN architectures has been improved, and the spectral efficiency of the operator's network can be increased. Furthermore, a number of RRHs can be installed at the cell sites by simply connecting with the BBU pool, which reduces the cost of deploying expensive baseband processing units nearer to the RRHs at the cell sites [4].

The 3rd generation partnership project (3GPP) [5] defined the eight functional splits for the C-RANs in order to connect the RRHs and the BBU pool. A chosen functional split uniquely defines the properties of the system design such as the complexity, costs, latency, data rates, throughputs, and achievable statistical multiplexing gains. In a low-level functional split, which has the maximum benefits of performance and cost efficiency, all the higher layer functions are moved into the BBU pool, and only the radio frequency (RF) functions are employed in the RRHs at a cell site [6]. This requires a high-speed communication link between the RRHs and the BBU pool, which is referred to as the mobile fronthaul (MFH). The fronthaul segment is connected to the core network through the backhaul links [7]. Figure 1 depicts the considered MFH networks of the C-RANs architecture.

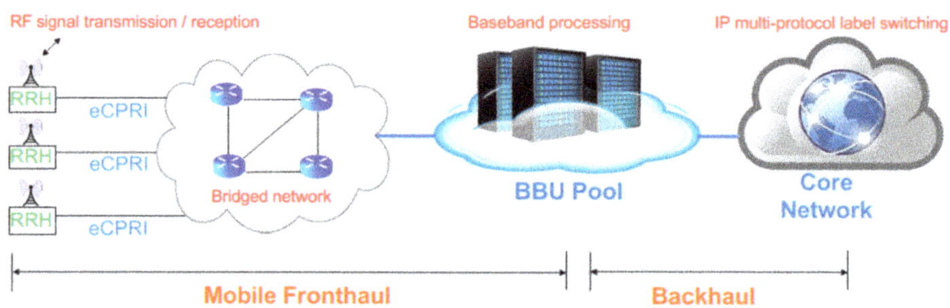

Figure 1. Cloud radio access networks (C-RANs) architecture for envisioned 5G cellular systems.

A MFH carries the digitized in-phase and quadrature (IQ) samples of baseband signals through the common public radio interface (CPRI) [8]. The CPRI is a non-packetized protocol that was specifically designed for the fronthaul networks by the leading telecom vendors, and cannot be integrated with other packetized transmissions unless a circuit (e.g., a wavelength) is reserved for it [9]. In this work, we consider a recent evolution of CPRI that is eCPRI (enhanced-CPRI) [10], which can packetize the IQ samples within Ethernet frames and is fully compatible with Ethernet systems. The eCPRI-based MFH networks are expected to deliver the throughputs of 10 Gbps, packet delay variations (PDVs) within 100 ns, and end-to-end (E2E) latency of less than 250 µs for all the RRHs streams [11,12]. The aforementioned requirements can only be fulfilled with the expensive and extremely high-capacity optical links. However, the cost of deploying such optical fronthaul bridged networks significantly increased the capital expenditure (CAPEX) and operational expenses (OPEX) for the network operators. As a result, the cost benefits of the C-RANs architecture could not be realized [13]. Thus, the implementation of low-cost MFH becomes a challenging issue in the C-RANs [14].

The 10 gigabit Ethernet systems can be more economical and remunerative solutions as compared to the optical networks to carry the eCPRI traffic in the MFH. Moreover, Ethernet supports the virtualization techniques and OAM (operations, administration, and management) functionalities, which make them more suitable to connect the RRHs with the BBU pool through the fronthaul bridged networks. However, the Ethernet networks that use the legacy quality of service (QoS)-aware routing [15] and packet forwarding schemes [16] yield the per-hop latency of about one millisecond and PDVs of up to hundreds of microseconds, which are much higher than the eCPRI traffic requirements [10,17]. A major cause in Ethernet systems for not complying with the eCPRI requirements is the non-consideration of frame-level queuing delays at the bridging nodes, which are caused by the globally synchronized eCPRI bursts flowing to and from the RRHs at different line rates. Despite the utilization of the QoS-aware forwarding schemes in the optical fronthaul bridged networks, the queuing delays among the burst of eCPRI streams are high, which increases the E2E latencies to an unacceptable level [18]. The queuing delays and E2E latencies become more critical in capacity-constraint Ethernet-based fronthaul bridged networks (EFBNs) [13]. For enhancing the capability of economically affordable and ubiquitously deployed Ethernet networks to transport the eCPRI streams at tolerable latencies, a novel packet forwarding mechanism that would efficiently utilize the limited Ethernet bandwidths and yield the low queuing delays at intermediate bridges is vital. Therefore, this paper proposes the E2E latency-aware path selection and packet transmission schemes that guarantee the low queuing delays in the EFBNs and retain the E2E latencies of maximum eCPRI streams closer to the threshold by taking advantage of the virtualization and OAM capabilities of the Ethernet systems. The proposed latency-aware path selection scheme assures the lowest E2E latencies in the fronthaul bridged networks by considering the frame-level queuing delays of eCPRI bursts. The proposed packet forwarding scheme mitigates the delays of eCPRI streams that experience high latency by slightly increasing the queuing delays of traffic streams that experience low latency. As a result, for all traffic streams, tolerable E2E latencies can be realized in the EFBNs. The computer simulations are undertaken on varied and realistic scenarios. The simulation results confirmed that the proposed schemes maximize the simultaneous transmissions of eCPRI streams of up to 100 Gbps without violating the QoS requirements as well as improve the link distances, whereas such performance is not guaranteed with the existing schemes. Moreover, the transmissions of time-sensitive eCPRI streams at tolerable latencies with the proposed schemes through the leveraging Ethernet links significantly reduced the CAPEX and OPEX of deploying extortionate optical links in fronthaul bridged networks. Hence, without hardware upgradations, the performance of low-cost Ethernet networks can be improved with the proposed schemes, which can be programmed at the intermediate nodes of the EFBNs using the OpenFlow protocols and software-defined networking (SDN) techniques [19].

The rest of the manuscript has the following organization. Section 2 introduces the state-of-the-art work on the CPRI and Ethernet-based fronthaul networks. Section 3 formulates the measurement procedures for the end-to-end latencies and distances in the EFBNs. Section 4 introduces the proposed transport schemes for the EFBNs. Section 5 presents the simulation setup for evaluating the proposed schemes. The results of the proposed schemes and comparison with the existing techniques are presented in Section 6. Finally, conclusions are provided in Section 7.

2. State of the Art—Ethernet Based Fronthaul Networks

Due to prior investments and backward compatibility issues, the existing standard bodies are focusing on the Ethernet-based fronthaul implementations. Currently, the IEEE 1914.3 [20] working group is investigating the techniques to encapsulate the CPRI samples within the Ethernet frames. This standard is also working on defining the techniques to improve the statistical multiplexing gains in the CPRI over Ethernet networks. The IEEE 802.1CM [21] is striving to standardize the Ethernet-based fronthaul systems for transporting the eCPRI streams in bridged networks. This standard is in the process of defining the default configurations, procedures, and profiles to carry the time-sensitive eCPRI streams in the MFH at low latencies. However, no mechanism has been defined explicitly in the standards to compensate the constraints of queuing delays at bridges and simultaneously transport the multiple eCPRI streams in Ethernet-based fronthaul bridged networks by satisfying the latency requirements.

Recently, due to the cost–benefits of Ethernet networks over the eCPRI-based optical switching networks, several academicals research proposals have been submitted to carry the time-sensitive fronthaul traffic in Ethernet networks. Most of the initial studies addressed the delay, jitter, and throughput challenges of the fronthaul networks by implementing the priority schemes [22] and packets scheduling [23] protocols, which were initially proposed for time-sensitive networking (TSN). The study [24] implemented the frames priority mechanism and traffic scheduling algorithm for the Ethernet-based MFH networks. This study shows that traffic priority can be effective in minimizing the delays in the Ethernet networks, but cannot fulfill the CPRI demands. The author claims that the queuing delays could be mitigated at switches with the traffic scheduling algorithm for a few CPRI streams. However, the parameters and configurations under which the results were obtained were not detailed in the paper. A performance analysis of passive optical networks (PON)-based fronthaul systems for long-range transmissions has been reported in [25]. This study shows that PON systems could not fully satisfy the eCPRI requirements without several hardware upgrades. This study also proposed to employ an expensive erbium-doped fiber amplifier (EDFA) at selected locations of the networks for achieving the QoS parameters within the eCPRI requirements. It is suggested to use an EDFA before the BBU pool for the considered fronthaul scenario to comply with the eCPRI demands. The studies [26] and [27] discussed the traffic scheduling and buffering techniques for Ethernet switching systems to improve the multiplexing gains in the fronthaul networks while retaining the delays and jitter within the CPRI limits. These studies showed that the CPRI traffic flows can be transported through at maximum one or two Ethernet switching nodes without violating the CPRI requirements. However, these schemes become ineffective for retaining the E2E latencies within the limits in more realistic fronthaul bridged networks, which comprises several intermediate bridging nodes between the RRHs and the BBU pool.

The constraint-based and QoS aware routing schemes would be alternative solutions to manage the latencies in multi-hop Ethernet networks. Since then, these routing schemes have been well studied in the past few years to tackle the network constraints such as the cost, capacity, throughputs, and delays. Dijkstra's shortest path algorithm that allocates the weights to different links, based on the available bandwidths and distances, has been studied in [28] for TSN networks, which could be an efficient solution for the delay constraint routing problems. Mutual constraint-based routing procedures have been discussed in [29] to fairly utilize the limited resources in the Ethernet bridged networks. In a study [30], multipath routing schemes were expedited for performing the dynamical load balancing and improving the congestions in TSNs by monitoring and maintaining the various performance indicating parameters. However, the existing QoS-aware routing schemes focus on searching the feasible routes subject to single or multiple QoS constraints, and select the best routes with no consideration of the frame level queuing delays, which is the case in the fronthaul bridged networks due to the simultaneous transmissions of the globally synchronized eCPRI bursts. Moreover, a high control messages overhead with the legacy routing schemes [31] due to the dynamic selection of the paths and continuous monitoring of the networking parameters further increases the delays, which makes them unsuitable for the E2E latency constraints of the EFBNs.

The study [18] proposes a queuing model to improve the delays in optical fronthaul bridged networks. In this study, Nakayama et al. proposed a low latency routing (LLR) mechanism based on the Markov chain Monte Carlo (MCMC) method, which searches the paths for individual streams by considering the traffic load on the different links. This study showed that the legacy QoS-aware routing schemes such as the constraint-based shortest path bridging (SPB) [32] become inefficient to satisfy the latency requirements in the MFH bridged networks, and schemes such as the LLR are required to satisfy the MFH requirements. However, the LLR scheme also yields intolerable E2E latencies, and becomes ineffective at complying with the eCPRI requirements once the traffic load increases to a certain level. This is because the LLR scheme uses a single path for transporting the time-sensitive flows, which are usually the shortest neighboring paths. Under the high load conditions, these alternative paths also get congested and cause intolerable E2E latencies for the eCPRI flows. Moreover, the LLR scheme lacks the mechanism to minimize the queuing delays of eCPRI streams of equal priority at the bridging nodes, which is highly desirable for alleviating the E2E latencies and successfully deploying the leveraging Ethernet-based fronthaul bridged networks in 5G cloud-RANs.

3. E2E Latency and Distance Formulation

The MFH networks require high multiplexing gains to efficiently implement the coordinated multiple point schemes in the C-RANs, which can be improved by increasing the aggregation of RRHs streams over the fronthaul links. However, the simultaneous transmissions of multiple eCPRI streams from the RRHs increased the E2E latencies to an unacceptable level, especially in the Ethernet networks [24–26]. Currently, the eCPRI-based Ethernet networks are gaining the attention of network operators and telecom vendors due to their low-cost and reconfigurability features, but achieving the required E2E latency is still a challenging issue [27] that requires further exploration. The E2E latency in such networks is the result of indeterministic and deterministic delays. The indeterministic delays such as the encapsulation delay Del_{Encp}, the queuing delay Del_{Que}, and the processing delay Del_{Proc} vary non-predictably in the EFBNs, while the deterministic delays such as the propagation delay Del_{Prop} and the transmission delay Del_{Tran} are pre-determined.

The indeterministic delays such as the Del_{Encp} correspond to the mapping delay of the eCPRI IQ samples within the Ethernet payload at the bridging nodes before transmitting them over the outgoing links. In this study, the Del_{Encp} for 10-gigabit Ethernet networks is estimated between the 19 μs to 2 μs for mapping the eCPRI flows from line rate option 1 to 7 [10], respectively within the Ethernet payload size of 1500 bytes [26]. However, the Del_{Encp} only increases when the traffic flows first enter the Ethernet networks. Therefore, the Del_{Encp} values are only considered at the first level of the bridging nodes within which the RRHs are physically connected.

The Del_{Que} is the length of time for which the packets of different streams wait in the buffers of the bridges before transmissions. When the f-th flow competes the g-th flow for simultaneous transmission over the n-th link, these flows are referred to as competitive flows [18]. Hence, the worst-case Del_{Que} experienced by the f-th flow over the n-th link is calculated from Equation (1):

$$Del_{Que} = \frac{m\lambda}{C} \tag{1}$$

where m is the maximum burst size of a flow, and C is the link capacity, while λ is the number of competitive flows [18].

The Del_{Proc} is the time that the switches take to decode the header of a packet to output them on the egress ports. The Del_{Proc} value, including the OpenFlow-based forwarding table lookup delays per bridging node experienced by the eCPRI streams, is estimated as 1.5 μs for the considered networks [27]. Thus, the indeterministic delays Del_{indet} in the EFBNs along a route can be calculated from Equation (2), and the deterministic delays Del_{det} can be calculated from Equation (3). Here, the deterministic delays such as Del_{Prop} are a measure of the time required for a frame to propagate from one node to the other. The Del_{Prop} is equal to d/s, where d is the link distance in km, and s is the signal propagation speed in copper (i.e., 2×10^8 ms^{-1}). The Del_{Tran} is the time taken by the physical layer at the source to transmit the packets over the link. In this study, the Del_{Tran} is estimated as 1.2 μs based on L/C, where L is the data packet length (i.e., 1500 bytes), and C is the channel data rate (i.e., 10 Gbps). Finally, the E2E latency of the n-th eCPRI traffic stream that is carried by a fronthaul link between the RRH and the BBU pool is formulated using Equation (4):

$$Del_{indet} = Del_{Encp} + [Del_{Que} + Del_{Proc}] \times Link_n \tag{2}$$

$$Del_{det} = Del_{Prop} + Del_{Tran} \tag{3}$$

$$E2E_{latency} = Del_{indet} + Del_{det} \tag{4}$$

Along with the latency, the distance of the fronthaul links between the RRHs and the BBU pool is another important performance indicating parameter for the EFBNs [24–26]. The E2E latency limits the maximum distances in the EFBNs. For commercially deploying the cloud-RANs, improvement in the link distances between the RRHs and the BBU pool is essential. The distance (km) of a fronthaul link in the considered scenario is calculated using Equation (5):

$$Distance = \frac{\left(Max_{threshold} - WE2E_{latency}\right)}{d/s} \tag{5}$$

where $Max_{threshold}$ is the maximum allowed E2E latency (i.e., 250 μs) for the fronthaul link, $WE2E_{latency}$ is the worst-case E2E latency experienced by a stream over the fronthaul link that can be computed from the Equation (4), and *d/s* is the propagation delay.

4. Description of the Proposed Transport System

Figure 2 illustrates the proposed EFBNs architecture to transport the multiple eCPRI traffic streams simultaneously between the RRHs and the BBU pool. A traffic stream is defined as a continuous bit rate transmission of the IQ packets at different eCPRI line rates that reserved the MFH bandwidth throughout the communications. However, with the improvement of bandwidth utilization in Ethernet-bridged networks, the E2E latencies for all the RRHs could not be minimized within the eCPRI limits [18]. This is because under high-load conditions and due to the rigorous requirements of the eCPRI, a few RRHs streams would experience the intolerable E2E latencies. Therefore, an optimized path selection algorithm with a low-latency packet forwarding procedure was highly desirable to retain the QoS parameters of EFBNs within the limits, and thus is proposed in this study.

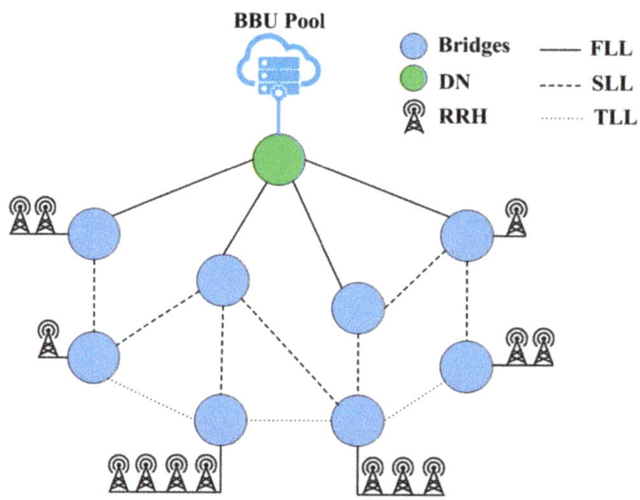

Figure 2. Proposed Ethernet-based fronthaul bridged networks.

4.1. E2E Latency-Aware Path Computation (LAPC) Procedure

In the considered fronthaul networks, a software-defined central controller (SDCC) [33] before the transmissions get a global view of the topology by collecting an information metric using the IS–IS (intermediate system to intermediate system) protocol, which was extended by IEEE 802.1Qca [34] for the time-sensitive networking. The IS–IS protocol provided the details about the connection of nodes, link capacities, costs of the links, and distances between the nodes of the topology. Typically, the numbers of RRHs connected to the nodes and accordingly configured eCPRI line rates of the RRHs are well determined in a fronthaul topology [26]. We assumed that all the nodes to which the RRHs are linked also connected with the SDCC, and a path computation element (PCE) is employed at the SDCC. The placement of the BBU pool is also predetermined in the topology, which is preferably connected to a node with which the maximum numbers of nodes are linked as shown in Figure 2. The PCE generates the initial paths for all the eCPRI-based RRHs streams in accordance with the k-th shortest paths algorithm [35] and maintains the paths in a global matrix. The SDCC selects the shortest paths for the individual traffic streams using the SPB algorithm based on the link costs, loads, and delays constraints. Then, for each stream, the SDCC calculates the E2E latency using equations (1) to (4) and updates the latencies in the global matrix along with the pre-selected shortest paths. Alternative paths are allocated to the traffic streams that could not satisfy the E2E latency requirements of the eCPRI, such as yield more than 250 μs. For this purpose, the first traffic load from the over-utilized adjacent links (ALs) of the destination node (DN) are offloaded, and then from the other links of previous ALs and so on. For simplicity, ALs are referred to as the first-level links (FLLs), the links that are adjacent to the FLLs are referred to as the second-level links (SLLs), next after the SLLs are referred to as the third-level links (TLLs), and so on, up to N-level links (NLLs), as illustrated in Figure 2. In this mechanism, the traffic loads from the FLLs to NLLs are systematically offloaded and uniformly distributed on the links. When a traffic load from a high-load link is shifted to a low-load link on each iteration, the according routes of the high-latency experiencing streams that were passing through the over-utilized high-load links are updated in the matrix. These routes were selected by searching from the k-th shortest paths by ensuring that it was not previously selected; then, E2E latencies on the new routes for the streams are recomputed. In the case when offloading the load from the high-load links results in E2E latencies that are greater than the threshold, under such scenarios, those paths are selected for the RRH traffic streams that yield the minimum E2E latencies for the maximum number of traffic streams. The above procedure is repeated until the E2E latencies of all the RRH streams reach the lowest level, and the further shifting of traffic loads between the links does not minimize the latencies. The global matrix contains the routes for each RRH traffic stream along with the E2E latencies computed with the proposed scheme. The output such as the global matrix of the low-latency paths computation algorithm will be used as the input of the low-latency packets transmission scheme, which is proposed in the next section. The proposed low-latency packet forwarding scheme cut down the queuing delays of those RRHs traffic streams, which could not comply with the stringent E2E latency requirements of the eCPRI, despite the low-latency paths computation procedure. The procedure of the proposed low-latency path computation scheme is described in Algorithm 1.

Algorithm 1 LAPC Algorithm

Input: m, c, λ, eCPRI line rates, link connections and distances
Output: Low-latency routes from the RRHs to the BBU pool with E2E latencies
1: SelectInitialPaths()
2: CalculateE2ELatency()
3: **procedure** BestPaths()
4: **if** (E2ELatencyStream$_i$ > 250 μs) **then**
5: **if** $\text{Load}^{\text{FLL}}_{\text{i to n}}$! isEqual **then**
6: Shift $\text{Load}^{\text{FLL}}_{\text{LowLatencyFLL}} \leftarrow \text{Load}^{\text{FLL}}_{\text{HighLatecnyFLL}}$
7: ReselectRoute()
8: UpdateGlobalMatrix()
9: Repeat **procedure**
10: **else if** $\text{Load}^{\text{SLL}}_{\text{i to n}}$! is Equal **then**
11: Shift $\text{Load}^{\text{SLL}}_{\text{LowLatencySLL}} \leftarrow \text{Load}^{\text{SLL}}_{\text{HighLatecnySLL}}$
12: ComputeNewRoute()
13: CalculateE2ELatency()
14: **if** $\text{E2ELatency}^{\text{NewRoute}}_{\text{stream}_i} < \text{E2ELatency}^{\text{PreviousRoute}}_{\text{stream}_i}$ **then**
15: SelectNewRoute()
16: UpdateGlobalMatrix()
17: **else**
18: RetainPreviousRoute()
19: **end if**
20: Repeat 10 to 20 for TLLs up to NLLs.
21: **end if**
22: **end if**
23: Repeat **procedure** for **all** streams
24: **return** GlobalMatrix
25: **end procedure**

4.2. Low-Latency Packet Forwarding (LLPF) Scheme

IEEE P802.1 Qbu [22] introduces a concept of frame preemption (FP) to transport the different priority traffics in time-sensitive Ethernet-bridged networks. In a frame preemption operation, the transmission of low-priority (preemptable) frames is suspended to allow the transmission of one or multiple high-priority (express) frames. After that, the transmission of low-priority frames from the same point is resumed. A low-priority frame that was in the process of transmission has not been completely transmitted, but is rather preempted and buffered until no high-priority frame is left in the queue. The preemptive queuing system enables the cut-through transmission of time-critical packets and minimizes the queuing delays. With this approach, the high-priority frames are transmitted through the switching nodes at low queuing delays without waiting in the queues for the complete transmission of low-priority frames. However, in fronthaul bridged networks, all the eCPRI streams have an equal priority, and simply employing the concepts of frame preemption is unfeasible for traffic streams of equal priority. Therefore, a mechanism that not only allocates the low-load routes to eCPRI traffic streams but also employs the sophisticated priority schemes such as frame preemption to minimize the queuing delays of equal priority eCPRI streams would be more feasible, as discussed below.

Under high traffic load conditions, a few RRH traffic streams are expected to experience the high queuing delays, despite the load balancing and efficiently utilizing the bandwidths [18], especially those that have long routes toward the BBU pool in term of delays and hop counts. As a result, the E2E latencies of such streams become intolerable. Thus, these high-latency experiencing RRHs streams (HERSs) can be given priority over the low-latency experiencing RRHs streams (LERSs) that have a margin of bearing slight queuing delays at intermediate nodes due to their physical or logical locality of being closer to the BBU pool. By considering the packets of HERSs as express frames and packets of LERSs as preemptable frames, the HERSs can be routed at low queuing delays through the nodes that the LERSs are connected to, and eventually, E2E latencies can be minimized. The global matrix of the LAPC algorithm provides the details about the high-latency experiencing RRHs streams over the most optimal paths. However, to manage the impact of preemption and retain the E2E latencies of the LERSs within limits, a frame of LERS should be delayed only a predefined number of times. The simulation result shows that a frame of the LERS can be delayed at maximum four to seven times under the considered scenarios, and after this threshold, a preempted frame of the LERS would be transmitted first instead of an express frame. In the case where a preempted frame is not transmitted according to a predetermined threshold, the E2E latencies of HERS might be low, but it will increase the E2E latencies of the LERS. The preempted frames are reassembled at the next nodes before processing so that these appear as valid Ethernet frames to the physical layer [22]. In the case, when more than one of the HERSs arrive at the bridges of a route, the HERSs that are connected with the most distant bridges, which is in terms of delays and hop counts from the BBU pool, would be given priority over the one with less distant connected bridges. As specified in the standard [22], packet tags are allocated to the preempted frames of the different streams that will be used at the receiver to reassemble the preempted frames. An example of the frames' arrival on the ingress ports, the preemption of frames, and transmissions of frames on the output port of the bridge are depicted in Figure 3. It can be observed that the HER streams of two different priorities (i.e., the priority of stream-3 is more than the priority of stream-2) and a LER stream of low priority (i.e. priority of stream-2 is more than the priority of stream-1) arrived at the ingress ports of the bridge. These priorities can be determined based on the information obtained from the global matrix of the proposed LAPC algorithm. It can be observed in Figure 3 that based on the priorities, the frames of high-priority HERS (i.e., stream-3) preempted the frames of low-priority HERS (i.e., stream-2); then frames of low-priority HERS (i.e., stream-2) preempted the frames of LERS (i.e., stream-1). The new and previously preempted frames of the LERS would be transmitted only when there is no high-priority frame left in the queues or a predefined threshold for the preempted frames is reached. As a result of the above procedure, the high-latency experiencing streams that were given more priority in the considered scenarios would experience the low queuing delays, and ultimately, the E2E latencies of all the eCPRI streams can be mitigated.

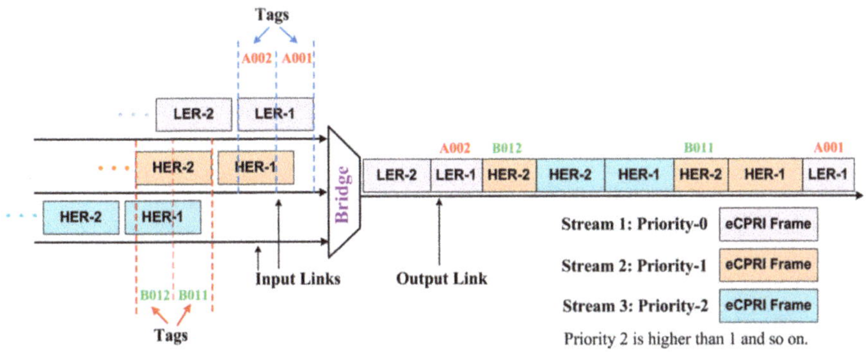

Figure 3. Frame preemption mechanism for enhanced common public radio interface (eCPRI) streams in the Ethernet-based fronthaul bridged networks (EFBNs).

The proposed schemes have certain implementation requirements to be deployed in the considered Ethernet-based fronthaul bridged networks. To compute the low-latency paths in the LAPC scheme, the values of link capacities, link distances, and link costs would be known to the algorithm, which can be determined with the IS–IS protocol. The number of connected RRHs, the configured eCPRI line rates on the RRHs, the size of eCPRI bursts, and the placement of the BBU pool in the topology are well determined in the fronthaul network [18]. The low-latency packet forwarding scheme mainly uses the output values of the LAPC scheme (i.e., global matrix) and requires the bridging nodes that can support the frame preemption operations. Both schemes require that the intermediate bridges should be reconfigurable so that the proposed schemes could be programmed. This functionality can be achieved in the fronthaul networks by using the worldwide available OpenFlow switches with the SDN techniques [19]. The proposed frame preemption-based mechanism is easy to implement and less complex as compared to the legacy packet forwarding schemes such as traffic scheduling [23]. The study [26] reported that due to the utilization of the traffic scheduling techniques in the fronthaul networks, the computational time increment exponentially with the increase of incoming traffic streams at the bridges, which makes it impractical for the multiple bridged networks. In the proposed EFBNs systems, the whole procedure of path selection, E2E latency computation, and accordingly enabling the preemption on the different routes is completed before the transmissions. During the transmissions, the controller periodically collects the topology information using the IS–IS protocol, and the above procedure is executed only when the number of connected RRHs, eCPRI line rates, or link capacities are modified. The proposed scheme yields the computation time up to 1.5 minutes for 100 Gbps of eCPRI traffic load in the considered fronthaul scenarios, which reduces with the reduction of traffic load or the number of connected RRHs in the topology. The path reconfiguration is executed only when the physical topology changes, which typically changes infrequently. Thus, the several minutes of computation time before the transmissions is not a problem or an issue of secondary importance as long as the latency parameters are within the limits, which are the main concern of fronthaul networks [18,36]. Moreover, the computational time can be further minimized by parallelizing the calculation for the proposed schemes.

5. Simulation Setup

The proposed schemes are evaluated with MATLAB programs. To verify the generality of the schemes, we employed the network topologies COST239 and ARPA2 [37], which are widely used topologies in evaluations of carrier networks [36].

IEEE 1914.3 [20] defines the two encapsulation techniques (namely structure-agnostic and structure-aware) to carry the CPRI/eCPRI streams by mapping in the Ethernet frames. In this work, we employed the structure-agnostic mapping scheme to encapsulate the eCPRI packets in the standard Ethernet frames. In this scheme, eCPRI packets are mapped in the Ethernet frames without the knowledge of framing protocols, while in structure-aware encapsulation, eCPRI packets are broken into antenna-carrier and control-data components to transport through the Ethernet switching nodes.

The RRHs connected to the bridging nodes are configured to support the different eCPRI line rates. The RRHs periodically transmit the eCPRI-based bursty traffic toward the BBU pool, which is independent of the end user's data rate and reserves the MFH bandwidth, even when no user is connected with the RRHs. Each RRH is configured to support a single eCPRI line rate randomly from the eCPRI line rate options 1 to 7 such as 0.61 Gbps, 1.22 Gbps, 2.45 Gbps, 3.07 Gbps, 4.91 Gbps, 6.14 Gbps, and 9.83 Gbps, respectively [10]. In contrast, an eCPRI line rate represents the number of IQ samples that an MFH link can transport between the RRH and the BBU pool [11]. Based on the configured line rates, the RRHs could transmit 50 to 100 Gbps of cumulative eCPRI traffic toward the BBU pool. However, for generality, loads are randomly connected to the bridging nodes. This is because the distribution of RRHs is determined by the demand distribution in reality [10,18]. In the considered EFBNs, each RRH sends a burst of 9000 bytes after every 2 ms toward the BBU pool based on [38] for the backward compatibility with the long term evolution-advanced (LTE-A) standards. The maximum size of the Ethernet payload is considered to be 1500 bytes.

The RRHs are connected to bridges through a link of 0.2 km length at the bandwidth of 10 Gbps. It is considered that all the bridges are located within an area of 20 km^2. The link distance between the consecutive bridges is randomly determined from 1 to 5 km similar to realistic networks [18]. The link bandwidth between two bridges is considered to be 10 Gbps. In each topology, the BBU pool is connected to a node where a maximum number of links are connected. In the considered scenarios, the BBU pool is connected to the bridging node-3 and to the bridging node-7 in COST239 and ARPA2 topologies, respectively. The link bandwidth between the BBU-DN is assumed as 100 Gbps and the link length is assumed as 0.2 km. Figure 4 shows the experimental setup to reflect the aforementioned configurations and parameters for simulating the fronthaul scenarios. As shown in Figure 4a,b, the RRHs are configured to support the cumulative eCPRI traffic of approximately 50 Gbps and 60 Gbps for COST239 and ARPA2 topologies, respectively. Analogous to this, the cumulative eCPRI traffic loads of 70 to 100 Gbps are formulated during the simulations for both topologies by randomly connecting the RRHs with different nodes at different line rates. The link length between the RRH bridges and DN-BBU pool is considered as 0.2 km. The link lengths between bridges and bridges would be determined randomly from 1 to 5 km. The bandwidth of all the links is considered to be 10 Gbps except for the link between the DN and the BBU pool, which has 100 Gbps of bandwidth.

It is assumed that all the bridges of the fronthaul networks are capable of implementing the frame preemption principles and policies, as discussed in Section 4.2. However, the frame preemption introduces a certain overhead for the preempted frames, which could impact the performance. Therefore, in order to realize the effect of preemption overhead in the EFBNs, the overhead per preemption is estimated as 124 ns, which is equivalent to the processing time of a 155-byte packet [39]. In order to obtain the average of the results, we repeated the simulations 10 times for different topologies and traffic loads. Furthermore, the proposed work has been compared with the LLR scheme [18]. This comparison is reasonable, because the LLR scheme is the most recent development on the given topic, and is considered to be one of the effective solutions to select the low-load paths in fronthaul bridged networks.

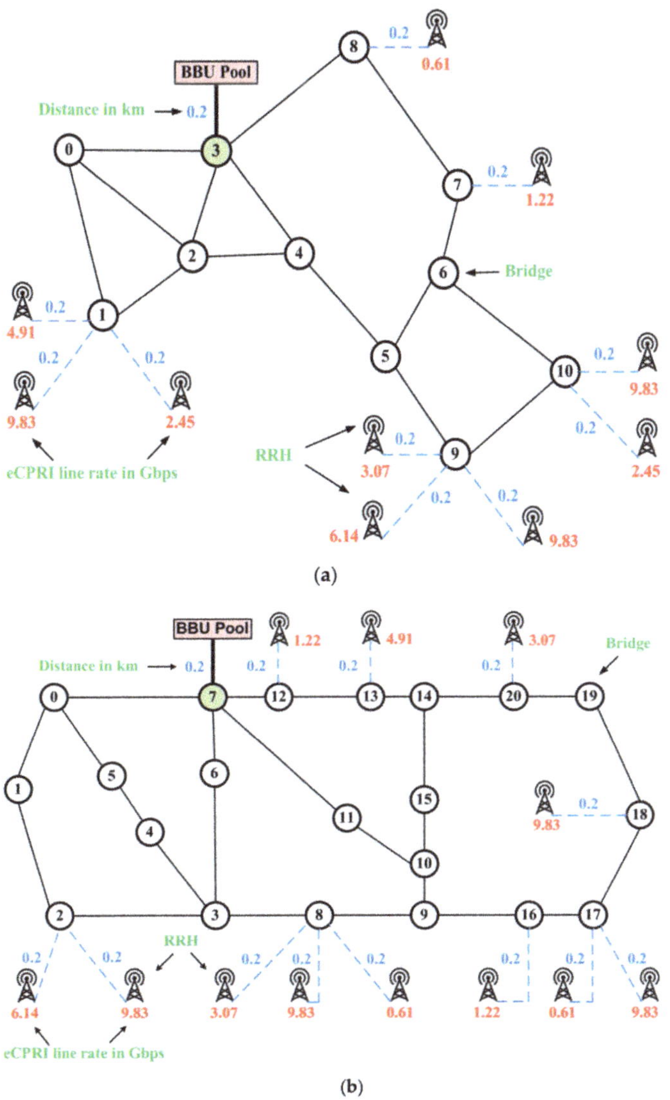

Figure 4. Considered network topologies for the evaluation of the proposed and existing schemes. (**a**) COST239; (**b**) ARPA2.

6. Results and Discussion

Figure 5a illustrates the average queuing delays experienced by the RRHs eCPRI streams for the COST239 topology while communicating with the BBU pool. The results showed that with the LLR scheme in the EFBNs, the average queuing delays increment as the incoming eCPRI data rates increase. This scheme resulted in the queuing delays up to 300 μs under the full load conditions, which are much higher than the required ranges of the fronthaul networks. The reason for the LLR scheme's inefficiency is that this scheme managed the delays by selecting the neighboring paths of the shortest paths, and under the high-load conditions, congestions at the neighboring routes were also high, which resulted in high queuing delays and ultimately increased the latencies to an unacceptable level. Our proposed low-latency path computation scheme more efficiently minimized the queuing delays, even under the high traffic load conditions. As shown in Figure 5a, the queuing delays with the LAPC scheme were reduced to 225 µs for 100 Gbps of eCPRI traffic. Further minimization of the queuing delays was realized by employing the proposed frame preemption concepts along with the low-latency path computation mechanism on the high-load paths. As shown in Figure 5a, the frame preemption-based LLPF scheme, which was named the LLPF-FP, more adroitly alleviated the queuing delays and enabled transmissions of up to 100 Gbps of traffic while retaining delays of less than 200 µs. With analogy to COST239, similar trends for ARPA2 were shown by the proposed schemes, as depicted by Figure 5b. The queuing delays for ARPA2 with the LAPC scheme were raised up to 265 µs for 100 Gbps of eCPRI traffic, while the LLR resulted beyond the 352 µs for the same load. This increment in delays is because of the increased in the propagation delays and the number of the bridges between the RRHs and the BBU in ARPA2 topology. However, the LLPF-FP scheme in ARPA2 topology retained the queuing delays for less than 248 µs, even when traffic load was high, which verifies the competency of the proposed mechanism to effectually minimize the queuing delays in time-sensitive EFBNs.

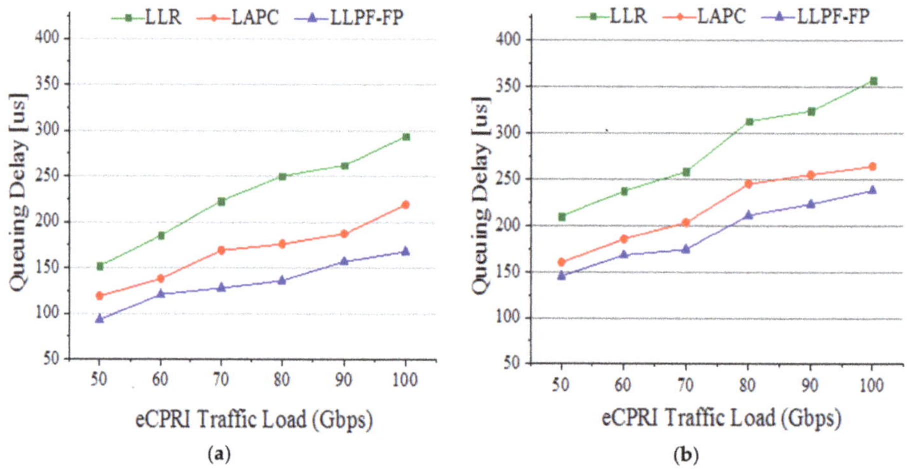

Figure 5. Performance results for average queuing delays for the (**a**) COST239; and (**b**) ARPA2.

Figure 6a depicts the worst-case E2E latencies experienced by the eCPRI streams with the LLR, LAPC, and LLPF-FP schemes. It can be observed that the LLR resulted in high E2E latencies for COST239, which exceeded beyond the 250 µs limits once the eCPRI traffic load increased to 60 Gbps. Whereas, for ARPA2 topology, as shown in Figure 6b, the LLR scheme lacked the mechanism to transport even 50 Gbps of traffic, and resulted in intolerable E2E latencies. Our proposed LAPC algorithm improved the E2E latency performance in the considered EFBNs topologies as compared to existing schemes such as the LLR. The simulation result showed that the LAPC can transport up to 82 Gbps and 65 Gbps of traffic in COST239 and ARPA2 topologies, respectively, by fully satisfying the E2E latency constraints of the fronthaul systems. The LLPF-FP scheme further improved the performance of the EFBNs and enabled the successful transmission of eCPRI traffic up to 100 Gbps for COST239 while retaining the E2E latencies within the threshold. In ARPA2 topology—which has comparatively long routes and more numbers of intermediate bridges with respect to COST239—the proposed LLPF-FP scheme transmitted up to 85 Gbps of traffic without violating the threshold. This verifies the feasibility of our proposed scheme to transport the time-sensitive eCPRI streams in low-cost Ethernet fronthaul networks without deploying the expensive optical fronthaul networks. Under the considered scenario and based on the results for ARPA2 topology, with a successful transmission of 85% of fronthaul traffic through the low-cost Ethernet infrastructure, a significant amount of CAPEX and OPEX, which would be required to deploy the complete optical fronthaul networks, can be saved. Hence, in this case, only 15% of the traffic required transmission through high-capacity and expensive optical links. For networks such as COST239, almost all of the traffic can be transported through the Ethernet systems, which would be highly cost-effective.

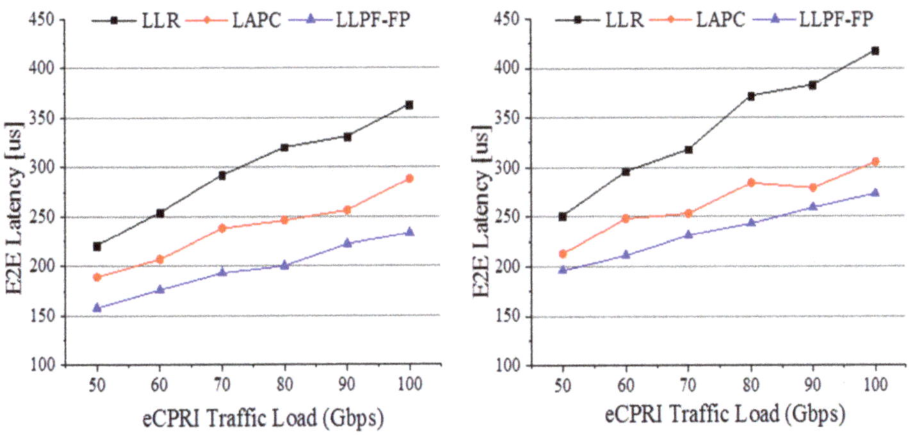

Figure 6. Performance results for worst-case end-to-end (E2E) latency for (**a**) COST239; and (**b**) ARPA2.

The results in Figure 7 show the distances that different schemes can support under different traffic load conditions based on the latency constraints. The distances are computed based on Equations (1)–(5). The proposed LAPC and LAPF-FP schemes outperform the LLR and considerably improve the distances supported by the fronthaul links. It can be observed that with the LAPF-FP scheme, the distances of the fronthaul segment can be increased up to 30 km for 50 Gbps of traffic, which reduced with the increase of traffic load, as shown in Figure 7a. In ARPA2 topology, as the number of intermediate bridges and distances between the RRHs and BBU pool has been increased as compared to COST239. As a result, the processing, transmission, queuing, and encapsulations delays at each bridging node would also be increased, which resulted in high E2E latencies based on Equation (4), and low fronthaul distances based on Equation (5). As shown in Figure 7b, the proposed LAPF-FP scheme can support fronthaul link distances from 3 km to 18 km for 50 Gbps to 100 Gbps of traffic, respectively. In contrast, the existing scheme can only support a maximum link distance of 4.5 km in Ethernet-based fronthaul bridged networks. The simulation results showed that in about 50% of cases, the proposed LAPF-FP scheme minimized the queuing delays and worst-case E2E latencies by more than 110 μs and 90 μs, respectively, compared to the LLR. In about 70% of cases, our scheme improved the link distances up to 35% compared to the LLR.

The results show that alone, delay-sensitive path computation schemes could not be enough to maximize the traffic transmissions in the EFBNs, and along with the path computation schemes, novel traffic forwarding concepts such as the one proposed in this study that enabled the frame preemption on high-load paths were required to fully exploit the advantages of leveraging Ethernet-based fronthaul networks. The simulation results showed a tradeoff between the maximum link distances and acceptable E2E latencies. By improving the E2E latencies in fronthaul networks, the length of Ethernet links can be increased. It can be concluded that with frame preemption in the proposed low-latency packet forwarding which is named the LAPF-FP scheme, the E2E latencies of maximum eCPRI streams can be retained closer to the threshold, and long-distance Ethernet links can be deployed between the RRHs and the BBU pool by fully satisfying the latency constraints of the fronthaul networks. However, this type of performance is not guaranteed with the existing schemes, which makes them less suitable to addressing the latency constraints of Ethernet-based fronthaul bridged networks.

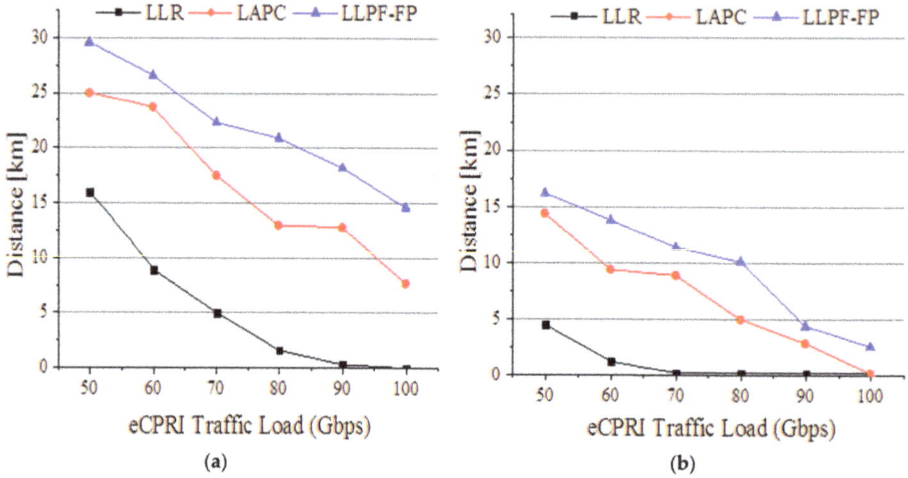

Figure 7. Fronthaul distances supported by the different schemes for the (**a**) COST239; and (**b**) ARPA2.

7. Conclusions

We proposed the end-to-end (E2E) latency-aware path computation (LAPC) scheme to improve the link utilization in capacity constraint Ethernet-based fronthaul bridged networks (EFBNs). The path computations were performed by primarily focusing on the E2E latency constraints of the eCPRI traffic streams. The latency-aware path selection scheme assures the lowest E2E latencies in the fronthaul bridged networks by considering the frame level queuing delays of eCPRI bursts. The performance of the EFBNs was further improved by exploiting the concepts of frame preemption in the proposed low-latency packet forwarding scheme, which is named the LAPF-FP. The LAPF-FP scheme reduces the queuing delays of high-latency experiencing RRHs (remote radio heads) streams by slightly increasing the queuing delays of the low-latency experiencing RRHs streams without the degradation of the overall system performance. The simulation results confirmed that the LAPF-FP scheme considerably reduced the E2E latencies and intensified the low-cost Ethernet systems to transport the time-sensitive eCPRI traffic up to 100 Gbps at tolerable latencies. Moreover, the low E2E latencies due to the LAPF-FP scheme significantly improved the link distances between the RRHs and the baseband unit (BBU) pool. This led to an expectation that the leveraging Ethernet systems could be a promising solution to implement the fronthaul bridged networks in the future cloud radio access networks (C-RANs). In summary, the following are the key findings from this study:

1. This study considered the queuing delays and end-to-end latencies for computing and selecting the optimal paths for individual eCPRI traffic streams in the EFBNs.
2. A novel packet forwarding mechanism for equal priority eCPRI streams has been proposed that maximizes the transmissions of multiple eCPRI streams at tolerable latencies as well as improves the link distances between the RRHs and the BBU pool.
3. The simulation results showed that in about 50% of cases, the proposed scheme minimized the queuing delays and worst-case E2E latencies by more than 110 μs and 90 μs compared to the LLR. In about 70% of cases, our scheme improved the link distances up to 35% as compared to the LLR. This level of performance is not guaranteed with the existing schemes.

4. With the proposed schemes, time-sensitive eCPRI streams can be transported through the leveraging Ethernet links, which significantly reduced the CAPEX and OPEX of deploying extortionate optical-based fronthaul bridged networks in envisioned 5G cloud-RANs.

Author Contributions: Conceptualization, M.W., and A.K.; methodology, M.W.; software, M.W.; validation, M.W., and A.K.; formal analysis, A.K.; investigation, M.W.; resources, A.K.; data curation, M.W.; writing—original draft preparation, M.W.; writing—review and editing, A.K.; visualization, A.K.; supervision, A.K.; project administration, A.K.; funding acquisition, A.K.

Funding: This work was supported in part by the National Research Foundation (NRF) and Ministry of Science, ICT and Future Planning (MSIP) of South Korea under grant number 2018R1D1A1B07049877.

Conflicts of Interest: The authors declare no conflict of interest.

References

1. Cisco Visual Networking Index, Global Mobile Data Traffic Forecast Update, 2017–2022, Cisco White Paper. 2019. Available online: https://www.cisco.com/c/en/us/solutions/collateral/service-provider/visual-networking-index-vni/vni-forecast-qa.pdf (accessed on 9 June 2019).
2. Luong, P.; Gagnon, F.; Despins, C.; Tran, L. Joint virtual computing and radio resource allocation in limited fronthaul green C-RANs. *IEEE Trans. Wirel. Commun.* **2018**, *17*, 2602–2617. [CrossRef]
3. Musumeci, F.; Silva, E.; Tornatore, M. Enhancing RAN throughput by optimized CoMP controller placement in optical metro networks. *IEEE J. Select. Areas Commun.* **2018**, *36*, 2561–2569. [CrossRef]
4. Das, S.; Ruffini, M. A variable rate fronthaul scheme for cloud radio access networks (C-RAN). *J. Lightwave Technol.* **2019**, *37*, 3153–3165. [CrossRef]
5. 3rd Generation Partnership Project (3GPP) Working Group, Tech. rep. 38.801. Study on New Radio Access Technology–Radio Access Architecture and Interfaces. Version 14.0.0. 2017. Available online: https://www.3gpp.org/ftp/Specs/archive/38_series/38.801/38801-e00.zip (accessed on 9 June 2019).
6. Zhou, Y.; Li, J.; Shi, Y.; Wong, V.W.S. Flexible functional split design for downlink C-RAN with capacity-constrained fronthaul. *IEEE Trans. Vehic. Technol.* **2019**. [CrossRef]
7. Frascolla, V.; Dominicini, C.; Paiva, M.; Caporossi, G.; Marotta, M.; Ribeiro, M.; Segatto, M.; Martinello, M.; Monteiro, M.; Both, C. Optimizing C-RAN backhaul topologies: A resilience-oriented approach using graph invariants. *App. Sci.* **2019**, *9*, 136. [CrossRef]
8. Common Public Radio Interface (CPRI), Interface Specification 2015, Version 7.0. Available online: http://www.cpri.info/downloads/CPRI_v_7_0_2015-10-09.pdf (accessed on 9 June 2019).
9. Molner, N.; Oliva, A.; Stavrakakis, I.; Azcorra, A. Optimization of an integrated fronthaul/backhaul network under path and delay constraints. *Ad Hoc Nets.* **2019**, *83*, 41–54. [CrossRef]
10. Common Public Radio Interface (CPRI), eCPRI Interface Specification 2019, Version 2.0. Available online: http://www.cpri.info/downloads/eCPRI_v_2.0_2019_05_10c.pdf (accessed on 9 June 2019).
11. Cai, M.; Liu, Q.; Jiang, H. A novel efficient wireless fronthaul (EWF) method for the common public radio interface (CPRI) signal transmission. In Proceedings of the 2018 IEEE 88th Vehicular Technology Conference (VTC-Fall), Chicago, IL, USA, 27–30 August 2018.
12. Nakayama, Y.; Hisano, D.; Kubo, T.; Fukada, Y.; Terada, J.; Otaka, A. TDD-based rapid fault detection and recovery for fronthaul bridged network. *IEEE Commun. Lett.* **2018**, *22*, 498–501. [CrossRef]
13. Gomes, N.J.; Sehier, P.; Thomas, H.; Chanclou, P.; Li, B.; Munch, D.; Assimakopoulos, P.; Dixit, S.; Jungnickel, V. Boosting 5G through Ethernet: How evolved fronthaul can take next-generation mobile to the next level. *IEEE Vehic. Technol. Magaz.* **2018**, *13*, 74–84. [CrossRef]
14. Peng, M.; Wang, Y.; Dang, T.; Yan, Z. Cost-efficient resource allocation in cloud radio access networks with heterogeneous fronthaul expenditures. *IEEE Trans. Wirel. Commun.* **2017**, *16*, 4626–4638. [CrossRef]
15. Kuipers, F.; Van Mieghem, P.; Korkmaz, T.; Krunz, M. An overview of constraint-based path selection algorithms for QoS routing. *IEEE Commun. Magaz.* **2002**, *40*, 50–55. [CrossRef]
16. Waqar, M.; Kim, A.; Yoon, J. A performance study of PON-Based 5G cloud-RANs with frame preemption and strict priority. In Proceedings of the International Conference on Green and Human Information Technology (ICGHIT), Kuala Lumpur, Malaysia, 16–18 January 2019.

17. Mountaser, G.; Mahmoodi, T.; Simeone, O. Reliable and low-latency fronthaul for tactile internet applications. *IEEE J. Select. Areas Commun.* **2018**, *36*, 2455–2463. [CrossRef]
18. Nakayama, Y.; Hisano, D.; Kubo, T.; Fukada, Y.; Terada, J.; Otaka, A. Low-latency routing scheme for a fronthaul bridged network. *J. Opt. Commun. Net.* **2018**, *10*, 14–23. [CrossRef]
19. Lin, F.P.; Tsai, Z. Hierarchical edge-cloud SDN controller system with optimal adaptive resource allocation for load-balancing. *IEEE Syst. J.* **2019**, *9*, 111–121. [CrossRef]
20. IEEE P1914.3, Draft Standard for Radio Over Ethernet Encapsulations and Mappings. Version 2.0. 2018. Available online: http://sites.ieee.org/sagroups-1914/files/2018/02/tf3_1914.3_D3_0_MEC_clean.pdf (accessed on 9 June 2019).
21. IEEE P802.1CM, Draft Standard for Local and Metropolitan Area Networks—Amendment: Time-Sensitive Networking for Fronthaul. Version 2.2. 2018. Available online: http://www.ieee802.org/1/files/private/cm-drafts/d2/802-1CM-d2-2.pdf (accessed on 9 June 2019).
22. IEEE P802.1Qbu, Draft Standard for Local and Metropolitan Area Networks—Bridges and Bridged Networks—Amendment: Enhancements for Frame Preemption. Version 3.0. 2015. Available online: http://www.ieee802.org/1/files/private/bu-drafts/d3/802-1Qbu-d3-0.pdf (accessed on 9 June 2019).
23. IEEE P802.1Qbv, Draft Standard for Local and Metropolitan Area Networks—Bridges and Bridged Networks—Amendment: Enhancements for Scheduled Traffic. Version 3.1. 2015. Available online: http://www.ieee802.org/1/files/private/bv-drafts/d3/802-1Qbv-d3-1.pdf (accessed on 9 June 2019).
24. Smith, P.A. CPRI Fronthaul Requirements Discussion with TSN. Tech. rep., San Diego, Huawei. 2015. Available online: http://www.ieee802.org/1/files/public/.../new-ashwood-tsn-cpri-fronthaul-0714-v03.pdf (accessed on 9 June 2019).
25. Waqar, M.; Kim, A.; Yoon, J.J. A performance analysis of 5G fronthaul networks for long-distance communications. In Proceedings of the 2019 Wireless Days (WD) Conference, Manchester, UK, 24–26 April 2019.
26. Chitimalla, D.; Kondepu, K.; Valcarenghi, L.; Tornatore, M.; Mukherjee, B. 5G fronthaul-latency and jitter studies of CPRI over Ethernet. *IEEE/OSA J. Opt. Commun. Net.* **2017**, *9*, 72–182. [CrossRef]
27. Waqar, M.; Kim, A.; Peter, K.C. A transport scheme for reducing delays and jitter in Ethernet-based 5G fronthaul networks. *IEEE Access.* **2018**, *6*, 446110–446121. [CrossRef]
28. Yahya, W.; Basuki, A.; Jiang, J.R. The extended Dijkstra's-based load balancing for OpenFlow network. *Int. J. Electr. Comp. Eng.* **2015**, *5*, 289–296.
29. Gurumoorthy, K.B.; Kumar, N. Mutual constraint-based GA suggested routing algorithm for improving QoS in clustered MANETS. *Wirel. Person. Commun.* **2018**, *98*, 2975–2991. [CrossRef]
30. Huo, C.; Yuan, J.; Song, G.; Shi, Z. Node reliability based multi-path routing algorithm of high-speed power line communication network. In Proceedings of the 2019 IEEE 4th International Conference on Cloud Computing and Big Data Analysis (ICCCBDA), Chengdu, China, 12–15 April 2019.
31. Guirguis, A.; Digham, F.; Seddik, K.G.; Ibrahim, M.; Harras, K.A.; Youssef, M. Primary user-aware optimal discovery routing for cognitive radio networks. *IEEE Trans. Mob. Comput.* **2019**, *18*, 193–206. [CrossRef]
32. IEEE 802.1aq, Draft Standard for Local and Metropolitan Area Networks—Bridges and Bridged Networks —Amendment: Shortest Path Bridging. Version 4.6. 2012. Available online: http://www.ieee802.org/1/files/private/aq-drafts/d4/802-1aq-D4-6.pdf (accessed on 9 June 2019).
33. Gutierrez, M.; Ademaj, A.; Steiner, W.; Dobrin, R.; Punnekkat, S. Self-configuration of IEEE 802.1 TSN networks. In Proceedings of the 2017 22nd IEEE International Conference on Emerging Technologies and Factory Automation (ETFA), Limassol, Cyprus, 12–15 September 2017.
34. IEEE 802.1Qca, Draft Standard for Local and Metropolitan Area Networks—Bridges and Bridged Networks—Amendment: IS-IS Path Control and Reservation. Version 2.1. 2015. Available online: http://www.ieee802.org/1/files/private/ca-drafts/d2/802-1Qca-d2-1.pdf (accessed on 9 June 2019).
35. Hershberger, J.; Maxel, M.; Suri, S. Finding the k shortest simple paths: A new algorithm and its implementation. *ACM Trans. Algos.* **2007**, *3*, 45. [CrossRef]
36. Hisano, D.; Nakayama, Y.; Kubo, T.; Uzawa, H.; Fukada, Y.; Terada, J. Decoupling of uplink user and HARQ response signals to relax the latency requirement for bridged fronthaul networks. *IEEE/OSA J. Opt. Commun. Net.* **2019**, *11*, B26–B36. [CrossRef]
37. Lee, D.; Lee, K.; Yoo, S. Efficient Ethernet ring mesh network design. *J. Lightweight Technol.* **2011**, *29*, 2677–2683. [CrossRef]

38. 3rd Generation Partnership Project (3GPP) Working Group, Tech. Rep. 36.213. Evolved Universal Terrestrial Radio Access (E-UTRA); Physical Layer Procedures. Version 14.6.0. 2018. Available online: https://www.etsi.org/deliver/etsi_ts/136200_136299/136213/14.06.00_60/ts_136213v140600p.pdf (accessed on 9 June 2019).
39. Park, T.; Samii, S.; Shin, K.G. Design optimization of frame preemption in real-time switched Ethernet. In Proceedings of the 2019 Design, Automation & Test in Europe Conference & Exhibition (DATE), Florence, Italy, 25–29 March 2019.

applied sciences

MDPI

Article

A Palm-Jacobaeus Loss Formula for Multi-Service Systems with Separated Resources

Mariusz Głąbowski *, Adam Kaliszan and Maciej Stasiak

Faculty of Computing and Telecommunications, Poznan University of Technology, ul. Polanka 3, 60-965 Poznan, Poland; adam.kaliszan@put.poznan.pl (A.K.); maciej.stasiak@put.poznan.pl (M.S.)
* Correspondence: mariusz.glabowski@put.poznan.pl

Received: 26 April 2020; Accepted: 5 June 2020; Published: 10 June 2020

Abstract: This article presents a method to determine the blocking probability (non-availability) for strictly determined multi-service resources that belong to a group of multi-service resources. The dependencies obtained during the process correspond to the Palm-Jacobaeus formula derived for the group of resources servicing single-service traffic. The approach to determining the blocking probability is based on the determination of the availability of resources at the occupancy level of allocation units. The analytical results are compared with the obtained results of the simulation experiments for a number of selected parameters of multi-service groups of resources. The results of the present study indicate a high accuracy of the proposed solutions. The elaborated extension of the Palm-Jacobaeus formula can be used in the modeling of separated wireless resources in cellular networks.

Keywords: Palm-Jacobaeus loss formula; blocking probability; multi-service traffic; separated resources; performance evaluation; cellular systems

1. Introduction

One of the main problems in service quality management in mobile networks is the problem of efficient resource allocation [1–3]. Resource allocation techniques are increasingly affecting the performance of subsequent cellular network technologies (3G, 4G, 5G) [4]. Among other things, this is directly related to two factors [5]. The first is the decreasing size of cells. This factor affects both the more frequent changes of cells due to user mobility and the increase in the number of cells (resources) that can handle a given request (call). The other factor, however, is the diversity of services provided in cellular networks in terms of, e.g., requested bit rates or acceptable delays.

In order to effectively use the resources of cellular networks, including in particular 4G and 5G multi-service networks, it became necessary to develop mechanisms that optimize the traffic distribution among neighboring cells. As a result of initial research [6], the concept of self-organizing (self-optimizing, self-configuring) networks (SONs) was developed, in which the load balancing of individual cells of cellular networks is a key element [7,8]. The developed concept of load balancing made it possible to take into account the changing load of individual cells over time—resulting, among others, from user mobility [9]. Such optimized traffic distribution in mobile networks allows mobile operators both to better use network resources and to improve service quality parameters (quality of experience) of multi-service traffic streams generated by end users [10,11].

Currently ongoing studies on self-optimizing cellular systems take into account not only the load-balancing criterion. The use of appropriate self-optimization techniques reduces the costs of the manual configuration of network elements, reduces the number of rejected connections, ensures better matching of resources to requests, and reduces energy costs [11]. SONs also support the implementation of the concept of end-to-end network slicing in 5G systems [12,13].

Regardless of the objective and technique used to optimize resource management, however, the area covered by a specific group of cells can be treated—from the point of view of traffic engineering—as a multi-service system in which multi-service requests are handled using "separated" resources (cells), creating network resources [10]. An analysis of the state of research on the analytical modeling of cellular systems indicates that multi-service (multi-rate) models are most often used to determine the traffic characteristics of such systems at the call level [14–20].

In most of the work on modeling cellular systems at the call level, only single cells [16–18,21] were taken into account. For the purpose of analyzing a group of cells (resources) as a single system, it is necessary to use more complex multi-service models, including the so-called limited-availability group model [22] and the non-full-availability group model [10]. In these studies [10,22], groups of cells with a connection handover mechanism for load balancing were analyzed. The developed methods take into account only information about the amount of occupied resources in a group of cells (resources), without the possibility of determining the occupancy states of specific (indicated) cells (resources) in the group. This assumption did not directly affect the determination of the probability of blocking in a group of cells (resources); however, it did not allow for taking into account separate call admission restrictions in individual cells.

In many new tasks posed to, among others, SON mechanisms and network slicing in 5G systems, the ability to determine the occupancy (non-availability) of precisely indicated (and not any) cells can be very important. The development of a general and effective analytical method for the determination of the non-availability of strictly determined/defined resources in a certain set of them for defined classes of traffic streams will make it possible for the method to be further parameterized, e.g., within the context of the aforementioned traffic distribution in a group of cells, the distribution of resources for network slicing, or the assessment of the number of necessary links in trunks (e.g., EtherChannels).

In the case of single-service systems, the Palm-Jacobaeus formula provided us with the ability to determine the probability of blocking the indicated resources within their group. A method for the determination of the blocking probability $H(x_d)$ of x_d strictly determined resources (from a set containing k resources) is proposed in [23], while in [24] this method is presented in the form of the following equation, known as the Palm-Jacobaeus loss formula:

$$H(x_d) = \frac{E_k(A)}{E_{k-x_d}(A)}, \tag{1}$$

where $E_k(A)$, the so-called Erlang B-formula, is the blocking probability in a single-service group composed of k resources to which traffic with an intensity of A Erlangs is offered. The parameter x_d defines the number of strictly determined resources to be investigated. In traffic engineering, Equation (1) has been used, directly or in the form of the so-called modified Palm-Jacobaeus loss formula [25,26], to model complex non-full-availability systems, single-service resource groups [25,27], and single-service multi-stage switching networks, e.g., [28–32].

A counterpart (corresponding equivalent) of the group of single-service resources is the group of separated multi-service resources—the limited availability group (LAG). The LAG is a system composed of k independent separated resources, each with the capacity of f adopted (assumed) AUs (allocation units). It is important to underline that the resources are independent of one another and are separated, which means that the LAG can service a given call only when at least one resource, from among all k resources, has sufficient resources needed to provide service to this call. Such a definition of service excludes any possibility of a division of resources demanded by a call between a given number of different resources.

The LAG is analyzed, for example, in [33–35]. In [36], a simple approximate LAG model for resources with different capacities is proposed, while [37] presents a LAG model for different resource capacities. In multi-service traffic engineering, the LAG is used to model and optimize mobile networks, cf. [10], as well as output groups of links (resources) in multi-service switching networks, e.g., [38–41].

We see the main application of the elaborated extension of the Palm-Jacobaeus formula in modeling of separated wireless resources in cellular networks. However, preliminary studies undertaken by us led to the conclusion that the determination of the blocking probability of strictly determined resources in the resource group will make it possible to simplify the modeling of a large number of complex multi-service systems in mobile and optical networks, including network nodes and switching networks.

This article is structured as follows. Section 2 presents the model of a group with limited availability. Section 3 includes a proposal of a method to determine the blocking probability for strictly defined resources. Section 4 provides a comparison of the results obtained in the analytical modeling with the results of the simulation experiments for a number of selected structures of multi-service resource groups. The last section summarizes the most important results of the study.

2. Model of the Group with Limited Availability

Consider a system called LAG (limited-availability group) in the literature, cf. [36,37]. The group is composed of k identical component resources, each with the capacity of f AUs (allocation units) (Allocation unit is a universal term describing the unit of resources required in a given system. An example of calculating the allocation unit in 5G systems with OFDM (orthogonal frequency division multiplexing)-based cells is presented in [42]), therefore defining the capacity of the system to be equal to $V = kf$ AUs. The system services a call only when the call can be serviced by the available resources of one of the component resources. In traffic engineering, AU is defined as the greatest common divisor of the bitrates of all the call classes offered to the system [43–47].

The assumption is that the LAG services a mixture M of independent Erlang traffic streams with the following intensities: $A_1, A_2, \ldots, A_M$. A call of class i requires t_i AUs to set up a connection. A multi-dimensional Markov process in the LAG can be approximated by a one-dimensional Markov chain and can be described by the following equation [48]:

$$n\,[P(n)]_V = \sum_{i=1}^{M} A_i t_i \xi_i(n-t_i)\,[P(n-t_i)]_V, \tag{2}$$

where $[P(n)]_V$ is the probability that the LAG is in a state of n busy AUs in the system, whereas $\xi_i(n)$ is the conditional transition probability for a traffic stream of class i between the adjacent (neighboring) states of the process. The blocking probability E_i for a stream of class i in the LAG model can be determined by the following equation:

$$E_i = \sum_{n=0}^{V} [P(n)]_V\,[1-\xi_i(n)]. \tag{3}$$

The conditional transition probability $\xi_i(n)$ in the LAG model [48] can be approximated by the following dependency:

$$\xi_i(n) = \frac{F(V-n,k,f,0) - F(V-n,k,t_i-1,0)}{F(V-n,k,f,0)}, \tag{4}$$

where $F(x,k,f,t)$ is the number of possible arrangements of x objects in k identical boxes, each of which is able to accommodate f objects. A further assumption is that there are at least t objects (from among all the x objects to be arranged) in each of the boxes:

$$F(x,k,f,t) = \sum_{i=0}^{\left\lfloor \frac{x-kt}{f-t+1} \right\rfloor} (-1)^i \binom{k}{i} \binom{x-k(t-1)-1-i(f-t+1)}{k-1}. \tag{5}$$

Note that Equation (4) applies to the arrangements of free (unoccupied) AUs in the LAG and can be interpreted as the ratio between the number of favorable arrangements, i.e., those in which at least one resource that has t_i free AUs can be found, and the number of all possible arrangements $V - n$ of free AUs.

3. Non-Availability Probability of Strictly Defined Resources

The non-availability probability $H(i, x_d)$ of x_d strictly determined resources determines the probability of an event in which each resource from the x_d selected resources does not have enough t_i free AUs to set up a connection of class i. This probability can be determined with respect to the occupancy of AUs, i.e., directly at the occupancy level of AUs. Thus, we have:

$$H(i, x_d) = \sum_{n=(f-t_i+1)x_d}^{V} H(i, x_d|n)[P(n)]_V, \tag{6}$$

where $H(i, x_d|n)$ is the conditional non-availability probability of a selected number of x_d resources for a call of class i, determined under the assumption that the total number of busy AUs in the LAG is equal to n. The lower limit of the sum in Equation (6) determines the minimum number of AUs that can induce non-availability for a selected number of x_d resources.

The conditional distribution $H(i, x_d|n)$ can be determined on the basis of the following reasoning. The selected x_d resources are unavailable to a call of class i if n_d AUs ($n_d \leq n$) are arranged in these x_d resources in such a way that none of them has t_i free AUs. On the basis of Equation (5), the number of such arrangements is $F(n_d, x_d, f, f - t_i + 1)$. The remaining $n - n_d$ busy AUs can be arranged (accommodated) in any way within the remaining $k - x_d$ resources. The number of such arrangements is $F(n - n_d, k - x_d, f, 0)$. Since the total number of arrangements of n AUs in k resources is $F(n, k, f, 0)$, the probability $H(i, x_d|n)$ can be determined in the following way:

$$H(i, x_d|n) = \sum_{n_d=(f-t_i+1)x_d}^{\min(n, fx_d)} \frac{F(n_d, x_d, f, f - t_i + 1)F(n - n_d, k - x_d, f, 0)}{F(n, k, f, 0)}. \tag{7}$$

Equation (6) derived in the present article for multi-service systems is the counterpart of the Palm-Jacobaeus loss formula (Equation (1)) derived for single-service systems [24]. The original Palm-Jacobaues formula is a substantial extension of the Erlang model for the full-availability group (Erlang B-formula [49]): The Erlang B-formula makes it possible to determine the blocking probability in a system with a capacity of k resources, whereas the Palm-Jacobaeus loss formula [24] allows us to determine the probability $H(x_d)$ of occupancy (non-availability) of x_d strictly determined resources (from a set containing k resources). In a similar way, Equation (6), developed in the present article, is a substantial extension of the model proposed in [48]. Equations (2) and (3) allow us to determine the occupancy distribution and the blocking probability in a multi-service system with a capacity of $V = kf$ AUs. The model proposed in the article makes it possible to determine the probability of an event in which the selected x_d (from among all the k resources) resources of a system can serve a call of a given class. Equation (6) proposed in the article is then in the same relation to the LAG model as the Palm-Jacobaeus formula is to the basic Erlang model and constitutes a substantial extension of the practical capabilities of the model in engineering applications.

4. Results and Discussion

4.1. General Assumptions

In the model that allows us to determine the probability of the non-availability of strictly determined resources for requests/calls (flows/streams) of particular traffic, we used the approach, generally accepted in the literature on the subject, according to which variable bitrates of real packet streams are replaced with constant bitrates [19,46,50,51]. Such an approach much simplifies the process of modeling, since it allows multi-rate (multi-service) systems to be analyzed in a Markovian way at the flow/stream/session level (call level). There are two approaches to the replacement of variable bitrates of packet streams with their equivalents with constant bitrates: They can be chosen on the basis of the maximum bitrates [42,51] of particular packet streams or, alternatively, on the basis of the so-called equivalent bandwidth determined for each packet stream [45,52]. An analysis at the flow level is the only approach to dimensioning, designing, and optimizing whole networks, which is much appreciated by network operators [19,46].

After the determination of constant bitrates (volume of required resources) for calls of individual classes serviced in the system, it is possible to determine the allocation unit (AU) for a given system. The AU is defined as the highest common divisor of the capacity of the system and all bitrates (equivalent bandwidths, volume of required resources) allocated to calls of individual classes. Subsequently, both the capacity of the system and the amount of resources required for a call of a given class to be serviced, are expressed in the number of AUs.

The value of a single AU can also depend on the applied technological solutions. Besides the convention of expressing AUs in bits per second, dominant in the literature, commonly used expressions include the so-called interference (noise) unit, used in WCDMA systems (wideband code division multiplexing) [19,53]. To maintain the maximum versatility of the developed model and taking the constant development of 5G systems for granted, we assume in our further research that it will be possible to express the volume of demands and the capacity of the system in the multiplicity of a certain AU, without specifying precisely the method for their determination. A suitable example of the determination of demands of individual traffic classes in 5G systems with OFDMA multiplying can be found in [42].

The proposed method for the determination of the non-availability probability of strictly determined resources is an approximate method, since the LAG model it uses is an approximate model for any analysis at the call level. It is possible to develop more accurate models, although an accurate and precise solution for the service process in the considered system would, however, involve its analysis at the level of the so-called microstates (a microstate is defined by the number of serviced calls of individual classes) whose sheer number would almost entirely prevent its solution. Such an approach is impractical and totally ineffective from an engineering point of view. Therefore, an approximation of real service processes with the LAG model (an analysis at the macrostate level, where the macrostate is determined by the number of occupied AUs) requires the validity of the adopted assumptions to be verified by simulation experiments.

4.2. Simulator

In order to determine the accuracy of the analytical method for determining the probability of the non-availability of strictly determined resources for calls of particular traffic classes, the results of the analytical calculations have been compared with the simulation data. The simulator was developed for the sole purpose of this article and was based on the process interaction approach, in line with the detailed and specific assumptions concerning modeling of multi-serve traffic streams presented in [54]. The simulator was implemented in C++. The input data for the simulator are

- the number k of resources,
- the capacity f of a single resource,
- the number M of traffic stream classes,
- the number t_i of demanded allocation units for a stream of class i ($1 \leq i \leq M$),
- the intensity μ_i of a service stream of class i,
- the proportions of offered traffic $A_1 t_1 : A_2 t_2 : \ldots : A_M t_M = x : y : \ldots : z$, i.e., the values of the parameters $x, y, \ldots, z$,

- the average value a of traffic offered to a single allocation unit in the system, where:

$$a = \frac{\sum_{i=1}^{M} A_i t_i}{kf}. \tag{8}$$

On the basis of the parameters $x, y, \ldots, z$, which define the proportions of offered traffic $A_1 t_1$: $A_2 t_: \ldots : A_M t_M$, the simulator determines the values of the traffic intensity A_i for each of the offered traffic classes ($1 \leq i \leq M$). Then, on the basis of the definition of traffic intensity,

$$A_i = \frac{\lambda_i}{\mu_i}, \tag{9}$$

the simulator determines the value of the parameter λ_i, which defines the intensity of generating traffic streams of class i. The parameters λ_i and μ_i are given to exponential distribution generators used in the program to determine the time intervals between the arrival of a next traffic stream of class i and the service time of a given call of class i.

To determine the probability of the non-availability of strictly determined resources for calls of particular traffic classes, the condition for the termination of a simulation experiment is the amount of elapsed time (duration time) for individual series necessary to generate a predefined number of calls of the class that is least active (most frequently, this is the class with the biggest number of demanded allocation units). The average result is calculated on the basis of 10 series. In practice, to obtain confidence intervals that are not greater than 5% of the average value of the simulation, it is necessary to generate about 100,000,000 calls of the least active class.

4.3. Accuracy of the Model

In line with the information given in Section 4.1, the process of determining the non-availability probability in the system is based on an approximation of the Markov process in the system. The simulator developed for the purpose of this article (presented in Section 4.2) was used to evaluate the influence of this approximation on the accuracy of the proposed analytical model. The study carried out by us included a wide range of systems. The influence of the number of resources (k), their capacity (f), the number of classes of offered traffic streams (M), the volume of demanded resources (t_i), the proportion of offered traffic $A_1 t_1 : A_2 t_2 : \ldots : A_M t_M$, and the intensity of service streams (μ_i) were all investigated. A a result of this study, three representative systems (from the perspective of the differences in the results) were chosen. The systems had the following parameters:

1. System 1: $k = 3$, $f = 20$ AUs, $V = 60$ AUs, $M = 3$, $t_1 = 1$ AU, $t_2 = 2$ AUs, $t_3 = 3$ AUs, $A_1 t_1 : A_2 : t_2 : A_3 t_3 = 1:1:1, \forall_{1 \leq i \leq M}\, \mu_i = 1$;
2. System 2: $k = 5$, $f = 20$ AUs, $V = 100$ AUs, $M = 3$, $t_1 = 1$ AU, $t_2 = 2$ AUs, $t_3 = 5$ AUs, $A_1 t_1 : A_2 : t_2 : A_3 t_3 = 1:1:1, \forall_{1 \leq i \leq M}\, \mu_i = 1$;
3. System 3: $k = 5$, $f = 50$ AUs, $V = 250$ AUs, $M = 3$, $t_1 = 1$ AU, $t_2 = 3$ AUs, $t_3 = 7$ AUs, $A_1 t_1 : A_2 : t_2 : A_3 t_3 = 1:1:1, \forall_{1 \leq i \leq M}\, \mu_i = 1$.

The obtained results of the non-availability for the strictly determined resources in the systems under investigation are presented in Figures 1–9. The determined confidence intervals for the simulation results are far lower than the markers used to indicate the results of the simulation experiments. For this reason they are not visible in the figures. In the specification of the examined systems, both the resources and the demands of the offered traffic streams are expressed in AUs. As a result, the notion of a small or large system is rather relative. If a system services call classes for which the AU is equal to 100 Mbps, then such a system can be considered to be large. If, however, the AU is 100 kbps, then the system can be regarded as small even though the number of demanded AUs is the same in both systems.

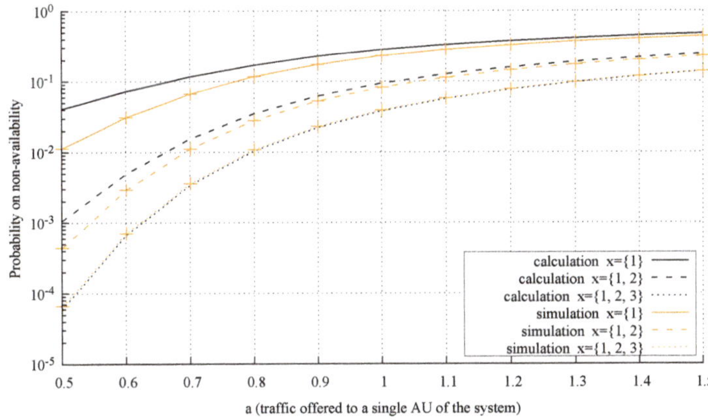

Figure 1. Probability of the non-availability of x strictly determined resources for class 1 calls ($t_1 = 1$) in System 1, where x is the set of strictly determined resources in the limited-availability group.

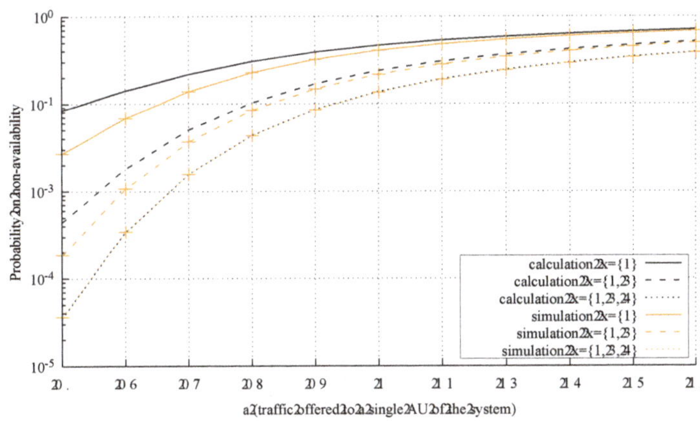

Figure 2. Probability of the non-availability of x strictly determined resources for class 2 calls ($t_2 = 2$) in System 1, where x is the set of strictly determined resources in the limited-availability group.

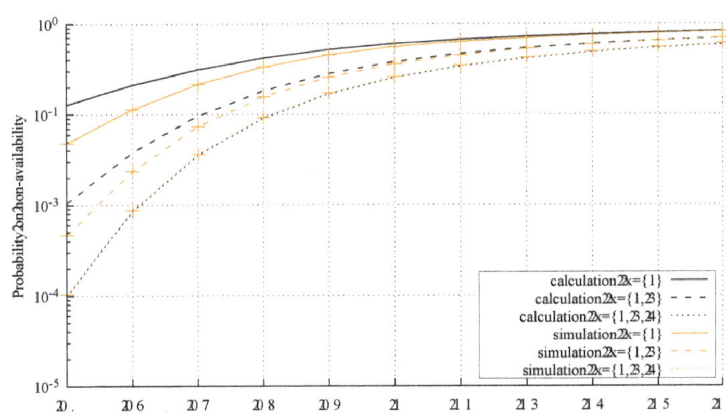

Figure 3. Probability of the non-availability of x strictly determined resources for class 3 calls ($t_3 = 3$) in System 1, where x is the set of strictly determined resources in the limited-availability group.

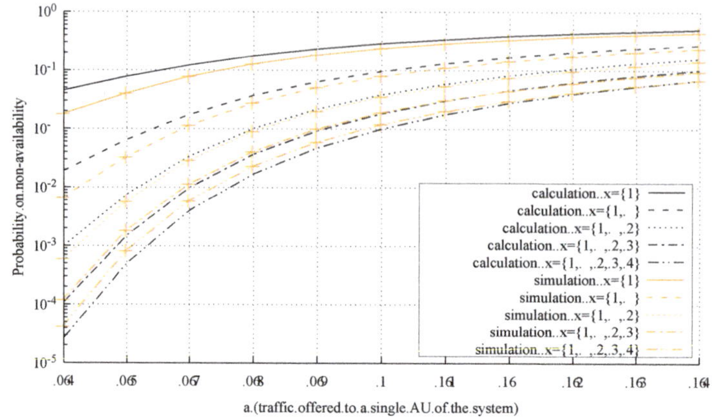

Figure 4. Probability of the non-availability of x strictly determined resources for class 1 calls ($t_1 = 1$) in System 2, where x is the set of strictly determined resources in the limited-availability group.

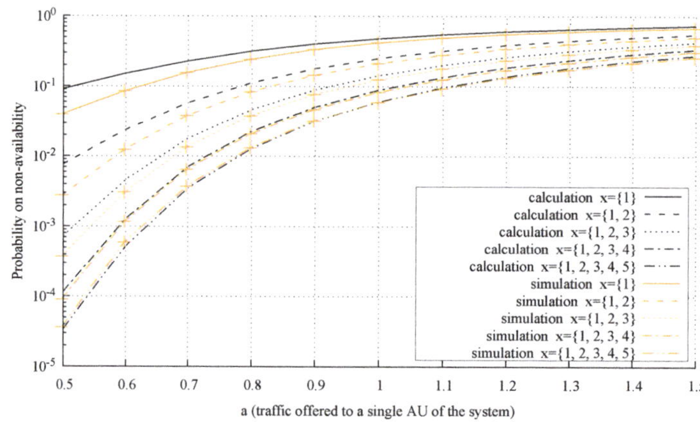

Figure 5. Probability of the non-availability of x strictly determined resources for class 2 calls ($t_2 = 2$) in System 2, where x is the set of strictly determined resources in the limited-availability group.

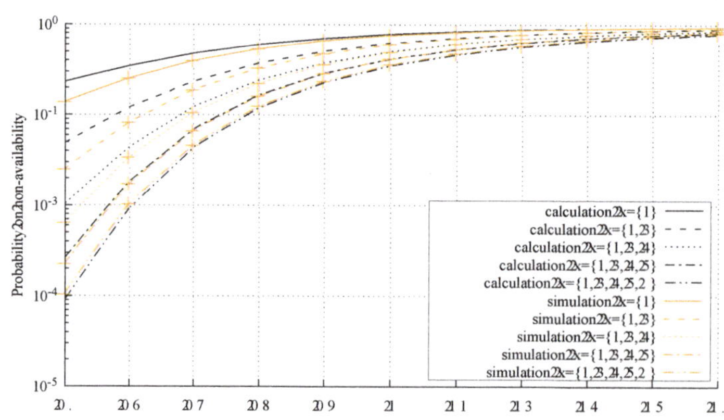

Figure 6. Probability of the non-availability of x strictly determined resources for class 3 calls ($t_3 = 5$) in System 2, where x is the set of strictly determined resources in the limited-availability group.

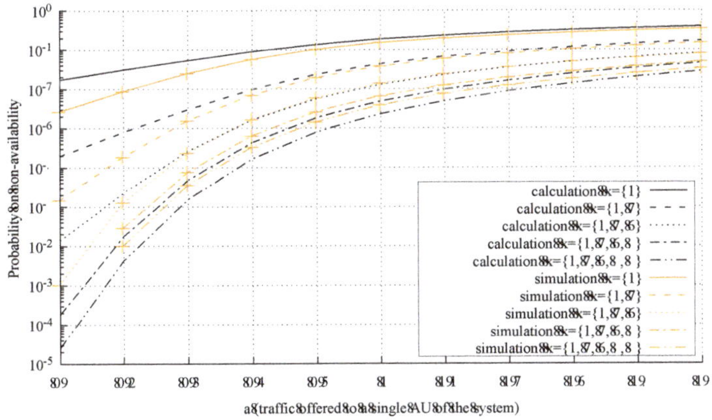

Figure 7. Probability of the non-availability of x strictly determined resources for class 1 calls ($t_1 = 1$) in System 3, where x is the set of strictly determined resources in the limited-availability group.

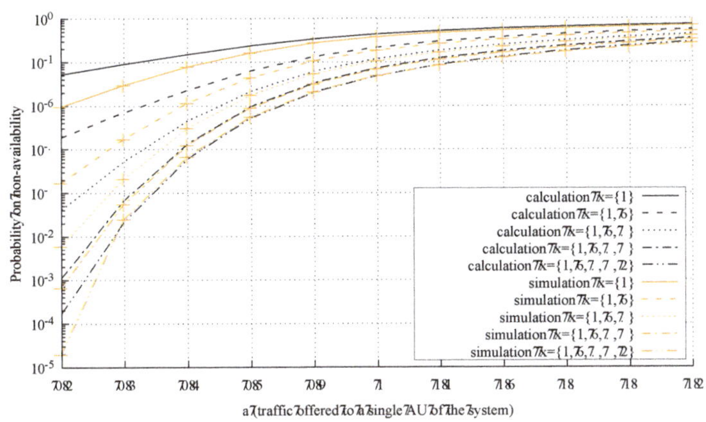

Figure 8. Probability of the non-availability of x strictly determined resources for class 2 calls ($t_2 = 3$) in System 3, where x is the set of strictly determined resources in the limited-availability group.

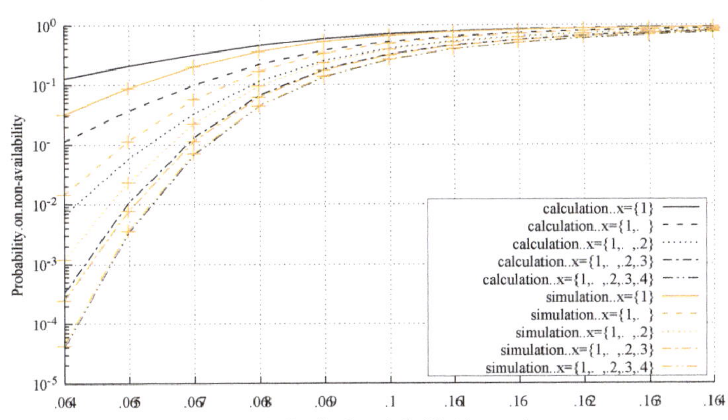

Figure 9. Probability of the non-availability of x strictly determined resources for class 3 calls ($t_3 = 7$) in System 3, where x is the set of strictly determined resources in the limited-availability group.

The comparative analysis of the obtained analytical and simulation results verifies that the developed analytical model, which allows us to determine the non-availability probability for strictly determined resources, is an approximate model. The main reasons for the inaccuracies, inherited from the basic model of the limited-availability group (which allows the probability to be determined in whole the system regardless of the occupancy distribution in individual resources), are the following:

- The analysis of the considered system is carried out from the microstate level (multi-dimensional Markov process) to the macrostate level (one-dimensional Markov process)—a detailed analysis of this problem is presented in [55].
- The conditional transition coefficients are determined in an approximate way (Equation (4)): in the process of determining the conditional transition probability (Equation (4)) in the occupancy distribution (Equation (2)) the division of n busy AUs between individual call classes is omitted.

On the basis of a large number of simulation experiments, it has been proven [48] that the influence of these two criteria on the inaccuracies in the obtained results stabilizes if the following condition is satisfied: $f > 5t_{\max}$, where $t_{\max}$ is the number of AUs required for a call with the maximum demands to be set up. In such cases, the basic LAG model offers the accuracy that is adequate for engineering applications.

The analysis of the results of the accuracy of the model proposed here shows that the accuracy of the model is not influenced by changes in the parameter μ_i, the proportions of the offered traffic, or the number of traffic classes. In addition, the influence of the number of demanded allocation units and the number of resources and their capacities is only slight. The results of the study, presented in Figures 1–9, allow us to evaluate the influence of any change in the value of these parameters. Any further increase in the number of resources, their capacity, and the demands of the individual traffic classes did not cause any increase in the inaccuracy of the systems under investigation. The main influence on the inaccuracy of the model is exerted by the number of resources x, whose non-availability is determined. The highest accuracy was obtained for a case in which we determined the non-availability of a large part of the resources, whereas the lowest accuracy was obtained when we determined the non-availability of only one resource. However, these dependencies are not critical for engineering applications related to the dimensioning of telecommunication networks, for which the possibility of servicing a given demand is critical. From the point of view of network operators, essential is also that the proposed method for the majority of traffic classes leads to overdimensioning of the system (i.e., the value of the non-availability probability is overestimated)—Figures 2, 3, 5, 6, 8 and 9—regardless of the number of unavailable resources. It is only for the traffic class with the lowest number of resources (Figures 1, 4 and 7) that the value of the non-availability probability is slightly underestimated. It is worthwhile to stress, though, that the classes with a larger number of required resources are mainly responsible for blocking.

To sum up, in the typical range of applications, i.e., for a traffic intensity per single AU between 0.7–1.0 Erlangs (Figures 1–9), the method ensures high accuracy and can be used in teletraffic engineering of multi-service communication systems.

Because the system is analyzed at the call (session) level, the direct determination of traffic parameters at the packet level is not possible. However, the possibility of expressing the amount of resources demanded at the packet level as the maximum bit rate or the equivalent bandwidth at the call level allows us to dimension the system (by determining its capacity) in such a way as to include, for example, acceptable packet delays, acceptable packet loss ratios, etc.

5. Conclusions

In this paper, a new method for calculating the probability of strictly determined multi-service resources in a group of resources has been proposed. The method allows us to elaborate a new formula for calculating the blocking probability of strictly determined resources in systems servicing multi-service traffic, e.g., in a group of cells in 4G and 5G systems. The new formula is an extension of the well-known Palm-Jacobaeus loss formula elaborated for systems with single-rate traffic. The new multi-service Palm-Jacobaeus formula opens new possibilities in the analytical modeling of communication systems with separated resources, such as multi-service switching networks [38,40,41,56–58], data centers [59–61], and systems with traffic overflow [62–65].

Author Contributions: Conceptualization, M.G. and M.S.; methodology, M.G. and M.S.; software, A.K.; validation, A.K. and M.G.; formal analysis, M.G. and M.S.;investigation, A.K.; resources, A.K., M.G., and M.S.; data curation, A.K.; writing—original draft preparation, M.G. and M.S.; writing—review and editing, M.G. and M.S.; visualization, A.K. and M.G.; supervision, M.G.; project administration, M.G.; funding acquisition, M.S. All authors have read and agreed to the published version of the manuscript.

Funding: This research was funded by the Polish Ministry of Science and Higher Education, Grant 08/82/SBAD/8229.

Conflicts of Interest: The authors declare no conflict of interest. The funders had no role in the design of the study; in the collection, analyses, or interpretation of data; in the writing of the manuscript, or in the decision to publish the results.

References

1. Oladejo, S.O.; Falowo, O.E. Latency-Aware Dynamic Resource Allocation Scheme for Multi-Tier 5G Network: A Network Slicing-Multitenancy Scenario. *IEEE Access* **2020**, *8*, 74834–74852. [CrossRef]
2. Jian, Z.; Muqing, W.; Ruiqiang, M.; Xiusheng, W. Dynamic Resource Sharing Scheme across Network Slicing for Multi-tenant C-RANs. In Proceedings of the 2018 IEEE/CIC International Conference on Communications in China (ICCC Workshops), Beijing, China, 16–18 August 2018. [CrossRef]
3. Kazi, B.U.; Wainer, G.A. Next generation wireless cellular networks: Ultra-dense multi-tier and multi-cell cooperation perspective. *Wirel. Netw.* **2018**, *25*, 2041–2064. [CrossRef]
4. Wang, C.X.; Haider, F.; Gao, X.; You, X.H.; Yang, Y.; Yuan, D.; Aggoune, H.; Haas, H.; Fletcher, S.; Hepsaydir, E. Cellular architecture and key technologies for 5G wireless communication networks. *IEEE Commun. Mag.* **2014**, *52*, 122–130. [CrossRef]
5. Jaber, M.; Imran, M.A.; Tafazolli, R.; Tukmanov, A. 5G Backhaul Challenges and Emerging Research Directions: A Survey. *IEEE Access* **2016**, *4*, 1743–1766. [CrossRef]
6. 3GPP. *Self-Configuring and Self-Optimizing Network (Son) Use Cases and Solutions (Rlease 9)*; Technical Report TR 36.902; 3GPP: Sophia Antipolis Cedex, France, 2010.
7. 3GPP. *TS32.500 V14.0.0: Telecommunication Management; Self-Organizing Networks (SON); Concepts and Requirements*; Technical Report; 3GPP: Sophia Antipolis Cedex, France, 2017.
8. 3GPP. *TS32.501 V14.0.0: Telecommunication Management; Self-Configuration of Network Elements; Concepts and Requirements*; Technical Report; 3GPP: Sophia Antipolis Cedex, France, 2017.
9. Bassoy, S.; Jaber, M.; Imran, M.A.; Xiao, P. Load Aware Self-Organising User-Centric Dynamic CoMP Clustering for 5G Networks. *IEEE Access* **2016**, *4*, 2895–2906. [CrossRef]
10. Głąbowski, M.; Hanczewski, S.; Stasiak, M. Modelling Load Balancing Mechanisms in Self-Optimising 4G Mobile Networks with Elastic and Adaptive Traffic. *IEICE Trans. Commun.* **2016**, *E99-B*, 1718–1726.
11. Ramirez-Perez, C.; Ramos, V. SDN meets SDR in self-organizing networks: Fitting the pieces of network management. *IEEE Commun. Mag.* **2016**, *54*, 48–57. [CrossRef]
12. Moysen, J.; Giupponi, L. From 4G to 5G: Self-organized network management meets machine learning. *Comput. Commun.* **2018**, *129*, 248–268. [CrossRef]
13. Xu, J.; Yao, J.; Wang, L.; Wu, K.; Chen, L.; Lou, W. Revolution of Self-Organizing Network for 5G MmWave Small Cell Management: From Reactive to Proactive. *IEEE Wirel. Commun.* **2018**, *25*, 66–73. [CrossRef]
14. Paik, C.; Suh, Y.S. Generalized Queueing Model for Call Blocking Probability and Resource Utilization in OFDM Wireless Networks. *IEEE Commun. Lett.* **2011**, *15*, 767–769. [CrossRef]
15. Moscholios, I.D.; Vassilakis, V.G.; Panagoulias, P.I.; Logothetis, M.D. On Call Blocking Probabilities and Resource Utilization in OFDM Wireless Networks. In Proceedings of the 2018 11th International Symposium on Communication Systems, Networks Digital Signal Processing (CSNDSP), Budapest, Hungary, 18–20 July 2018; pp. 1–5.
16. Panagoulias, P.I.; Moscholios, I.D.; Sarigiannidis, P.G.; Logothetis, M.D. Congestion probabilities in OFDM wireless networks with compound Poisson arrivals. *IET Commun.* **2020**, *14*, 674–681. [CrossRef]
17. Efstratiou, P.; Moscholios, I.D. User Mobility in a 5G Cell with Quasi-Random Traffic under the Complete Sharing and Bandwidth Reservation Policies. *Autom. Control Comput. Sci.* **2019**, *53*, 376–386. [CrossRef]

18. Panagoulias, P.I.; Moscholios, I.D. Congestion probabilities in the X2 link of LTE networks. *Telecommun. Syst.* **2019**, *71*, 585–599. [CrossRef]
19. Stasiak, M.; Głąbowski, M.; Wiśniewski, A.; Zwierzykowski, P. *Modeling and Dimensioning of Mobile Networks*; Wiley: Chichester, UK, 2011.
20. Chousainov, I.; Moscholios, I.D.; Kaloxylos, A.; Logothetis, M.D. Performance Evaluation of a C-RAN Supporting Quasi-Random Traffic. In Proceedings of the 2019 International Conference on Software, Telecommunications and Computer Networks, SoftCOM 2019, Split, Croatia, 19–21 September 2019; Begusic, D., Rozic, N., Radic, J., Saric, M., Eds.; IEEE: New York, NY, USA, 2019; pp. 1–6. [CrossRef]
21. Zhang, Y.; Xiao, Y.; Chen, H. Queueing analysis for OFDM subcarrier allocation in broadband wireless multiservice networks. *IEEE Trans. Wirel. Commun.* **2008**, *7*, 3951–3961. [CrossRef]
22. Głąbowski, M.; Sobieraj, M.; Stasiak, M. Blocking Probability Calculation in UMTS Networks with Bandwidth Reservation, Handoff Mechanism and Finite Source Population. In Proceedings of the 7th International Symposium on Communications and Information Technologies, Sydney, Australia, 17–19 October 2007; pp. 433–438. [CrossRef]
23. Palm, C. Nagra foljdsatser urde Erlang'ska formlerna. *Tekn. Medd. Kungl. Telegr.* **1943**.
24. Jacobaeus, C. A study on congestion in link-systems. *Ericsson Tech.* **1950**, *48*, 1–68.
25. Lotze, A. History and development of grading theory. In Proceedings of the 5th International Teletraffic Congress, New York, NY, USA, 14–20 June 1967; pp. 148–161.
26. Syski, R. *Introduction to Congestion Theory in Telephone Systems*; Studies in Telecommunication, North Holland: Amsterdam, The Netherlands, 1986.
27. Herzog, U. *Adaption of the MPJ Loss Formula to Gradings of Various Types (with Special Regard to O'Dell Gradings and Standard Gradings with Slipping)—4. Bericht über verkehrstheoretische Arbeiten*; Technical Report; Institut für Nachrichtenvermittlung und Datenverarbeitung, University of Stuttgart: Stuttgart, Germany, 1967.
28. Binida, N.; Wend, W. Die Effektive Erreichbarkeit für Abnehmerbundel hinter Zwischenleitungsanungen. *Nachrichtentechnische Ztg. (NTZ)* **1959**, *11*, 579–585.
29. Charkiewicz, A. An approximate method for calculating the number of junctions in a crossbar system exchange. *Elektrosvyaz* **1959**, *2*, 55–63.
30. Lotze, A. *Bericht uber Verkehrtheoretische Untersuchungen CIRB*; Technical Report; Inst. für Nachrichten-Vermittlung und Datenverarbeitung der Technischen Hochschule, University of Stuttgart: Stuttgart, Germany, 1963.
31. Bazlen, D.; Kampe, G.; Lotze, A. On the Influence of Hunting Mode and Link Wiring on the Loss of Link Systems. In Proceedings of the 7th International Teletraffic Congress, Stockholm, Swedenm, 13–20 June 1973; Organizing Committee: Stockholm, Swedenm, 1973; pp. 232/1–232/12.
32. Rothmaier, K.; Scheller, R. Design of Economic PCM Arrays with a Prescribed Grade of Service. *IEEE Trans. Commun.* **1981**, *29*, 925–935. [CrossRef]
33. Conradt, J.; Buchheister, A. Considerations on Loss Probability of Multi-Slot Connections. In Proceedings of the 11th International Teletraffic Congress, Kyoto, Japan, 4–11 September 1985; North-Holland: Kyoto, Japan, 1985.
34. Karlsson, J. Loss performance in trunk groups with different capacity demands. In Proceedings of the 13th International Teletraffic Congress, Copenhagen, Denmark, 19–26 June 1991; North-Holland: Copenhagen, Denmark, 1991; Volume Discussion Circles; pp. 201–212.
35. Ramaswami, V.; Rao, K. Flexible time slot assignment—A performance study for the integrated services digital network. In Proceedings of the 11th International Teletraffic Congress, Kyoto, Japan, 4–11 September 1985; North-Holland: Kyoto, Japan, 1985.
36. Stasiak, M. Blocking probability in a limited-availability group carrying mixture of different multichannel traffic streams. *Ann. Télécommun.* **1993**, *48*, 71–76.
37. Głąbowski, M.; Stasiak, M. Multi-rate Model of the Group of Separated Transmission Links of Various Capacities. In *Telecommunications and Networking—ICT 2004*; Lecture Notes in Computer Science; de Souza, J., Dini, P., Lorenz, P., Eds.; Springer: Berlin/Heidelberg, Germany, 2004; Volume 3124, pp. 1101–1106. [CrossRef]
38. Głąbowski, M.; Stasiak, M.; Stasiak, M.D. QoS Criteria for Energy-Aware Switching Networks. In *Quality, Reliability, Security and Robustness in Heterogeneous Systems*; Duong, T.Q., Vo, N.S., Phan, V.C., Eds.; Springer International Publishing: Cham, Switzerland, 2019; pp. 137–147.

39. Głąbowski, M.; Stasiak, M.D. Modelling of Multiservice Switching Networks with Overflow Links for Any Traffic Class. *IET Circuits Devices Syst.* **2014**, *8*, 358–366. [CrossRef]
40. Głąbowski, M.; Stasiak, M.D. Multiservice Switching Networks with Overflow Links and Resource Reservation. *Math. Probl. Eng.* **2016**, *2016*, 4090656. [CrossRef]
41. Głąbowski, M.; Sobieraj, M. Analytical modelling of multiservice switching networks with multiservice sources and resource management mechanisms. *Telecommun. Syst.* **2017**, *66*, 559–578. [CrossRef]
42. Panagoulias, P.I.; Moscholios, I.D.; Sarigiannidis, P.G.; Głąbowski, M.; Logothetis, M.D. An Analytical Framework in OFDM Wireless Networks Servicing Random or Quasi-Random Traffic. *Appl. Sci.* **2019**, *9*, 5376. [CrossRef]
43. Roberts, J. (Ed.) *Performance Evaluation and Design of Multiservice Networks, Final Report COST 224*; Commission of the European Communities: Brussels, Belgium, 1992; pp. 36–44.
44. Pras, A.; Nieuwenhuis, L.; van de Meent, R.; Mandjes, M. Dimensioning network links: A new look at equivalent bandwidth. *IEEE Netw.* **2009**, *23*, 5–10. [CrossRef]
45. Hui, J.Y. Resource Allocation in Broadband Networks. *J. Sel. Areas Commun.* **1988**, *6*, 1598–1608. [CrossRef]
46. Moscholios, I.; Logothetis, M. *Efficient Multirate Teletraffic Loss Models Beyond Erlang*; Wiley: Chichester, UK, 2019.
47. Moscholios, I.D.; Logothetis, M.D.; Vardakas, J.S.; Boucouvalas, A.C. Congestion probabilities of elastic and adaptive calls in Erlang-Engset multirate loss models under the threshold and bandwidth reservation policies. *Comput. Netw.* **2015**, *92 Pt 1*, 1–23. [CrossRef]
48. Stasiak, M. Combinatorial considerations for switching systems carrying multi-channel traffic streams. *Ann. Télécommun.* **1996**, *51*, 611–625.
49. Erlang, A. Solution of some problems in the theory of probabilities of significance in automatic telephone exchanges. *Elektrotechnikeren* **1917**, *13*, 5.
50. Roberts, J.; Mocci, V.; Virtamo, I. (Eds.) *Broadband Network Teletraffic, Final Report of Action COST 242*; Commission of the European Communities; Springer: Berlin, Germany, 1996.
51. Bonald, T.; Roberts, J.W. Internet and the Erlang Formula. *ACM SIGCOMM Comput. Commun. Rev.* **2012**, *42*, 23–30. [CrossRef]
52. Kelly, F. Effective bandwith at multi-class queues. *Queueing Syst.* **1991**, *9*, 5–15. [CrossRef]
53. Kallos, G.A.; Vassilakis, V.G.; Moscholios, I.D.; Logothetis, M.D. Performance Modelling of W-CDMA Networks supporting Elastic and Adaptive Traffic. In Proceedings of the 4th International Working Conference on Performance Modelling and Evaluation of Heterogeneous Networks (HET-NETs '06), Ilkley, UK, 11–13 September 2006; Organizing Committee: Ilkley, UK, 2006.
54. Głąbowski, M.; Kaliszan, A. Simulator of Full-Availability Group with Bandwidth Reservation and Multi-Rate Bernoulli-Poisson-Pascal Traffic Streams. In Proceedings of the Eurocon 2007, Warsaw, Poland, 9–12 September 2007; pp. 2271–2277. [CrossRef]
55. Głąbowski, M.; Kaliszan, A.; Stasiak, M. Modeling product-form state-dependent systems with BPP traffic. *Perform. Eval.* **2010**, *67*, 174–197. [CrossRef]
56. Żal, M. Energy-efficient optical switching nodes based on banyan-type switching fabrics. *Opt. Switch. Netw.* **2019**, *33*, 95–102. [CrossRef]
57. Abdulsahib, M.; Michalski, M.; Kabaciński, W. Optimization of wide-sense nonblocking elastic optical switches. *Opt. Switch. Netw.* **2019**, *33*, 85–94. [CrossRef]
58. Hanczewski, S.; Sobieraj, M.; Stasiak, M.D. The Direct Method of Effective Availability for Switching Networks with Multi-Service Traffic. *IEICE Trans. Commun.* **2016**, *E99-B*, 1291–1301. [CrossRef]
59. Yang, H.; Zhang, J.; Ji, Y.; Tian, R.; Han, J.; Lee, Y. Performance evaluation of multi-stratum resources integration based on network function virtualization in software defined elastic data center optical interconnect. *Opt. Express* **2015**, *23*, 31192–31205. [CrossRef] [PubMed]
60. Dizaji, L.G.; Rahbar, A.G. PAHON: Power-Aware Hybrid Optical Network. *J. Parallel Distrib. Comput.* **2018**, *117*, 1–16. [CrossRef]
61. Kabaciński, W.; Michalski, M.; Rajewski, R.; Żal, M. Optical datacenter networks with elastic optical switches. In Proceedings of the 2017 IEEE International Conference on Communications (ICC), Paris, France, 21–25 May 2017; pp. 1–6. [CrossRef]
62. Głąbowski, M.; Kmiecik, D.; Stasiak, M. Modelling of Multiservice Networks with Separated Resources and Overflow of Adaptive Traffic. *Wirel. Commun. Mob. Comput.* **2018**, *2018*, 7870164. [CrossRef]

63. Głąbowski, M.; Kmiecik, D.; Stasiak, M. Overflows in Multiservice Systems. *IEICE Trans. Commun.* **2019**, *E102.B*, 958–969. [CrossRef]
64. Głąbowski, M.; Hanczewski, S.; Stasiak, M. Modelling of Cellular Networks with Traffic Overflow. *Math. Probl. Eng.* **2015**, *2015*, 286490. [CrossRef]
65. Keramidi, I.P.; Moscholios, I.D.; Sarigiannidis, P.G.; Logothetis, M.D. Call Blocking Probabilities in a Two-Link Loss Model with Restricted Accessibility. In Proceedings of the 15th International Conference on Telecommunications (ConTEL 2019), Graz, Austria, 3–5 July 2019; pp. 1–7. [CrossRef]

applied sciences

MDPI

Article

An Analytical Framework in OFDM Wireless Networks Servicing Random or Quasi-Random Traffic

Panagiotis I. Panagoulias [1], Ioannis D. Moscholios [1,*], Panagiotis G. Sarigiannidis [2], Mariusz Głąbowski [3] and Michael D. Logothetis [4]

1 Department of Informatics & Telecommunications, University of Peloponnese, Tripolis 221-31, Greece; panagoulias@uop.gr

2 Department of Electrical and Computer Engineering, University of Western Macedonia, Kozani 501-00, Greece; psarigiannidis@uowm.gr

3 Faculty of Electronics and Telecommunications, Poznan University of Technology, 609-65 Poznan, Poland; mariusz.glabowski@put.poznan.pl

4 Department of Electrical and Computer Engineering, University of Patras, Patras 265-04, Greece; mlogo@upatras.gr

* Correspondence: idm@uop.gr

Received: 15 November 2019; Accepted: 4 December 2019; Published: 9 December 2019

Abstract: We consider the downlink of an orthogonal frequency division multiplexing (OFDM)-based cell that services calls from many service-classes. The call arrival process is random (Poisson) or quasi-random, i.e., calls are generated by an infinite or a finite number of sources, respectively. In order to determine congestion probabilities and resource utilization, we model the cell as a multirate loss model. Regarding the call admission, we consider the restricted accessibility, the bandwidth reservation (BR), and the complete sharing (CS) policies. In a system of restricted accessibility, a new call may be blocked even if resources do exist. In a BR system, subcarriers can be reserved in favor of calls of high subcarrier requirements. Finally, in a CS system, a new call is blocked due to resource unavailability. In all three policies, we show that there exist recursive formulas for the determination of the various performance measures. Based on simulation, the accuracy of the proposed formulas is found to be quite satisfactory.

Keywords: OFDM; congestion; random; quasi-random; recursive; restricted; reservation; complete sharing

1. Introduction

The determination and evaluation of the main quality of service (QoS) parameters, such as call blocking probabilities (CBP) and network resources utilization, is a complex task in contemporary networks, due to the growth of network traffic and the high traffic stream diversity [1]. The latter necessitates research on teletraffic loss or queueing models, either at call-level or at packet-level [2–7]. Such models not only assist in network optimization and dimensioning procedures but they may also be used in combination with machine learning techniques [8,9] or as an input to computational intelligent techniques, such as the fuzzy analytical hierarchy process techniques [10,11]. In this paper, we concentrate on call-level teletraffic loss models.

The simplest call-level loss models adopt as a call arrival process the classical Poisson process, which leads to analytically tractable formulas for the determination of performance measures, such as CBP and resources utilization. The origination of calls in the Poisson process results from an infinite number of users (or traffic sources). This means that the Poisson process cannot capture the case of calls generated via a finite number of users. The latter can be well described by the quasi-random arrival process. This call arrival process depends on the number of idle users (that are capable of generating traffic) and is smoother than the Poisson process. Because of this, the CBP in a system that carries quasi-random traffic are much lower compared to the corresponding CBP of a system that carries random (Poisson) traffic. For recent applications of the quasi-random process in loss systems, the interested reader may resort to [12–16].

We consider the cases of random and quasi-random traffic in the downlink of an orthogonal frequency division multiplexing (OFDM)-based cell which provides service to calls from many service-classes. OFDM is a dominant technology in 4th generation (4G) networks and can also be considered as a candidate technology in 5th generation (5G) networks [17–21] and in cognitive radio networks [22]. The analysis of this OFDM-based cell relies on the loss models of [23–26]. The case of a batch arrival process and two types of calls, narrow-band and wide-band, has been studied in [27] under two different batch blocking disciplines: (a) the complete and (b) the partial batch blocking discipline. A possible extension of [27] that may result in efficient formulas for the CBP determination can be based on the works of [28–30] (for the partial batch blocking discipline) and [31] (for the complete batch blocking discipline).

In [23], Paik and Suh (P-S) consider an OFDM-based cell that accommodates different service-classes whose calls follow a Poisson process. The P-S system is described as a loss system, i.e., calls are blocked and lost in the case of resource unavailability. This means that the adopted policy for resource sharing in [23] is the complete sharing (CS) policy. The CS policy is the default policy in teletraffic loss models, but it may result in an unfair resource allocation among calls [1,2].

Contrary to [24,25], where a new call is accepted in the OFDM-based cell only if the required subcarriers are available, in the P-S model the admission of a new call is based on the availability of both subcarriers and power. Apart from this significant modification in the call admission, it is important to mention that the model of [23] has a product form solution (PFS) for the steady-state probabilities. The existence of a PFS is significant in loss/queueing models, since it results in efficient algorithms (of recursive or convolutional form) for the performance measures calculation [32–37]. However, in the P-S model, the calculation of CBP and resource utilization is based on complicated algorithms which are unattractive for network planning engineers. To overcome this problem of complexity that appears in [23], a recursive formula is proposed in [26] for the calculation of CBP and resource utilization.

In this paper, firstly, we extend the model of [23] by incorporating restricted accessibility. We name this model P-S/res. In such a system, each state is related to a pre-specified blocking probability and therefore a call may be blocked and lost even if subcarriers do exist at the time of the call's arrival [38]. For the proposed model, we show that the calculation of CBP and resource utilization can be based on recursive formulas. In addition, we show the relationship of the P-S/res model with the P-S model under the bandwidth reservation (BR) policy (P-S/BR model). This policy permits the subcarriers' reservation so as to favor those calls that have high subcarrier requirements. In that sense, the BR policy provides QoS to certain service-classes. Secondly, we propose the quasi-random P-S model with restricted accessibility (qr-P-S/res model) and provide recursive formulas for the determination of time and call congestion (TC and CC, respectively) probabilities as well as resource utilization which are the main performance measures. TC probabilities, for a particular service-class, can be calculated via the proportion of time the system has no available resources for this service-class. On the other hand, CC probabilities can be determined via the proportion of arriving calls that find no available resources in the system. Note that in the P-S model, TC and CC probabilities coincide (and named CBP) due to the Poisson process. Thirdly, we show the relationship between the qr-P-S/res model and the qr-P-S/BR, qr-P-S models, where the arrival process remains quasi-random but the adopted policy is the BR and the CS policy, respectively. The calculation of all performance measures in the qr-P-S/BR and the qr-P-S models can also be based on recursive formulas. Based on simulation results, the accuracy of all proposed formulas is quite satisfactory.

This paper is organized as follows: In Section 2, we review the P-S model and present the formulas for the CBP determination and resource utilization. In Section 3, we propose the P-S/res model and present recursive formulas for the determination of the performance measures. In addition, we show the relationship between the P-S/res model and the P-S/BR model. In Section 4, we study the case of quasi-random traffic, propose the qr-P-S/res model, and present recursive formulas for the determination of the performance measures. In addition, we show how the qr-P-S/res model and the qr-P-S/BR, qr-P-S models are related. In Section 5, we compare the simulation with analytical results for the P-S, the P-S/res and the qr-P-S models. The comparison verifies the accuracy of all formulas. We conclude in Section 6.

2. The P-S Multirate Loss Model—A Review

To describe the P-S multirate loss model, consider the downlink of an OFDM-based cell that has M subcarriers and let B, P, and R be the system's bandwidth, the cell's available power and the average data rate per subcarrier, respectively. We consider that the range of channel gains (i.e., signal to noise ratios per unit power) is partitioned into K intervals that are consecutive but do not overlap and denote the average channel gain of each interval k as γ_k, $k = 1, \ldots, K$. In addition, we assume L subcarrier requirements which result in KL service-classes. Upon its arrival, a service-class (k,l) call ($k = 1, \ldots$, K and $l = 1, \ldots, L$) requires a total of b_l subcarriers. This means that the requirement of each new call in terms of data rate is b_lR. In addition, the new call has an (average) channel gain γ_k. If the b_l subcarriers are not available, then call blocking occurs and the call is lost without affecting the system (i.e., a blocked call cannot retry to be accepted in the cell with the same or less subcarrier requirements). Otherwise, the call is accepted in the cell and remains for a service time which is generally distributed with mean μ^{-1}. To achieve the data rate R of a subcarrier assigned to a call (with channel gain γ_k), we determine the power p_k via the Shannon theorem: $R = (B/M)\log_2(1+\gamma_k p_k)$.

Assuming that service-class (k,l) calls arrive in the cell according to a Poisson process with arrival rate λ_{kl} and that n_{kl} is the number of in-service calls of service-class (k,l), then the steady-state probabilities $\pi(\boldsymbol{n})$ are given via a PFS [23]:

$$\pi(\boldsymbol{n}) = G^{-1}\left(\prod_{k=1}^{K}\prod_{l=1}^{L}\alpha_{kl}^{n_{kl}}/n_{kl}!\right), \tag{1}$$

where $\boldsymbol{n} = (n_{11},\ldots,n_{k1},\ldots,n_{K1},\ldots,n_{1L},\ldots,n_{kL},\ldots,n_{KL})$ $G = \sum\limits_{\boldsymbol{n}\in\boldsymbol{\Omega}}\left(\prod\limits_{k=1}^{K}\prod\limits_{l=1}^{L}\alpha_{kl}^{n_{kl}}/n_{kl}!\right)$ $\boldsymbol{\Omega}$ is the system's state space with $\boldsymbol{\Omega} = \left\{\boldsymbol{n} : 0 \le \sum\limits_{k=1}^{K}\sum\limits_{l=1}^{L} n_{kl}b_l \le M, 0 \le \sum\limits_{k=1}^{K}\sum\limits_{l=1}^{L} p_k n_{kl} b_l \le P\right\}$ and $\alpha_{kl} = \lambda_{kl}/\mu$ refers to the offered traffic-load (in erlang) of service-class (k,l) calls.

To derive (1), both P and p_k should be integers. This is achieved by multiplying them by a constant so as to obtain an equivalent representation of $\sum\limits_{k=1}^{K}\sum\limits_{l=1}^{L} p'_k n_{kl} b_l \le P'$, where P' and p'_k are integers. Thus, it can be assumed that P and p_k are integers, without loss of generality.

According to [23], all performance metrics are based on (1). As a specific example, consider the CBP $B(k,l)$ of service-class (k,l) calls which is determined via

$$B(k,l) = 1 - G(P - p_k b_l, M - b_l)/G(\boldsymbol{\Omega}). \tag{2}$$

However, since $\boldsymbol{\Omega}$ grows as $(MP)^{KL}$, the applicability of (1) (and consequently of (2)) can only be limited to moderate size systems and therefore is not adequate for procedures related to network planning.

In [23], the determination of $G(P - p_k b_l, M - b_l)$ (and consequently the determination of CBP via (2)) is based on the algorithms of [39,40]. These algorithms were initially proposed for the CBP calculation in circuit-switched networks [41,42]. The algorithms of [39] are based on mean-value analysis and z-transforms. The algorithm of [40] is based on the quite complex approach of numerical inversion of generating functions [43]. Both algorithms are applied to loss models that have a PFS and are impractical compared to the formula of Kaufman–Roberts (K–R) [44,45]. The K–R formula is recursive and therefore leads to an efficient way for the CBP calculation in a loss system that services multirate Poisson traffic. Because of this, the interested reader can find many applications of the K–R formula in PFS and non-PFS models [46–58].

The complexity problem of (1) can be circumvented via a recursive formula, proposed in [26], that resembles the K–R formula. The presentation of this formula, requires the following notation: $j_1 = \sum_{k=1}^{K}\sum_{l=1}^{L} n_{kl}b_l$ refers to the occupied subcarriers, i.e., $j_1 = 0, \ldots, M$ and $j_2 = \sum_{k=1}^{K}\sum_{l=1}^{L} p_k n_{kl} b_l$ refers to the occupied cell's power, i.e., $j_2 = 0, \ldots, P$. Furthermore, let $q(\vec{j}) = q(j_1, j_2)$ be the occupancy distribution, denoted as:

$$q(\vec{j}) = q(j_1, j_2) = \sum_{n \in \boldsymbol{\Omega}_{\vec{j}}} \pi(\boldsymbol{n}), \tag{3}$$

where $\boldsymbol{\Omega}_{\vec{j}}$ refers to those states where the occupied subcarriers and power is j_1 and j_2, respectively.

The recursive calculation of $q(j_1, j_2)$'s is based on (4), whose complexity is $O(MPKL)$:

$$q(j_1, j_2) = \begin{cases} 1, \quad \text{for } j_1 = j_2 = 0 \\ \frac{1}{j_1} \sum_{k=1}^{K}\sum_{l=1}^{L} \alpha_{kl} b_l q(j_1 - b_l, j_2 - p_k b_l) \\ \qquad \text{for } j_1 = 1, \ldots, M \text{ and } j_2 = 1, \ldots, P \end{cases} . \tag{4}$$

Having obtained the unnormalized values of $q(j_1, j_2)$, we can calculate the CBP $B(k,l)$ of service-class (k,l) via:

$$B(k,l) = \sum_{\{(j_1+b_l>M) \,\cup\, (j_2+p_k b_l>P)\}} G^{-1} q(j_1, j_2), \tag{5}$$

and the mean number of service-class (k,l) in-service calls, $E(k,l)$, via:

$$E(k,l) = \alpha_{kl}(1 - B(k,l)), \tag{6}$$

where G is the normalization constant, given by $G = \sum_{j_1=0}^{M}\sum_{j_2=0}^{P} q(j_1, j_2)$.

Based on $E(k,l)$, we can determine the blocking probability (BP) of the entire system, the subcarrier and the power utilization (SU and PU, respectively) via:

$$\text{BP} = \sum_{k=1}^{K}\sum_{l=1}^{L} B(k,l)\lambda_{k,l}/\Lambda, \quad \Lambda = \sum_{k=1}^{K}\sum_{l=1}^{L} \lambda_{k,l}, \tag{7}$$

$$\text{SU} = \sum_{k=1}^{K}\sum_{l=1}^{L} E(k,l) b_l / M, \tag{8}$$

$$\text{PU} = \sum_{k=1}^{K}\sum_{l=1}^{L} p_k E(k,l) b_l / P. \tag{9}$$

3. The P-S Multirate Loss Model under Restricted Accessibility

3.1. The Analytical Model

We consider again the P-S model and apply the notion of restricted accessibility. To this end, let each state $j_1 > 0$ be associated with a blocking probability factor, $pb_{k,l}(j_1)$. Note that if the available subcarriers for service-class (k,l) calls are not enough (i.e., when $j_1 \geq M - b_l + 1$), then $pb_{k,l}(j_1) = 1$. Similarly, if the system is empty, then $pb_{k,l}(0) = 0$.

The admission mechanism for a call of service-class (*k,l*) in the P-S/res model is the following: (a) if $(M - j_1 \geq b_l) \cap (j_2 + p_k b_l \leq P)$ then the service-class (*k,l*) call is accepted in the cell with probability $1 - pb_{k,l}(j_1)$ and remains for a service-time which is generally distributed with mean μ^{-1}, (b) if $(M - j_1 < b_l) \cup (j_2 + p_k b_l > P)$ then the call is blocked due to subcarriers' unavailability.

The proof of a recursive formula for the unnormalized values of $q(j_1, j_2)$'s requires the existence of local balance between states $(j_1 - b_l, j_2 - p_k b_l)$ and (j_1, j_2). These two states differ only by one in-service call of service-class (*k,l*). The form of local balance equation is as follows:

$$\alpha_{kl}\Big(1 - pb_{k,l}(j_1 - b_l)\Big)q(j_1 - b_l, j_2 - p_k b_l) = y_{kl}(j_1, j_2)q(j_1, j_2), \tag{10}$$

where $y_{kl}(j_1, j_2)$ refers to the mean number of service-class (*k,l*) calls in state (j_1, j_2).

Multiplying both sides of (10) by b_l and summing over *k* and *l* we obtain the following formula for the recursive calculation of $q(j_1, j_2)$'s in the P-S/res model:

$$\sum_{k=1}^{K}\sum_{l=1}^{L}\alpha_{kl}\Big(1 - pb_{k,l}(j_1 - b_l)\Big)q(j_1 - b_l, j_2 - p_k b_l) = j_1 q(j_1, j_2), \tag{11}$$

or

$$q(j_1, j_2) = \begin{cases} 1, \quad \text{for } j_1 = j_2 = 0 \\ \frac{1}{j_1}\sum\limits_{k=1}^{K}\sum\limits_{l=1}^{L}\alpha_{kl}b_l\Big(1 - pb_{k,l}(j_1 - b_l)\Big)q(j_1 - b_l, j_2 - p_k b_l) \\ \qquad \text{for } j_1 = 1, \ldots, M \text{ and } j_2 = 1, \ldots, P \end{cases}. \tag{12}$$

Based on the unnormalized values of $q(j_1, j_2)$, we can determine the CBP *B*(*k,l*) of service-class (*k,l*) via:

$$B(k,l) = \sum_{\{(j_1 + b_l > M) \cup (j_2 + p_k b_l > P)\}} G^{-1}q(j_1, j_2)pb_{k,l}(j_1), \tag{13}$$

while *E*(*k,l*), BP, SU, and PU are calculated via (6)–(9), respectively.

3.2. The Case of the BR Policy (P-S/BR Model)

A proper selection of the values of $pb_{k,l}(j_1)$ results in the classical BR policy. In that policy, a call of service-class (*k,l*) has a reservation parameter t_l and a requirement of b_l subcarriers. The parameter t_l denotes the number of subcarriers reserved to benefit calls of all service-classes except for (*k,l*). Since the BR policy is used to favor calls of high subcarrier requirements, it is obvious that it provides QoS to certain service-classes.

The call admission mechanism in the case of the BR policy (P-S/BR model) consists of the following two cases: (a) if $(M - j_1 - t_l \geq b_l) \cap (j_2 + p_k b_l \leq P)$ then the service-class (*k,l*) call is accepted for service in the system, (b) if $(M - j_1 - t_l < b_l) \cup (j_2 + p_k b_l > P)$ then call blocking occurs.

By assuming that $pb_{k,l}(j_1) = 0$ when $j_1 \leq M - b_l - t_l$ and $pb_{k,l}(j_1) = 1$ when $j_1 > M - b_l - t_l$, then the BR policy is incorporated in the model.

In the P-S/BR model, the determination of $q(j_1, j_2)$'s is based on (14), whose complexity is *O*(*MPKL*) [56]:

$$q(j_1, j_2) = \begin{cases} 1, \quad \text{for } j_1 = j_2 = 0 \\ \frac{1}{j_1}\sum\limits_{k=1}^{K}\sum\limits_{l=1}^{L}\alpha_{kl}(j_1 - b_l)b_l q(j_1 - b_l, j_2 - p_k b_l) \\ \qquad \text{for } j_1 = 1, \ldots, M \text{ and } j_2 = 1, \ldots, P \end{cases}, \tag{14}$$

where $\alpha_{kl}(j_1 - b_l) = \begin{cases} \alpha_{kl}, & \text{for} j_1 \leq M - t_l \\ 0, & \text{for} j_1 > M - t_l \end{cases}$.

Having obtained the unnormalized values of $q(j_1, j_2)$, the calculation of the CBP $B(k,l)$ of service-class (*k*,*l*) can be based on:

$$B(k,l) = \sum_{\{(j_1+b_l+t_l>M)\cup(j_2+p_kb_l>P)\}} G^{-1}q(j_1, j_2), \tag{15}$$

while $E(k,l)$, BP, SU, and PU are calculated via (6)–(9), respectively.

4. The Quasi-Random P-S Multirate Loss Model with Restricted Accessibility

4.1. The Analytical Model

Consider again the model of [23] that provides service to *KL* service-classes. In the proposed qr-P-S/res model, new calls of service-class (*k*,*l*) are generated from a finite number of sources N_{kl}. The mean arrival rate of idle service-class (*k*,*l*) sources is given by $\lambda_{kl,\text{idle}} = (N_{kl} - n_{kl})v_{kl}$, where n_{kl} is the number of in-service calls of service-class (*k*,*l*) and v_{kl} is the per idle source arrival rate. Based on the above, the per idle source offered traffic-load of service-class (*k*,*l*) is determined by $\alpha_{kl,\text{idle}} = v_{kl}/\mu$ (in erlang). If $N_{kl} \to \infty$ for all service-classes and the total offered traffic-load is constant, then we have the P-S/res model (since the arrival process becomes Poisson).

Upon its arrival, a service-class (*k*,*l*) call ($k = 1, \ldots, K$ and $l = 1, \ldots, L$) requires b_l subcarriers. Let the occupied subcarriers and power in the cell be j_1 and j_2, respectively, when the new call arrives. Then, the admission mechanism for the new call is as follows: (a) if $(M - j_1 \geq b_l) \cap (j_2 + p_kb_l \leq P)$ then the service-class (*k*,*l*) call is accepted in the cell with probability $1 - pb_{k,l}(j_1)$. In that case, the service-time is generally distributed with mean μ^{-1}, (b) if $(M - j_1 < b_l) \cup (j_2 + p_kb_l > P)$ then the call is blocked due to subcarriers' unavailability.

Let $q_{\text{fin}}(\vec{j}) = q_{\text{fin}}(j_1, j_2)$ be the occupancy distribution in the proposed qr-P-S/res model. To prove a formula for the unnormalized values of $q_{\text{fin}}(j_1, j_2)$'s, we assume that local balance exists between the states $(j_1 - b_l, j_2 - p_kb_l)$ and (j_1, j_2). The form of local balance equation is as follows:

$$\big(N_{kl} - y_{kl,\text{fin}}(j_1 - b_l, j_2 - p_kb_l)\big)\alpha_{kl,\text{idle}}\big(1 - pb_{k,l}(j_1 - b_l)\big)q_{\text{fin}}(j_1 - b_l, j_2 - p_kb_l) = y_{kl,\text{fin}}(j_1, j_2)q_{\text{fin}}(j_1, j_2), \tag{16}$$

where $y_{kl,\text{fin}}(j_1 - b_l, j_2 - p_kb_l)$ and $y_{kl,\text{fin}}(j_1, j_2)$ refer to the mean number of service-class (*k*,*l*) calls in states $(j_1 - b_l, j_2 - p_kb_l)$ and (j_1, j_2), respectively.

Multiplying both sides of (16) by b_l and summing over *k* and *l* we obtain:

$$\sum_{k=1}^{K}\sum_{l=1}^{L} \big(N_{kl} - y_{kl,\text{fin}}(j_1 - b_l, j_2 - p_kb_l)\big)\alpha_{kl,\text{idle}}b_l\big(1 - pb_{k,l}(j_1 - b_l)\big)q_{\text{fin}}(j_1 - b_l, j_2 - p_kb_l) = j_1q_{\text{fin}}(j_1, j_2). \tag{17}$$

The value of $y_{kl,\text{fin}}(j_1 - b_l, j_2 - p_kb_l)$ in (17) is unknown. To determine it, the following lemma is necessary [59]: Two stochastic systems will be equivalent and result in the same congestion probabilities, if they have (a) the same traffic parameters $(K, L, N_{kl}, \alpha_{kl,\text{idle}})$, where $k = 1, \ldots, K$, $l = 1, \ldots, L$ and (b) are the same states.

Therefore, the purpose is to find a stochastic system, whereby the values of $y_{kl,\text{fin}}(j_1 - b_l, j_2 - p_kb_l)$ can be determined. The subcarriers' requirements of calls of all service-classes and the values of *M* and *P* in the new system are chosen so that both conditions (a) and (b) are valid and the occupancy (j_1, j_2) of each state $\vec{j}$ is unique.

In that case, state $(\vec{j}) = (j_1, j_2)$ is reached only via state $(j_1 - b_l, j_2 - p_kb_l)$. Thus, $y_{kl,\text{fin}}(j_1 - b_l, j_2 - p_kb_l) = n_{kl} - 1$. Based on the above, (17) can be written as:

$$q_{\text{fin}}(j_1, j_2) = \begin{cases} 1, & \text{for} j_1 = j_2 = 0 \\ \frac{1}{j_1}\sum_{k=1}^{K}\sum_{l=1}^{L} (N_{kl} - n_{kl} + 1)\alpha_{kl,\text{idle}}b_l\big(1 - pb_{k,l}(j_1 - b_l)\big)q_{\text{fin}}(j_1 - b_l, j_2 - p_kb_l) & \\ \quad \text{for} j_1 = 1, \ldots, M \text{ and } j_2 = 1, \ldots, P \end{cases}. \tag{18}$$

Note that if $N_{kl} \to \infty$ for all service-classes and the total offered traffic-load is constant, then we obtain (12) of the proposed P-S/res model.

4.2. Performance Measures Calculation

Having obtained the unnormalized values of $q_{\text{fin}}(j_1, j_2)$, we can calculate the TC probabilities of service-class (k,l) calls, $B_{TC}(k,l)$, via:

$$B_{TC}(k,l) = \sum_{\{(j_1+b_l>M) \cup (j_2+p_k b_l>P)\}} G^{-1} q_{\text{fin}}(j_1, j_2) pb_{k,l}(j_1), \tag{19}$$

and the CC probabilities of service-class (k,l) calls via (19) but for a cell with $N_{kl} - 1$ sources.

Furthermore, we can determine the average number of service-class (k,l) in-service calls, $E_{\text{fin}}(k,l)$, via:

$$E_{\text{fin}}(k,l) = \sum_{j_1=1}^{M} \sum_{j_2=1}^{P} G_{\text{fin}}^{-1} y_{kl,\text{fin}}(j_1, j_2) q_{\text{fin}}(j_1, j_2), \tag{20}$$

where $G_{\text{fin}} = \sum\limits_{j_1=0}^{M} \sum\limits_{j_2=0}^{P} q_{\text{fin}}(j_1, j_2)$ and $y_{kl,\text{fin}}(j_1 - b_l, j_2 - p_k b_l)$ is the mean number of service-class (k,l) calls in state $(j_1 - b_l, j_2 - p_k b_l)$ calculated via:

$$y_{kl,\text{fin}}(j_1, j_2) = \frac{(N_{kl} - n_{kl} + 1)\alpha_{kl,\text{idle}}\left(1 - pb_{k,l}(j_1 - b_l)\right) q_{\text{fin}}(j_1 - b_l, j_2 - p_k b_l)}{q_{\text{fin}}(j_1, j_2)}. \tag{21}$$

In addition, we can determine the entire system BP based on the TC probabilities of all service-classes, BP_{TC}, the SU_{fin}, and the PU_{fin}, via:

$$\text{BP}_{\text{TC}} = \sum_{k=1}^{K} \sum_{l=1}^{L} B_{TC}(k,l) N_{kl} v_{kl} / \Lambda_{\text{fin}}, \quad \Lambda_{\text{fin}} = \sum_{k=1}^{K} \sum_{l=1}^{L} N_{kl} v_{kl}, \tag{22}$$

$$\text{SU}_{\text{fin}} = \sum_{k=1}^{K} \sum_{l=1}^{L} E_{\text{fin}}(k,l) b_l / M, \tag{23}$$

$$\text{PU}_{\text{fin}} = \sum_{k=1}^{K} \sum_{l=1}^{L} p_k E_{\text{fin}}(k,l) b_l / P. \tag{24}$$

In order to determine the values of $q_{\text{fin}}(j_1, j_2)$ according to (18), the unknown values of n_{kl} are required. These values can be obtained via a stochastic system, with the same parameters and the same states as already described for the proof of (18). However, the state space determination of this system becomes complex due to the large number of service-classes. To this end, we propose an algorithm which is simpler and easy to implement:
(a) Determine the values of $q(j_1, j_2)$ according to (12) (i.e., via the P-S/res model).
(b) Determine the values of $y_{kl}(j_1, j_2)$ via the formula:

$$y_{kl}(j_1, j_2) = \frac{\alpha_{kl}\left(1 - pb_{k,l}(j_1 - b_l)\right) q(j_1 - b_l, j_2 - p_k b_l)}{q(j_1, j_2)}. \tag{25}$$

(c) Modify (18) to the following formula, where $y_{kl}(j_1, j_2)$ is given by (25):

$$q_{\text{fin}}(j_1, j_2) = \begin{cases} 1, & \text{for} j_1 = j_2 = 0 \\ \frac{1}{j_1} \sum\limits_{k=1}^{K} \sum\limits_{l=1}^{L} (N_{kl} - y_{kl}(j_1 - b_l, j_2 - p_k b_l)) \alpha_{kl,\text{idle}} b_l \left(1 - pb_{k,l}(j_1 - b_l)\right) q_{\text{fin}}(j_1 - b_l, j_2 - p_k b_l) \\ \quad \text{for} j_1 = 1, \ldots, M \text{and} j_2 = 1, \ldots, P \end{cases}. \tag{26}$$

(d) Determine $E_{\text{fin}}(k,l)$ via (20), where the values of $y_{kl,\text{fin}}(j_1, j_2)$ are given by:

$$y_{kl,\text{fin}}(j_1, j_2) = \frac{(N_{kl} - y_{kl}(j_1 - b_l, j_2 - p_k b_l))\alpha_{kl,\text{idle}}\big(1 - pb_{k,l}(j_1 - b_l)\big)q_{\text{fin}}(j_1 - b_l, j_2 - p_k b_l)}{q_{\text{fin}}(j_1, j_2)}. \quad (27)$$

(e) Determine: (1) the TC probabilities of service-class (k,l) calls, $B_{\text{TC}}(k,l)$, via (19), and (2) the BP_{TC}, the SU_{fin}, and the PU_{fin}, via (22)–(24), respectively.

4.3. The Case of the BR Policy (qr-P-S/BR Model)

The admission mechanism in the qr-P-S/BR model is the same with that of the P-S/BR model. Since the local balance is destroyed (due to the BR policy), the recursive formulas presented in this subsection are approximate. Following the previous analysis of Section 4, we propose an algorithm for the calculation of performance measures in the qr-P-S/BR model:
(a) Determine the values of $q(j_1, j_2)$ according to (14) (i.e., via the P-S/BR model).
(b) Determine the values of $y_{kl}(j_1, j_2)$, for $j_1 \le M - t_l$, via the formula:

$$y_{kl}(j_1, j_2) = \frac{\alpha_{kl} q(j_1 - b_l, j_2 - p_k b_l)}{q(j_1, j_2)}. \quad (28)$$

(c) Modify (18) to the following formula where $y_{kl}(j_1, j_2)$ has been calculated via (28):

$$q_{\text{fin}}(j_1, j_2) = \begin{cases} 1, & \text{for} j_1 = j_2 = 0 \\ \frac{1}{j_1}\sum\limits_{k=1}^{K}\sum\limits_{l=1}^{L} (N_{kl} - y_{kl}(j_1 - b_l, j_2 - p_k b_l))\alpha'_{kl,\text{idle}} b_l q_{\text{fin}}(j_1 - b_l, j_2 - p_k b_l) & , \\ \quad \text{for} j_1 = 1, \ldots, M \text{and} j_2 = 1, \ldots, P \end{cases} \quad (29)$$

where $\alpha'_{kl,\text{idle}} \equiv \alpha'_{kl,\text{idle}}(j_1 - b_l) = \alpha_{kl,\text{idle}}$ for $j_1 \le M - t_l$.
(d) Determine the average number of in-service calls of service-class (k,l), $E_{\text{fin}}(k,l)$ via (20), where $y_{kl,\text{fin}}(j_1, j_2)$ is given by, for $j_1 \le M - t_l$:

$$y_{kl,\text{fin}}(j_1, j_2) = \frac{(N_{kl} - y_{kl}(j_1 - b_l, j_2 - p_k b_l))\alpha_{kl,\text{idle}} q_{\text{fin}}(j_1 - b_l, j_2 - p_k b_l)}{q_{\text{fin}}(j_1, j_2)}. \quad (30)$$

(e) Determine the TC probabilities of service-class (k,l) calls, $B_{\text{TC}}(k,l)$, via:

$$B_{\text{TC}}(k,l) = \sum_{\{(j_1 + b_l + t_l > M) \cup (j_2 + p_k b_l > P)\}} G^{-1} q_{\text{fin}}(j_1, j_2), \quad (31)$$

and the BP_{TC}, the SU_{fin}, and the PU_{fin}, via (22)–(24), respectively.

4.4. The Case of the CS Policy (qr-P-S Model)

In the qr-P-S multirate loss model, a new service-class (k,l) call requires b_l subcarriers. Assuming that the occupied subcarriers and power in the cell are j_1 and j_2, respectively, then, the new call: (a) is accepted for a generally distributed service-time with mean μ^{-1}, if $(M - j_1 \ge b_l) \cap (j_2 + p_k b_l \le P)$ and (b) is blocked and lost, if $(M - j_1 < b_l) \cup (j_2 + p_k b_l > P)$.

Following Section 4, it can be proved that the (unnormalized) values of $q_{\text{fin}}(j_1, j_2)$ in the qr-P-S model are given by:

$$q_{\text{fin}}(j_1, j_2) = \begin{cases} 1, & \text{for} j_1 = j_2 = 0 \\ \frac{1}{j_1}\sum\limits_{k=1}^{K}\sum\limits_{l=1}^{L} (N_{kl} - n_{kl} + 1)\alpha_{kl,\text{idle}} b_l q_{\text{fin}}(j_1 - b_l, j_2 - p_k b_l) & . \\ \quad \text{for} j_1 = 1, \ldots, M \text{and} j_2 = 1, \ldots, P \end{cases} \quad (32)$$

Note that if $N_{kl} \rightarrow \infty$ for all service-classes and the total offered traffic-load is constant, then we have (4) of the P-S model.

In order to overcome the equivalent stochastic system determination required for the calculation of $q_{\mathrm{fin}}(j_1, j_2)$ via (32), we present the following algorithm for the calculation of the various performance measures in the qr-P-S model:
(a) Determine the values of $q(j_1, j_2)$ according to (4) (i.e., via the P-S model).
(b) Determine the values of $y_{kl}(j_1, j_2)$ via the formula:

$$y_{kl}(j_1, j_2) = \frac{\alpha_{kl} q(j_1 - b_l, j_2 - p_k b_l)}{q(j_1, j_2)}. \tag{33}$$

(c) Modify (32) to the following formula, where $y_{kl}(j_1, j_2)$ has been determined via (33):

$$q_{\mathrm{fin}}(j_1, j_2) = \begin{cases} 1, & \mathrm{for} j_1 = j_2 = 0 \\ \frac{1}{j_1} \sum\limits_{k=1}^{K} \sum\limits_{l=1}^{L} (N_{kl} - y_{kl}(j_1 - b_l, j_2 - p_k b_l)) \alpha_{kl,\mathrm{idle}} b_l q_{\mathrm{fin}}(j_1 - b_l, j_2 - p_k b_l) & . \\ \quad \mathrm{for} j_1 = 1, \ldots, M \mathrm{and} j_2 = 1, \ldots, P \end{cases} \tag{34}$$

(d) Determine $E_{\mathrm{fin}}(k, l)$ via (20), where $y_{kl,\mathrm{fin}}(j_1, j_2)$ is given by:

$$y_{kl,\mathrm{fin}}(j_1, j_2) = \frac{(N_{kl} - y_{kl}(j_1 - b_l, j_2 - p_k b_l)) \alpha_{kl,\mathrm{idle}} q_{\mathrm{fin}}(j_1 - b_l, j_2 - p_k b_l)}{q_{\mathrm{fin}}(j_1, j_2)}. \tag{35}$$

(e) Determine the TC probabilities of service-class (*k*,*l*) calls, $B_{\mathrm{TC}}(k, l)$, via:

$$B_{\mathrm{TC}}(k, l) = \sum_{\{(j_1 + b_l > M) \cup (j_2 + p_k b_l > P)\}} G^{-1} q_{\mathrm{fin}}(j_1, j_2), \tag{36}$$

and the $\mathrm{BP_{TC}}$, the $\mathrm{SU_{fin}}$, and the $\mathrm{PU_{fin}}$, via (22)–(24), respectively.

5. Performance Evaluation

In this section, we consider a cell that accommodates *KL* service-classes and provide simulation and analytical congestion probabilities results for the P-S, the P-S/res, and the qr-P-S models. In addition, we provide simulation and analytical SU and PU results for the P-S and the P-S/res models. The required input for these models is: $B = 20$ MHz, $M = 256$, $P = 25$ Watt, $R = 329.6$ kbps, $L = 64$, $b_l = l$, $l = 1, \ldots, 64$, while we assume that b_l is uniformly distributed. Due to this assumption, a new call has an average subcarrier requirement $\hat{g} = 32.5$. In addition, let $K = 3$, which means that the cell accommodates $KL = 192$ service-classes. In the case of the qr-P-S model, we assume that $N_{kl} = 20$ sources for all service-classes. Let the integer representations of p_k ($k = 1, 2, 3$) and P be: $p'_1 = 6$, $p'_2 = 10$, $p'_3 = 16$, $P\prime = 2500$. The values of p'_k require that: $p_1 \approx 0.06$, $p_2 \approx 0.01$, $p_3 \approx 0.16$ achieved via the following values of the average channel gain γ_k ($k = 1, 2, 3$): $\gamma_1 = 24.679$ dB, $\gamma_2 = 22.460$ dB, $\gamma_3 = 20.419$ dB. In addition, an arriving call has an average channel gain γ_k with a probability that is determined via: (1) set 1: $r_k = 1/3$ ($k = 1, 2, 3$) and (2) set 2: $r_1 = 1/4$, $r_2 = 1/4$, $r_3 = 1/2$. Furthermore, let $\lambda_{kl} = \Lambda r_k / L$, where Λ is the total arrival rate given by $\Lambda = \alpha M \mu / \hat{g}$, α is the cell's traffic intensity and μ is the service rate of calls with $\mu = 0.00625$. In the case of the P-S/res model, we assume that $pb_{k,l}(j_1) = 0$ when $j_1 \leq M - b_l - t_l$ and $pb_{k,l}(j_1) = 1$ when $j_1 > M - b_l - t_l$. Due to this assumption, we have the P-S/BR model. The values of t_l are $t_l = 64 - l$, $l = 1, \ldots, 64$, and they are chosen in such a way so that $b_1 + t_1 = \ldots = b_{64} + t_{64}$.

Regarding the simulation results, they are based on Simscript III [60]. Each simulation run is based on 10 million generated calls while the results presented herein are mean values of 7 runs. Furthermore, the blocking events of the first 3% of these calls are not taken into account in the results, in order to account for a warm-up period. In all figures of this section, the analytical and simulation results are quite close. Note that in the x-axis of Figures 1–5 the value of α increases from 0.2 to 1.0 in steps of 0.1, while in the x-axis of Figures 6 and 7 the value of α increases from 0.05 to 0.2 in steps of 0.025.

In Figures 1 and 2, we consider the P-S and the P-S/BR models and present the simulation and analytical CBP of service-classes (3, 64), (2, 64), (1, 64) (Figure 1) and service-classes (3, 48), (2, 48), (1, 48) (Figure 2). Note that service-classes (3, 64), (2, 64) and (1, 64) have the highest requirement in terms of subcarriers ($l = 64$). Regarding the average channel gain we consider set 1 ($r_k = 1/3$ ($k = 1, 2, 3$)). In Figure 1, it is obvious that the BR policy decreases the CBP values of service-classes (3, 64), (2, 64) and (1, 64) compared to the corresponding CBP values of the P-S model. In Figure 2, the BR policy increases (in most of the cases) the CBP values of service-classes (3, 48), (2, 48), (1, 48) compared to the corresponding CBP values of the P-S model. A similar behavior appears in most of the service-classes whose calls have a requirement of less than 64 subcarriers. In addition, the same behavior (in terms of CBP) appears when we consider set 2 for the average channel gain.

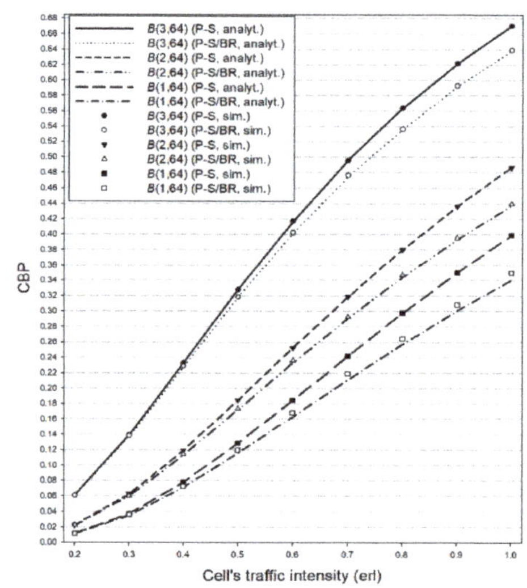

Figure 1. CBP—Service-classes (3, 64), (2, 64), and (1, 64).

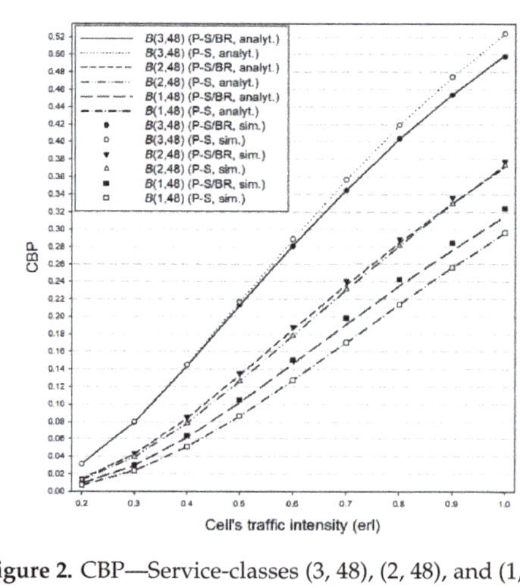

Figure 2. CBP—Service-classes (3, 48), (2, 48), and (1, 48).

In Figure 3, we present the entire system BP for both sets of r_k. We observe that the BP increases for both sets of r_k when the BR policy is considered. This is because the values of t_l parameters are chosen to increase the CBP of service-classes with low subcarrier requirements and benefit service-classes with high subcarrier requirements. The increase of BP in the case of the P-S/BR model results in a slight decrease of the SU and PU (for both sets of r_k) compared to the P-S model, as we show in Figures 4 and 5, respectively. A similar behavior appears in the case of the corresponding quasi-random models.

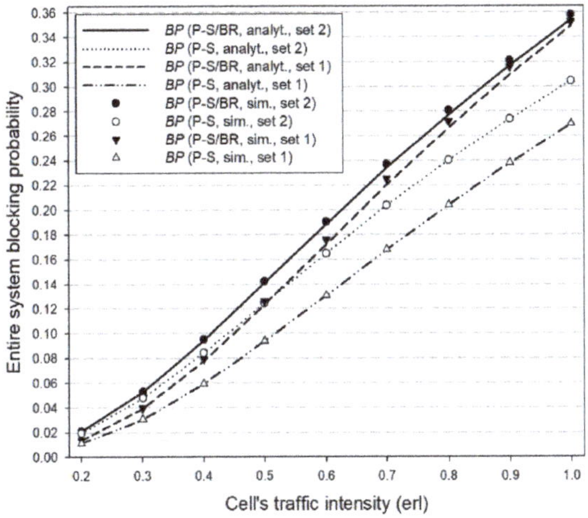

Figure 3. BP of the entire system.

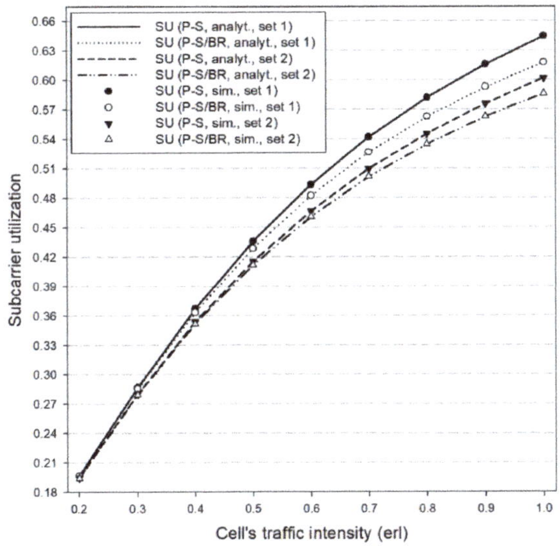

Figure 4. Subcarrier utilization.

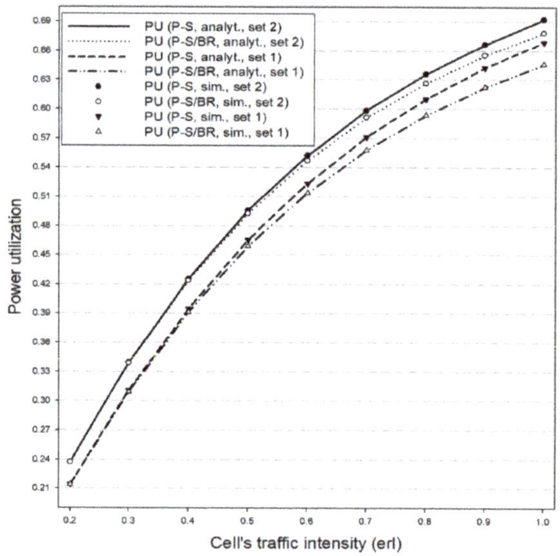

Figure 5. Power utilization.

In Figures 6 and 7, we consider the PS and the qr-P-S models for both sets of r_k. Figures 6 and 7 show the simulation and analytical TC probabilities of service-classes (3, 16) and (3, 64), respectively. We observe that: (1) in the qr-P-S model the TC probabilities are lower compared to those obtained in the P-S model, which is due to the quasi-random process which is smoother than the Poisson process, and (2) the selection of set 2 for the values of r_k, increases the TC probabilities since the power assigned to calls in the case of set 2 is larger compared to set 1.

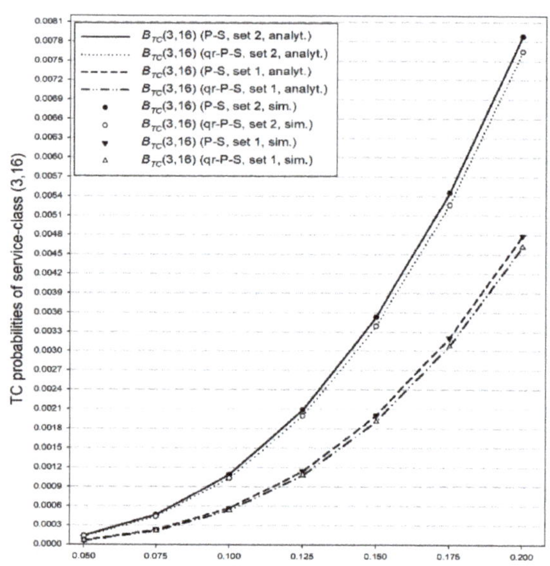

Figure 6. TC probabilities service-class (3, 16).

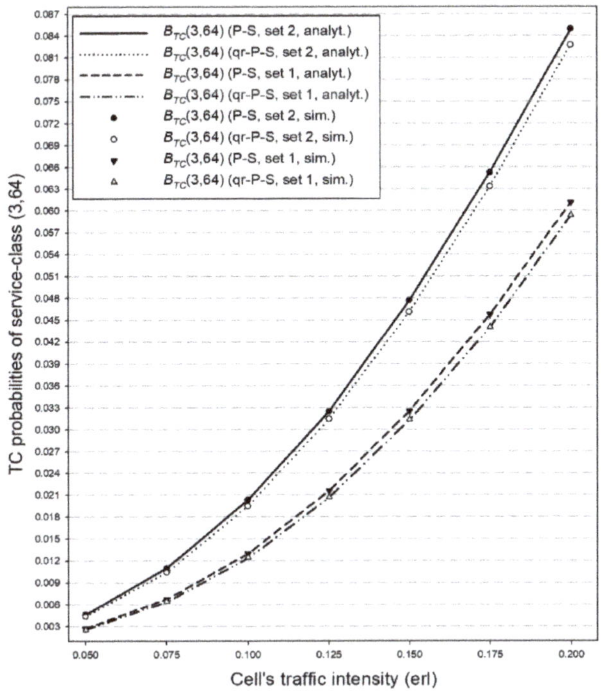

Figure 7. TC probabilities service-class (3, 64).

6. Conclusions

We propose loss models for the analysis of the downlink of an OFDM cell that accommodates random or quasi-random generated calls from different service-classes under the restricted accessibility, the BR and the CS policies. The cell is analysed as a loss system, i.e., calls are blocked in case of resource unavailability. To determine the main performance measures, such as congestion probabilities and resource utilization, we propose approximate but recursive formulas. All formulas are quite accurate, compared to simulation, and can be applied to network planning and dimensioning procedures.

Author Contributions: Conceptualization, all authors; methodology, all authors; software, P.I.P., I.D.M.; validation, P.I.P., I.D.M.; writing—original draft preparation, all authors; writing—review and editing, all authors.

Funding: This research received no external funding.

Conflicts of Interest: The authors declare no conflict of interest.

References

1. Moscholios, I.D.; Logothetis, M.D. *Efficient Multirate Teletraffic Loss Models beyond Erlang*; Wiley IEEE Press: Hoboken, NJ, USA, 2019. [CrossRef]
2. Stasiak, M.; Głąbowski, M.; Wisniewski, A.; Zwierzykowski, P. *Modeling and Dimensioning of Mobile Networks: From GSM to LTE*; Wiley: Hoboken, NJ, USA, 2011. [CrossRef]
3. Shioda, S. Fundamental trade-offs between resource separation and resource share for quality of service guarantees. *IET Netw.* **2014**, *3*, 4–15. [CrossRef]
4. Bouloukakis, G.; Moscholios, I.D.; Georgantas, N.; Issarny, V. Performance modeling of the middleware overlay infrastructure of mobile things. In Proceedings of the 2017 IEEE International Conference on Communications (ICC), Paris, France, 21–25 May 2017. [CrossRef]
5. Tadayon, N.; Kaddoum, G. Packet-level modeling of cooperative diversity: A queueing network approach. *IEEE Access* **2018**, *6*, 35223–35242. [CrossRef]

6. Benson, K.; Bouloukakis, G.; Issarny, V.; Mehrotra, S.; Moscholios, I.; Venkatasubramanian, N. FIREDEX: A prioritized IoT data exchange middleware for emergency response. In Proceedings of the ACM/IFIP/USENIX 19th International Middleware Conference, Rennes, France, 10–14 December 2018. [CrossRef]
7. Hanczewski, S.; Stasiak, M.; Weissenberg, J. Non-full-available queueing model of an EON node. *Opt. Switch. Netw.* **2019**, *33*, 131–142. [CrossRef]
8. Xue, J.; Yan, F.; Riska, A.; Smirni, E. Scheduling data analytics work with performance guarantees: Queuing and machine learning models in synergy. *Clust. Comput.* **2016**, *19*, 849–864. [CrossRef]
9. Ataie, E.; Gianniti, E.; Ardagna, D.; Movaghar, A. A combined analytical modeling machine learning approach for performance prediction in mapreduce jobs in cloud environment. In Proceedings of the 18th International Symposium on Symbolic and Numeric Algorithms for Scientific Computing (IEEE SYNASC), Timisoara, Romania, 24–27 September 2016. [CrossRef]
10. Abbasi, A.; Jin, H.; Wu, S. A software-defined cloud resource management framework. In Proceedings of the Asia-Pacific Services Computing Conference, Bangkok, Thailand, 7–9 December 2015. [CrossRef]
11. Abbasi, A.; Jin, H. v-Mapper: An application-aware resource consolidation scheme for cloud data centers. *Future Internet* **2018**, *10*, 90. [CrossRef]
12. Vardakas, J.; Zorba, N.; Verikoukis, C. Power demand control scenarios for smart grid applications with finite number of appliances. *Appl. Energy* **2016**, *162*, 83–98. [CrossRef]
13. Moscholios, I.; Vassilakis, V.; Logothetis, M.; Boucouvalas, A. State-dependent bandwidth sharing policies for wireless multirate loss networks. *IEEE Trans. Wirel. Commun.* **2017**, *16*, 5481–5497. [CrossRef]
14. Głąbowski, M.; Kmiecik, D.; Stasiak, M. Modelling of multiservice networks with separated resources and overflow of adaptive traffic. *Wirel. Commun. Mob. Comput.* **2018**, *2018*, 17. [CrossRef]
15. Abd El-atty, S.; Gharsseldien, Z.; Lizos, K. Predictive reservation for handover optimization in two-tier heterogeneous cellular networks. *Wirel. Pers. Commun.* **2018**, *98*, 1637–1661. [CrossRef]
16. Efstratiou, P.; Moscholios, I. User mobility in a 5G cell with quasi-random traffic under the complete sharing and bandwidth reservation policies. *Autom. Control Comp. Sci.* **2019**, *53*, 376–386. [CrossRef]
17. Vahidi, V.; Saberinia, E. OFDM high speed train communication systems in 5G cellular networks. In Proceedings of the 15th IEEE Annual Consumer Communications Networking Conference (IEEE CCNC), Las Vegas, NV, USA, 12–15 January 2018. [CrossRef]
18. Zhang, L.; Xiao, A.; Molu, M.; Tafazolli, R. Filtered OFDM systems, algorithms, and performance analysis for 5G and beyond. *IEEE Trans. Commun.* **2018**, *66*, 1205–1218. [CrossRef]
19. Yang, G.; Liang, Y.; Zhang, R.; Pei, Y. Modulation in the air: Backscatter communication over ambient OFDM carrier. *IEEE Trans. Commun.* **2018**, *66*, 1219–1233. [CrossRef]
20. Na, Z.; Wang, Y.; Xiong, M.; Liu, X.; Xia, J. Modeling and throughput analysis of an ADO-OFDM based relay-assisted VLC system for 5G networks. *IEEE Access* **2018**, *6*, 17586–17594. [CrossRef]
21. Lu, W.; Fang, S.; Hu, S.; Liu, X.; Li, B.; Na, Z.; Gong, Y. Energy efficiency optimization for OFDM based 5G wireless networks with simultaneous wireless information and power transfer. *IEEE Access* **2018**, *6*, 75937–75946. [CrossRef]
22. Sultana, A.; Fernando, X.; Zhao, L. Power allocation using geometric water filling for OFDM-based cognitive radio networks. In Proceedings of the 2016 IEEE 84th Vehicular Technology Conference (IEEE VTC-Fall), Montreal, QC, Canada, 18–21 September 2016. [CrossRef]
23. Paik, C.; Suh, Y. Generalized queueing model for call blocking probability and resource utilization in OFDM wireless networks. *IEEE Commun. Let.* **2011**, *15*, 767–769. [CrossRef]
24. Pla, V.; Martinez-Bauset, J.; Casares-Giner, V. Comments on call blocking probability and bandwidth utilization of OFDM subcarrier allocation in next-generation wireless networks. *IEEE Commun. Lett.* **2008**, *12*, 349. [CrossRef]
25. Chen, J.; Chen, W. Call blocking probability and bandwidth utilization of OFDM subcarrier allocation in next-generation wireless networks. *IEEE Commun. Lett.* **2006**, *10*, 82–84. [CrossRef]
26. Moscholios, I.; Vassilakis, V.; Panagoulias, P.; Logothetis, M. On call blocking probabilities and resource utilization in OFDM wireless networks. In Proceedings of the 2018 11th International Symposium on Communication Systems, Networks Digital Signal Processing (CSNDSP), Budapest, Hungary, 18–20 July 2018. [CrossRef]
27. Zhang, Y.; Xiao, Y.; Chen, H. Queueing analysis for OFDM subcarrier allocation in broadband wireless multiservice networks. *IEEE Trans. Wirel. Commun.* **2008**, *7*, 3951–3961. [CrossRef]

28. Kaufman, J.; Rege, K. Blocking in a shared resource environment with batched Poisson arrival processes. *Perf. Eval.* **1996**, *24*, 249–263. [CrossRef]
29. Moscholios, I.; Logothetis, M. The Erlang multirate loss model with batched Poisson arrival processes under the bandwidth reservation policy. *Comput. Commun.* **2010**, *33*, S167–S179. [CrossRef]
30. Moscholios, I.; Vassilakis, V.; Sarigiannidis, P. Performance modelling of a multirate loss system with batched Poisson arrivals under a probabilistic threshold policy. *IET Netw.* **2018**, *7*, 242–247. [CrossRef]
31. Ezhilchelvan, P.; Mitrani, I. Multi-class resource sharing with batch arrivals and complete blocking. In Proceedings of the International Conference on Quantitative Evaluation of Systems (QEST), Berlin, Germany, 5–7 September 2017; Lecture Notes in Computer Science. Springer: Cham, Switzerland, 2017; Volume 10503. [CrossRef]
32. Iversen, V. The exact evaluation of multi-service loss system with access control. *Teleteknik* **1987**, *31*, 56–61.
33. Moscholios, I.; Logothetis, M.; Koukias, M. An ON-OFF multirate loss model of finite sources. *IEICE Trans. Commun.* **2007**, *90*, 1608–1619. [CrossRef]
34. Głąbowski, M.; Kaliszan, A. Convolution algorithm for overflow calculation in integrated services networks. In Proceedings of the 17th Asia Pacific Conference on Communications (APCC), Sabah, Malaysia, 2–5 October 2011. [CrossRef]
35. Stasiak, M.; Parniewicz, D.; Zwierzykowski, P. Traffic engineering for multicast connections in multiservice cellular network. *IEEE Trans. Ind. Inform.* **2013**, *9*, 262–270. [CrossRef]
36. Moscholios, I.; Vassilakis, V.; Logothetis, M.; Boucouvalas, A. A probabilistic threshold-based bandwidth sharing policy for wireless multirate loss networks. *IEEE Wirel. Commun. Lett.* **2016**, *5*, 304–307. [CrossRef]
37. Głąbowski, M.; Sobieraj, M. Analytical modelling of multiservice switching networks with multiservice sources and resource management mechanisms. *Telecom. Syst.* **2017**, *66*, 559–578. [CrossRef]
38. Iversen, V. Modelling restricted accessibility for wireless multi-service systems. *LNCS* **2006**, *3883*, 93–102. [CrossRef]
39. Pinsky, E.; Conway, A. Computational algorithms for blocking probabilities in circuit-switched networks. *Ann. Oper. Res.* **1992**, *35*, 31–41. [CrossRef]
40. Choudhury, G.; Leung, K.; Whitt, W. An algorithm to compute blocking probabilities in multi-rate multi-class multi-resource loss models. *Adv. Appl. Prob.* **1995**, *27*, 1104–1143. [CrossRef]
41. Caro, F.; Simchi-Levi, D. Optimal static pricing for a tree network. *Ann. Oper. Res.* **2012**, *196*, 137–152. [CrossRef]
42. Wang, M.; Li, S.; Won, E.; Zukerman, M. Performance analysis of circuit-switched multi-service multi-rate networks with alternative routing. *J. Light. Technol.* **2014**, *32*, 179–200. [CrossRef]
43. Beard, C.; Frost, V. Prioritized resource allocation for stressed networks. *IEEE/ACM Trans. Netw.* **2001**, *9*, 618–633. [CrossRef]
44. Kaufman, J. Blocking in a shared resource environment. *IEEE Trans. Commun.* **1981**, *29*, 1474–1481. [CrossRef]
45. Roberts, J. *A service system with heterogeneous user requirements. Performance of Data Communications Systems and Their Applications*; North Holland: Amsterdam, The Netherlands, 1981; pp. 423–431.
46. Vardakas, J.; Moscholios, I.; Logothetis, M.; Stylianakis, V. Performance analysis of OCDMA PON configuration supporting multirate bursty traffic with retrials and QoS differentiation. *Opt. Switch. Netw.* **2014**, *13*, 112–123. [CrossRef]
47. Huang, Y.; Rosberg, Z.; Ko, K.; Zukerman, M. Blocking probability approximations and bounds for best-effort calls in an integrated service system. *IEEE Trans. Commun.* **2015**, *63*, 5014–5026. [CrossRef]
48. Vassilakis, V.; Moscholios, I.; Logothetis, M. Uplink blocking probabilities in priority-based cellular CDMA networks with finite source population. *IEICE Trans. Commun.* **2016**, *99*, 1302–1309. [CrossRef]
49. Casares-Giner, V. Some teletraffic issues in optical burst switching with burst segmentation. *Electr. Lett.* **2016**, *52*, 941–943. [CrossRef]
50. Głąbowski, M.; Kaliszan, A.; Stasiak, M. Modelling overflow systems with distributed secondary resources. *Comp. Netw.* **2016**, *108*, 171–183. [CrossRef]
51. Vassilakis, V.; Moscholios, I.; Logothetis, M. Efficient radio resource allocation in SDN/NFV based mobile cellular networks under the complete sharing policy. *IET Netw.* **2018**, *7*, 103–108. [CrossRef]
52. Sagkriotis, S.; Pantelis, S.; Moscholios, I.; Vassilakis, V. Call blocking probabilities in a two-link multi rate loss System for Poisson traffic. *IET Netw.* **2018**, *7*, 233–241. [CrossRef]

53. Vakilinia, S.; Cheriet, M. Preemptive cloud resource allocation modeling of processing jobs. *J. Supercomput.* **2018**, *74*, 2116–2150. [CrossRef]
54. Panagoulias, P.; Moscholios, I. Congestion probabilities in the X2 link of LTE Networks. *Telecommun. Syst.* **2019**, *71*, 585–599. [CrossRef]
55. Francisco, C.; Martins, L.; Medhi, D. Dynamic multicriteria alternative routing for single-and multi-service reservation-oriented networks and its performance. *Ann. Telecommun.* **2019**, *74*, 697–715. [CrossRef]
56. Panagoulias, P.; Moscholios, I.; Logothetis, M. Performance metrics in OFDM wireless networks under the bandwidth reservation policy. In Proceedings of the International Conference on Image Processing and Communications, Bydgoszcz, Poland, 11–13 September 2019. [CrossRef]
57. Hanczewski, S.; Horiushkina, A.; Stasiak, M.; Weissenberg, J. The analytical model of 5G networks. In Proceedings of the International Conference on Image Processing and Communications, Bydgoszcz, Poland, 11–13 September 2019. [CrossRef]
58. Chousainov, I.-A.; Moscholios, I.; Kaloxylos, A.; Logothetis, M. Performance evaluation of a C-RAN supporting quasi-random traffic. In Proceedings of the 2019 International Conference on Software, Telecommunications and Computer Networks (Softcom), Split, Croatia, 19–21 September 2019. [CrossRef]
59. Stamatelos, G.; Hayes, J. Admission control techniques with application to broadband networks. *Comp. Commun.* **1994**, *17*, 663–673. [CrossRef]
60. Simscript III. Available online: http://www.simscript.com (accessed on 8 December 2019).

applied sciences

MDPI

Article

A Flexible FPGA-Based Channel Emulator for Non-Stationary MIMO Fading Channels

Qiuming Zhu [1,*], Wei Huang [1], Kai Mao [1], Weizhi Zhong [2], Boyu Hua [1], Xiaomin Chen [1] and Zikun Zhao [1]

1 The Key Laboratory of Dynamic Cognitive System of Electromagnetic Spectrum Space, College of Electronic and Information Engineering, Nanjing University of Aeronautics and Astronautics, Nanjing 211106, China; hw_val@nuaa.edu.cn (W.H.); maokai@nuaa.edu.cn (K.M.); byhua@nuaa.edu.cn (B.H.); chenxm402@nuaa.edu.cn (X.C.); zhaozikun@nuaa.edu.cn (Z.Z.)

2 The Key Laboratory of Dynamic Cognitive System of Electromagnetic Spectrum Space, College of Astronautics, Nanjing University of Aeronautics and Astronautics, Nanjing 211106, China; zhongwz@nuaa.edu.cn

* Correspondence: zhuqiuming@nuaa.edu.cn

Received: 13 May 2020; Accepted: 13 June 2020; Published: 17 June 2020

Abstract: In this paper, a discrete non-stationary multiple-input multiple-output (MIMO) channel model suitable for the fixed-point realization on the field-programmable gate array (FPGA) hardware platform is proposed. On this basis, we develop a flexible hardware architecture with configurable channel parameters and implement it on a non-stationary MIMO channel emulator in a single FPGA chip. In addition, an improved non-stationary channel emulation method is employed to guarantee accurate channel fading and phase, and the schemes of other key modules are also illustrated and implemented in a single FPGA chip. Hardware tests demonstrate that the output statistical properties of proposed channel emulator, i.e., the probability density function (PDF), cross-correlation function (CCF), Doppler power spectrum density (DPSD), and the power delay profile (PDP) agree well with the corresponding theoretical ones.

Keywords: channel emulator; non-stationary MIMO channel; discrete channel model; field-programmable gate array (FPGA) platform

1. Introduction

Multiple-input multiple-output (MIMO) technologies have played an important role in the fifth generation (5G) and previous communication systems [1–3], as they can boost channel capacity and improve spectral efficiency without increasing transmitting power or system bandwidth [4,5]. It is inevitable to evaluate and validate the performance of MIMO communication devices during the development. The most realistic method is field testing, but it is uncontrollable, unrepeatable, and expensive. Channel emulators can reproduce the real propagation scenario in a controllable way and is a good alternative so far [6].

There are several commercial channel emulators such as Agilent's N5106A PXB, Keysight's Propsim F32 [7], and Azimuth's ACE 400WB [8]. However, these emulators are very large, expensive, and complicated, and mainly developed for the standard channel models, which are all based on the wide-sense stationary (WSS) assumption. Meanwhile, various academic researches on hardware emulation can be found in [9–18], which were focused on the emulation of stationary channel models [9–12]. However, recent measurements have proved that the stationary channel model is not suitable for certain propagation scenarios [13–18], such as high-speed train (HST) [16,17], vehicle-to-vehicle (V2V) [13–15], and unmanned aerial vehicle (UAV) channels [18].

There are very limited non-stationary channel emulators reported in the literatures [19–27]. A hardware emulator for the discrete-time triply selective fading channel was developed in [19]. The channel coefficients were calculated by software dynamically, which cannot support real-time updating. The authors in [20,21] proposed an improved sum-of-sinusoid (SoS) method to generate channel fading, and implemented it into a 2×2 non-stationary MIMO channel emulator. A 4×4 MIMO channel emulator was designed in [22], but the authors did not give the details of implementation. In [23,24], two specific MIMO channel emulators for high speed WLAN 802.ac and LTE-A channels were developed, respectively. In [25], the authors divided the non-stationary channel into several stationary channel segments and adopted the traditional stationary channel emulation method. The authors in [26] designed a channel emulator based on software defined radios (SDR) platform, but the emulator can only be applied to vehicular communications. To the best of our knowledge, the aforementioned channel emulators still adopted traditional stationary channel models and considered the non-stationary aspect by updating parameters periodically. However, we have found that the output fading phases of this kind of method are not accurate, which leads to the output Doppler power spectrum density (DPSD) not fitting well with the theoretical ones [28]. To overcome this shortcoming, an improved 3D non-stationary geometry-based stochastic model (GBSM) was proposed in [27] and implemented in a 2×2 MIMO channel emulator. However, the developed hardware was only suitable for the corresponding channel model and the structure was not general and flexible. This paper proposes a discrete non-stationary channel model with accurate channel fading and phase. The channel parameters such as power, delay, and Doppler frequency are all time-variant in order to take the non-stationarity into account. Furthermore, a flexible hardware architecture is proposed and implemented in a single FPGA chip. Finally, we validate the correctness of the proposed channel model as well as the hardware emulator. The major contributions are summarized as follows.

- Based on the improved GBSM with the accurate channel fading phase and Doppler frequency in [27], this paper proposes a discrete non-stationary MIMO channel model, which is suitable to implement on the FPGA-based hardware platforms. Meanwhile, a flexible hardware architecture tailored for the proposed model is developed, in which the channel size and parameters can easily be reconfigured.
- An improved emulation method of channel fading, namely, sum-of-frequency-modulated-signals (SoFM), is employed to guarantee the accurate channel fading and phase. In addition, the architectures of other key modules, i.e., the delay module, fading generation module, and interpolator module, are developed and implemented on a single Xilinx XC7VX690T FPGA.
- For the developed channel emulator, the output statistical properties, i.e., the probability density function (PDF), cross-correlation function (CCF), and DPSD are tested and verified by the theoretical results. The power delay profile (PDP) is also validated by the measurement data.

The rest of this paper is organized as follows. In Section 2, a discrete non-stationary MIMO channel model is briefly introduced. Section 3 proposes the hardware architecture of channel emulator as well as the channel fading emulation algorithm. In addition, the detailed implementation of key modules are also presented. In Section 4, the developed channel emulator is tested and validated. Finally, some conclusions are drawn in Section 5.

2. Discrete Non-Stationary MIMO Channel Model

Considering a MIMO channel with S transmitting antennas and U receiving antennas, the channel can be defined by a complex channel matrix. Moreover, the input–output relationship in the discrete time domain can be expressed by a convolution operation as

$$\mathbf{y}(l) = \mathbf{H}(l,\zeta) \otimes \mathbf{x}(l) \tag{1}$$

where $\mathbf{x}(l) = [x_1(l), x_2(l), \cdots, x_S(l)]^T$ is the transmitted signal vector; $\mathbf{y}(l) = [y_1(l), y_2(l), \cdots, y_U(l)]^T$ is the received signal vector; l and ζ are the discrete time indexes in the time domain and delay domain, respectively; and $(\cdot)^T$ denotes the transpose operator of a matrix or vector. In (1), the channel matrix $\mathbf{H}(l,\zeta)$ can be further defined as

$$\mathbf{H}_{U\times S}(l,\zeta) = \begin{bmatrix} h_{1,1}(l,\zeta) & h_{1,2}(l,\zeta) & \cdots & h_{1,S}(l,\zeta) \\ h_{2,1}(l,\zeta) & h_{2,2}(l,\zeta) & \cdots & h_{2,S}(l,\zeta) \\ \vdots & \vdots & \ddots & \vdots \\ h_{U,1}(l,\zeta) & h_{U,2}(l,\zeta) & \cdots & h_{U,S}(l,\zeta) \end{bmatrix} \tag{2}$$

where $h_{u,s}(l,\zeta)$ denotes the channel impulse response (CIR) of the sub-channel between the uth $(u = 1,2,\cdots,U)$ receiving antenna and the sth $(s = 1,2,\cdots,S)$ transmitting antenna, and it can be modeled in the discrete time domain as [20]

$$h_{u,s}(l,\zeta) = \sum_{n=1}^{N(l)} \sqrt{P_n(l)}\tilde{h}_{u,s,n}(l)\delta(\zeta - \lfloor\tau_n(l)\rfloor_{Ts}) \tag{3}$$

where $P_n(l)$ and $N(l)$ are the path power and valid path number at time instant l, respectively; $\tilde{h}_{u,s,n}(l)$ is the channel coefficient with the normalized power; T_s is the sampling interval; and $\lfloor\tau_n(l)\rfloor_{Ts}$ denotes the discrete time delay. It should be noticed that the channel parameters in (3), such as $P_n(l)$, $N(l)$, $\lfloor\tau_n(l)\rfloor_{Ts}$, and $\tilde{h}_{u,s,n}(l)$, are all time-variant, which can take into account the non-stationary aspects of real MIMO channels.

3. Flexible Hardware Architecture and Implementation

3.1. System Architecture

The flexible architecture of our proposed channel emulator is presented in Figure 1. It includes two primary units: the config unit and the signal processing unit. The config unit consists of user-defined scenario module and channel parameters calculation module. It provides an interactive interface for setting environment related parameters, and then calculates the channel parameters, i.e., the path number, delay, power, Doppler frequency, and phase. These channel parameters are passed through by the peripheral component interconnect express (PCIE) bus to the signal processing unit. Each signal processing unit has a four-channel structure with the analog-to-digital converters (ADC), digital-to-analog converters (DAC), and FPGA. Thus, a single signal processing unit can implement a 4×4 MIMO channel emulation. It should be noted that the proposed system architecture is flexible and theoretically supports arbitrary scaled MIMO channels within the limitation of transmission rate of PCIE.

The signal processing unit in Figure 1 is the most important and difficult part and it generates and superposes the multiple channel fading in real-time. Due to the flexibility and parallelism, FPGA is adopted as the core operation chip in the signal processing unit. It includes three modules: delay module (DM), generation module (GM), and superposition module (SM). The first module realizes the predefined delay of each propagation path, the second module generates channel fading coefficients, and the last one carries out the superposition operation and outputs the signal. As we can see, the final output can be expressed as

$$\begin{bmatrix} y_1(l) \\ y_2(l) \\ y_3(l) \\ y_4(l) \end{bmatrix} = \begin{bmatrix} \sum_{n=1}^{N(l)} \sqrt{P_n(\zeta)}\tilde{h}_{1,1,n}(\zeta)x_1(\zeta - [\tau_n(l)]_{Ts}) + \cdots + \sum_{n=1}^{N(l)} \sqrt{P_n(\zeta)}\tilde{h}_{1,4,n}(\zeta)x_4(\zeta - [\tau_n(l)]_{Ts}) \\ \sum_{n=1}^{N(l)} \sqrt{P_n(\zeta)}\tilde{h}_{2,1,n}(\zeta)x_1(\zeta - [\tau_n(l)]_{Ts}) + \cdots + \sum_{n=1}^{N(l)} \sqrt{P_n(\zeta)}\tilde{h}_{2,4,n}(\zeta)x_4(\zeta - [\tau_n(l)]_{Ts}) \\ \sum_{n=1}^{N(l)} \sqrt{P_n(\zeta)}\tilde{h}_{3,1,n}(\zeta)x_1(\zeta - [\tau_n(l)]_{Ts}) + \cdots + \sum_{n=1}^{N(l)} \sqrt{P_n(\zeta)}\tilde{h}_{3,4,n}(\zeta)x_4(\zeta - [\tau_n(l)]_{Ts}) \\ \sum_{n=1}^{N(l)} \sqrt{P_n(\zeta)}\tilde{h}_{4,1,n}(\zeta)x_1(\zeta - [\tau_n(l)]_{Ts}) + \cdots + \sum_{n=1}^{N(l)} \sqrt{P_n(\zeta)}\tilde{h}_{4,4,n}(\zeta)x_4(\zeta - [\tau_n(l)]_{Ts}) \end{bmatrix} \tag{4}$$

which is equivalent with the theoretical result obtained from (1)–(3).

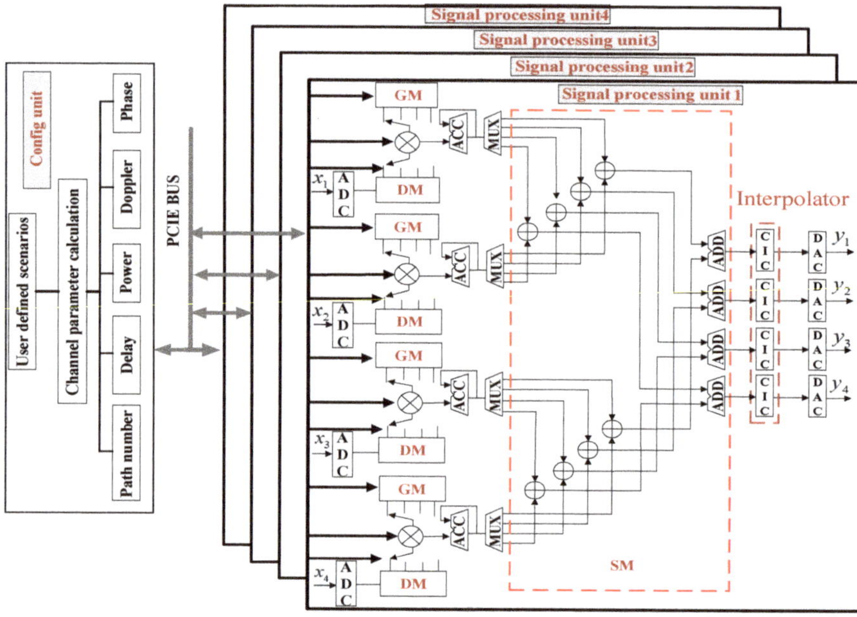

Figure 1. System architecture of the proposed emulator.

3.2. Channel Fading Generation

Several methods for generating the channel fading coefficients, i.e., SoC method, Doppler filter method, AR method, and their derivatives can be addressed in [10,11,28,29]. However, these methods can only be used for stationary channels with fixed channel parameters. In this paper, we upgrade the traditional SoC method to the non-stationary channel fading generation. In order to guarantee the continuity of output fading phase, we use an improved method to generate non-stationary channel as shown in Figure 2. The non-stationary fading coefficient can be generated based on the summation of several linear frequency modulated signals as

$$\tilde{h}_{u,s,n}(l) = \sum_{m=1}^{M} c_{n,m}[l]\, e^{j\left(2\pi \sum_{k=0}^{l} T_s f_{n,m}[k] + \theta_{n,m}\right)} \tag{5}$$

where l is the discrete time index, M is the number of frequency modulated signal, $c_{n,m}$ denotes the sub-path gain, and $f_{n,m}$ and $\theta_{n,m}$ are the discrete Doppler frequency and initial phase, respectively.

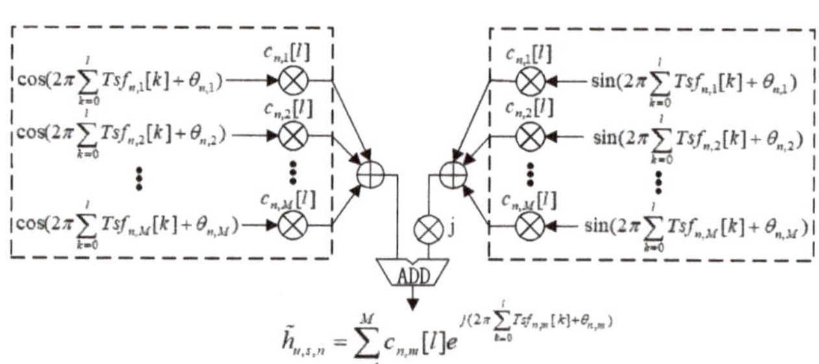

Figure 2. Sum-of-frequency-modulated-signals (SoFM)-based channel fading generator.

Note that the initial random phase of each branch is uniformly distributed over $[-\pi, \pi)$ and time-invariant. Considering the complexity of hardware implementation, it is assumed that the sub-path gain has the same value and does not change over time. Hold the condition of normalized path power, the sub-path gain of each branch equals to $\sqrt{1/N}$. As the time-variant discrete Doppler frequency would increase the complexity and uncertainty, it is very important to find an efficient way to update the Doppler frequency parameter over time. The theoretical Doppler frequency of the mth sub-path within the nth path can be defined by [27]

$$f_{n,m}(l) = k\frac{\vec{v}_{MS}\hat{r}_{MS,n,m}(l)}{2\pi} \tag{6}$$

where $k = 2\pi f_c/c$ denotes the wave number, f_c is the carrier frequency, c refers to the speed of light, $\vec{v}_{MS}$ denotes the vector of the mobile station (MS) velocity, and $\hat{r}_{MS,n,m}$ is the arrival angle unit vector of the mth sub-path within the nth path. As the Doppler frequency is usually much smaller than the system sampling rate, it is assumed that the statistical properties maintain unchanged within several sampling intervals, i.e., stationary interval T_u, which ranges from several millisecond to dozens of millisecond. The Doppler frequency of the mth sub-path within the uth interval, denoted as $f^u_{n,m}$, can be obtained by (6). In addition, we assume the discrete frequency parameters following the linear change within each interval T_u. Then, the Doppler frequency within the uth stationary interval can be expressed as

$$f^u_{n,m}(l) = a^u_{n,m} + b^u_m(l - (u-1)T_u) + \Delta^u_{n,m}(l) \tag{7}$$

where $a^u_{n,m}$ denotes the initial value of the mth sub-path within the nth path, b^u_m is the slope of the mth sub-path, $\Delta^u_{n,m}(l)$ is the small random offset of the frequency parameter, $a^u_{n,m}$ is random variable and distributes uniformly over $[F^1_{m-1}, F^1_m)$ when $u = 1$, and $a^u_{n,m}$ stays the value at the end of previous interval when $u = 2, 3, \cdots$. Finally, the slope b^u_m can be calculated by

$$b^u_m = L\frac{(F^u_m - F^u_{m-1}) + (F^{u+1}_m - F^{u+1}_{m-1})}{2T_u} \tag{8}$$

where L denotes the total number of the slope changes within each interval. In order to improve the performance, the following conditions for discrete Doppler frequency should be fulfilled [28],

$$\begin{aligned} &f_{n,m} \neq 0, \ \forall n, m \\ &f_{n,m} \neq f_{n,q}, \ \forall n \, and \, \forall m \neq q \end{aligned} \tag{9}$$

3.3. FPGA-Based Implementation

3.3.1. Delay Module

The delay module plays an important role in the channel emulation. It should be noted that the realization of multiple path delay is mainly based on the random access memory (RAM) or first input first output (FIFO). This method is easy to implement in FPGA, but cannot achieve the long-time delay, i.e., the aerial communication case, and high-precision delay, i.e., the indoor communication case. Especially, if the delay is relatively large, this method consumes a large amount of storage resources, which makes it impossible to realize in FPGA. To overcome this shortcoming, we adopt an external double-data-rate three synchronous dynamic random access memory (DDR3) and an interpolation filter to our scheme as shown in Figure 3. It includes three primary parts: DDR3, RAM, and high-precision interpolation filter. Take the advantage of large storage space of DDR3, it can achieve the large delay. Moreover, the data from RAM is multiplied by the coefficients of interpolation filter to achieve a high-precision delay. Thus, this scheme can adapt to a wide range of communication channels.

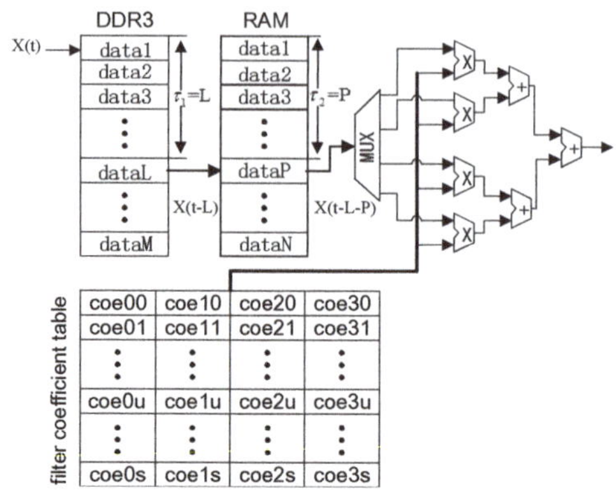

Figure 3. The implementation scheme of delay module.

In order to validate the proposed scheme of delay module, we run the module by modelsim software under the scenario of 3 GPP modified vehicular-A channel (MVA) [30]. In the simulation, the system sampling clock is 100 MHz, that is to say, the clock period is 10 ns. As the delay resolution in MVA is 5 ns, the interpolation filter is designed as a two-time interpolator. Figure 4 shows the corresponding output signal when a pulse signal passes through the delay module. Taking the first path as the reference path, the relative delay of each path is set as 375 ns, 750 ns, 1125 ns, 1750 ns, and 250 ns, respectively. It can be seen that the simulated results are consistent with the desired ones, which validates the effectiveness of this method.

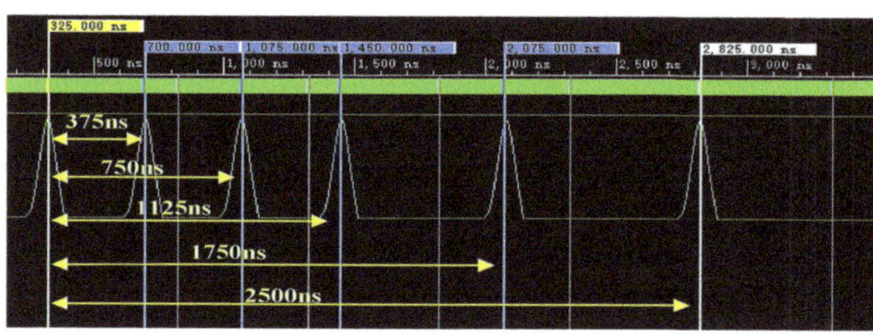

Figure 4. Hardware simulation of delay module.

3.3.2. Fading Generation Module

For an arbitrary $U \times S$ MIMO channel, the number of channel fading generation module should be $U \times S \times N$ and they could consume huge of hardware resources. As the maximum Doppler frequency is usually much smaller than the system sampling rate, in this paper we use a low initial sampling rate $f_s{}'$ to generate the channel fading, which can greatly reduce the hardware consumption. The implementation scheme of channel fading generation is showed in Figure 5. First, the parameter module updates the Doppler frequency and phase in real time. Then, it passes them to the subtractor (SUB), accumulator (ACC), multiplier, and adder (ADD) to complete the corresponding integral operations and generate a look-up table (LUT) address. The values of the cosine function stored in the cosine table can be found by the LUT address, and they are superimposed by the accumulator to obtain the channel fading coefficient. Finally, the cascaded integrator comb (CIC) filter is used to interpolate and match the data rate.

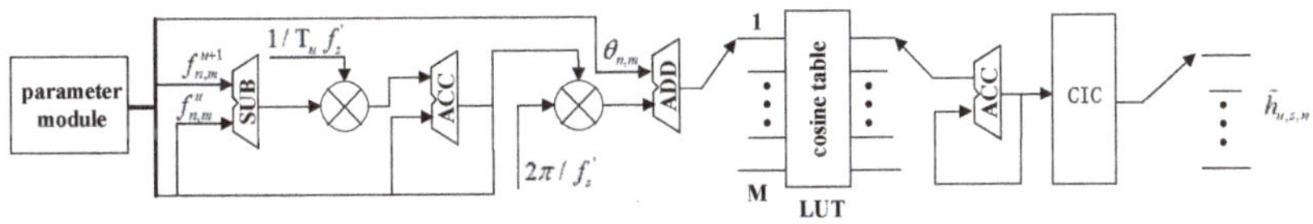

Figure 5. The implementation scheme of channel fading module.

According to the central limit theorem, the larger the number of sub-path, the closer the output channel fading is to the theoretical distribution. Considering a trade-off between the resource consumption and complexity, the number of sub-path is set as $N = 64$. The data width of LUT is set to 16 bits and the data depth is set to 12 bits. We use the idea of serial and time-division multiplexing to find the phase address in the LUT efficiently. Considering the symmetry of cosine function, only a quarter of cosine period needs to be stored, and thus the data width and depth are 15 bits and 10 bits, respectively. Note that this can significantly save the RAM resource when the sub-path number becomes large. Figure 6 shows the simulation result of hardware implementation. In this figure, only the first three sub-paths, i.e., three FM signals, and the superposition of 64 branches are given. As we can see that the output fading envelope is random fluctuation and it should approximate to the Rayleigh distribution according to the central limit theorem. For the latency of hardware, with the help of integrated logical analyzer (ILA) debugging tool, we find that it takes three clock cycles to reach the steady state and 16 clock cycles totally to output the first valid channel data. As the system clock is 100 MHz, the latency of proposed hardware emulator is about $16 \times 10^9/(100 \times 10^6) = 160$ ns.

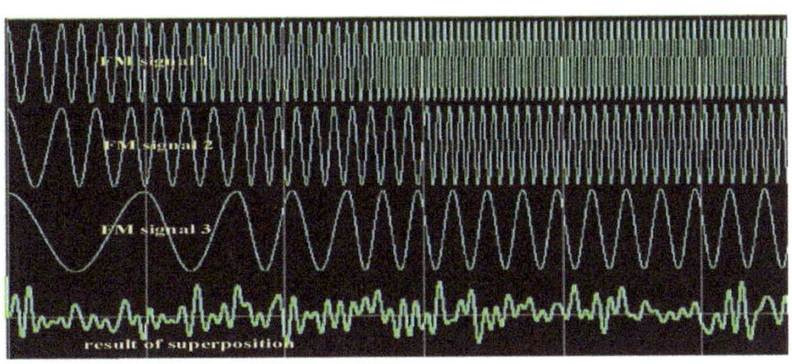

Figure 6. Hardware simulation of channel fading module.

3.3.3. Interpolator Module

The interpolator module performs a linear interpolation by I times to match the data rate between the channel fading and the input signal. The channel sampling rate $f_s{}'$ is much smaller than the system sampling rate f_s , so the channel fading rate should be interpolated to $f_s = I \times f_s{}'$. Let us denote two adjacent channel fading samples as $h[mI]$ and $h[(m+1)I]$, then the linear interpolation can be realized as

$$h[(mI+k)] = \frac{(h[(m+1)I] - h[mI])k}{I} + h[mI] \tag{10}$$

where $k = 0, 1, \cdots I-1$. The scheme of interpolator module in this paper includes one SUB, one multiplier, and one ADD shown in Figure 7. In this figure, two input ports of subtractor represent the adjacent channel fading samples, and the difference value is multiplied by the weight coefficient k/I . Finally, the output of multiplier and the first channel fading sample are summed up by an adder to obtain the interpolated channel fading sample.

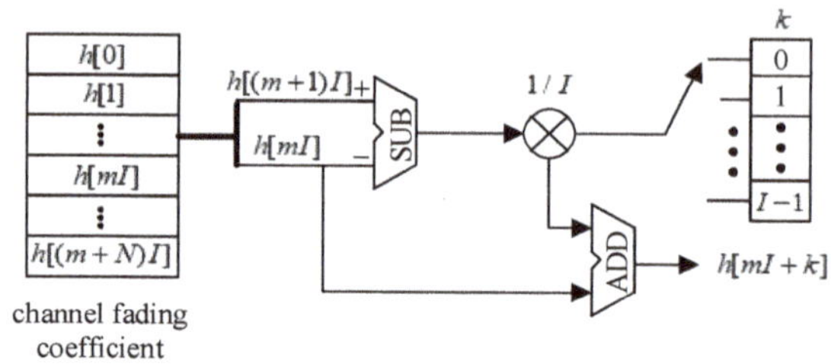

Figure 7. The implementation scheme of interpolation module.

4. Resource Consumption and Measurements

4.1. Resource Consumption

In this section, we take a 2×2 MIMO channel as an example to be implemented in one FPGA chip (Virtex-7). It should be noted that a single path generation needs 64 sub-paths or FM signals as shown in (5). Thus, for a single channel with M multiple paths, the traditional parallel method theoretically needs to prestore $128 \times M$ cosine tables. In this paper, we implement the channel fading module by adopting a serial scheme or time division idea as shown in Figure 6, which only needs $2 \times M$ cosine tables. Table 1 compares the hardware resources usages of a 2×2 MIMO channel emulator in [22] and a 2×2 MIMO channel emulator generated by the proposed method. It shows that the proposed method is more efficient than the one in [22]. The selected FPGA (Xilinx XC7VX690TFFG1927-2) consists of about 433,200 Slice LUTs, 1470 Block RAMs, and 3600 digital signal processors (DSPs). Considering the resource consumption of other modules and the efficiency of FPGA layout, it can be estimated that a 32×32 MIMO channel can be emulated on this single chip.

Table 1. Hardware resource usage of a 2×2 MIMO channel emulator.

	The Method in [22]	The Proposed Method
System sample rate	100 M	256 M
Slice LUTs	152,337	25,800
Block RAMs	191	116
DSPs	768	160

4.2. Measured Results and Analysis

In order to verify the output channel of proposed emulator, we consider that both of the base station (BS) and MS are equipped with normalized omnidirectional antennas, the carrier frequency is $f_c = 2.4$ GHz, and the scatterers are randomly distributed around the BS and MS. The number of paths and sub-paths are six and sixty-four, respectively, i.e., $N = 6, M = 64$. Moreover, all these six paths are assumed to be valid over the simulation period. The initial distance between the BS and MS is 318 m. The absolute speed, azimuth angle, and elevation angle of the moving MS are 40 km/h, $10°–8° \cdot t$, and $10°–0.1° \cdot t$, respectively. Other emulation parameters are as follows, $T_u = 25$ ms, $L = 10$.

Based on the parameter calculation method of GBSM in [27], we can obtain the theoretical time-variant PDP under the above scenario as shown in Figure 8a. As we can see, as the MS has an initial distance of 318 m from the BS, the initial time delay of line-of-sight (LOS) path equals to $318/(3 \times 10^8) = 1.06 \times 10^{-6}$ s. The time delay of non-line-of-sight (NLOS) paths can also be calculated and shown in Figure 8a with the dotted line. By using the ILA software, we store and export the data from the hardware emulator, and then analyze the data with Matlab. Finally, the measured time-variant PDP of emulator is given in Figure 8b, which clearly shows that it is consistent well with the theoretical one.

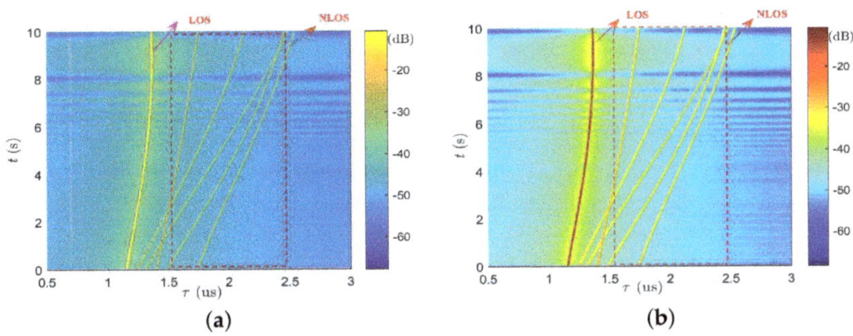

Figure 8. (**a**) The theoretical results of time-variant power delay profile (PDP) and (**b**) the measured time-variant PDP of proposed emulator.

Under the same condition, the time-variant DPSD is also tested and verified. With the help of (22) in [27], the theoretical time-variant DPSD is firstly calculated and shown in Figure 9a. For comparison purposes, we also give the simulated time-variant DPSD based on the model in [17] in Figure 9b. It is clearly showed that the part around circles is different from the theoretical one. The main reason is the output Doppler phase of that model is discontinuous which results in the output Doppler frequency or DPSD not accurate. In order to observe the output DPSD directly, a 2.4 GHz cosine signal generated by a Agilent E4438C is adopted as the input signal. Then, the measured DPSD of proposed emulator can be obtained by a spectrum analyzer of ROHDE&SCHWARZ FSV. The measured result is shown in Figure 9c. Due to the randomness and distortion caused by the fixed point process, the measured result can only be qualitatively compared with the theoretical one. Figure 9a,c show that the shape and trend of two DPSDs have a good approximation, which also validates the effectiveness of proposed emulator.

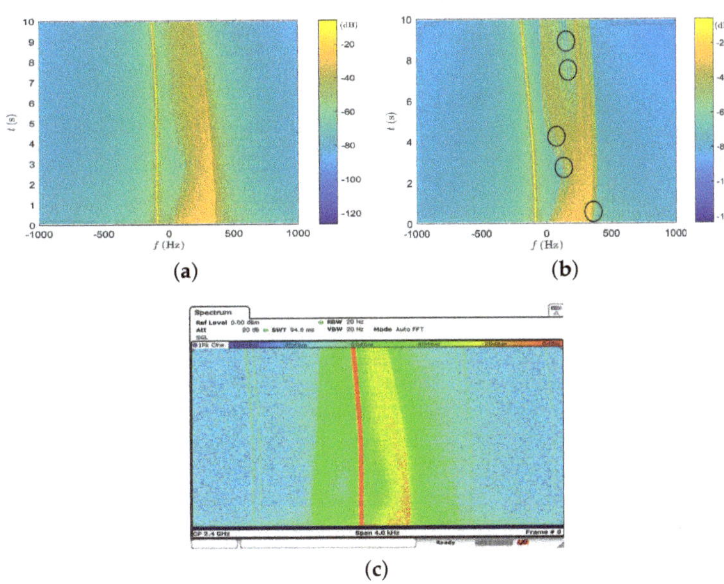

Figure 9. (**a**) The theoretical results of time-variant Doppler power spectrum density (DPSD), (**b**) the simulated time-variant DPSD of the model in [17], and (**c**) the measured time-variant DPSD of proposed emulator.

Without loss of generality, only the fading envelope PDF of first NLOS path for the first sub-channel is tested and validated. First, the theoretical time-variant PDF of channel fading is derived and shown in Figure 10a. It is apparently showed that the PDF changes over time due to the time-variant channel conditions. Similarly, with the help of Xilinx software development tool, we export the data of output fading envelope from the hardware, and then analyze the distribution by Matlab. Figure 10b gives the measured PDFs at three different time instants $t = 0$ s, 4 s, and 8 s. For comparison purpose, the corresponding theoretical results are also extracted from Figure 10a and showed in Figure 10b, which also fit well with the measured ones. In addition, we configure the channel parameters by referring to [31] as follow. The height of BS is 30 m, and the initial distance between the MS and BS is 90 m. The MS is moving towards the BS at a speed of 10 m per second. By using the similar method as above, we can obtain the measured PDF as shown in Figure 10c. It is shown that the measured PDF of proposed channel emulator is close to the result of field test in [31].

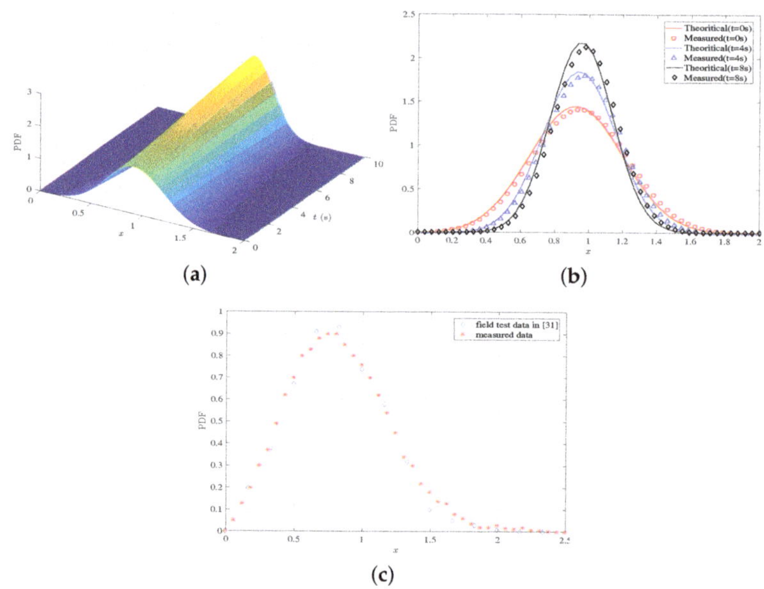

Figure 10. (**a**) The theoretical results of time-variant PDFs, (**b**) the measured time-variant PDFs of proposed emulator at different time instants, and (**c**) the measured PDF of proposed emulator and the PDF of field test.

Based on the theoretical expressions of (28)–(30) in [32] and (9)–(10) in [33], the absolute values of time-variant CCF of firt two paths are calculated and given in Figure 11. In the figure, we assume that the antenna spaces of the BS and MS are the same, and equal to twice the wavelength of carrier. Then, the measured CCF of proposed emulator can be obtained in a similar way as mentioned above and given in Figure 11 for comparison purpose. As can be seen from the figure, the CCF changes over time due to the movement of the MS. Again, the measured CCF aligns well with the theoretical one, proving the correctness of output correlation properties.

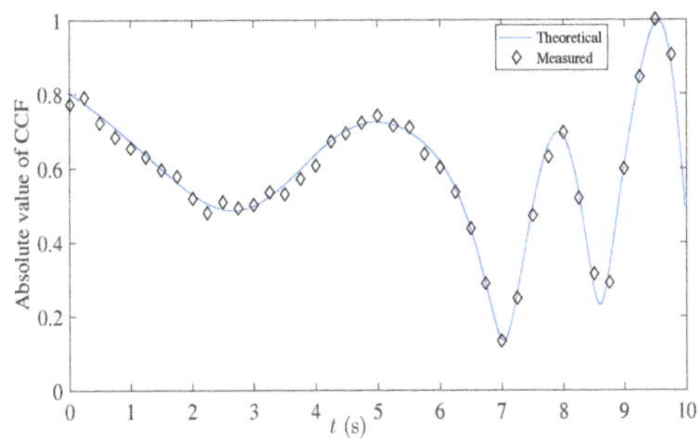

Figure 11. The absolute values of the theoretical and measured cross-correlation functions (CCFs).

5. Conclusions

This paper has proposed a discrete non-stationary MIMO channel model, which is suitable to realize on the FPGA-based platform. A tailored hardware architecture of channel emulator with flexible size and parameters has also been developed. In addition, the hardware implementation of key modules have been illustrated in details and applied in a single FPGA chip. Finally, the PDP and other statistical properties of proposed channel emulator have been tested. The measured results have shown that the output PDP, DPSD, PDF, and CCF are consistent well with the corresponding theoretical ones. Therefore, the proposed non-stationary channel emulator can be applied to evaluate and validate the performance of MIMO communication devices in the future.

Author Contributions: Conceptualization, Q.Z. and W.H.; methodology, Q.Z. and K.M.; software, Z.Z.; validation, W.H. and Z.Z.; writing—original draft preparation, Q.Z. and W.H.; writing—review and editing, all authors; supervision, W.Z. and B.H. All authors have read and agreed to the published version of the manuscript.

Funding: This work was supported in part by the Fundamental Research Funds for the Central Universities (No. NS2020026 and No. NS2020063), in part by the Aeronautical Science Foundation of China (No. 201901052001), and in part by the National Key Scientific Instrument and Equipment Development Project under Grant (No. 61827801).

Conflicts of Interest: The authors declare no conflict of interest.

References

1. Agiwal, M.; Roy, A.; Saxena, N. Next generation 5G wireless networks: A comprehensive survey. *Commun. Surv.* **2016**, *18*, 1617–1655. [CrossRef]
2. Wang, C.X.; Bian, J.; Sun, J. A survey of 5G channel measurements and models. *Commun. Surv.* **2018**, *20*, 3142–3168. [CrossRef]
3. Zhang, J.; Shafi, M.; Molisch, A.F.; Tufvesson, F.; Wu, S.; Kitao, K. Channel models and measurements for 5G. *IEEE Commun. Mag.* **2018**, *56*, 12–13. [CrossRef]
4. Kamga, G.; Xia, M.; Aissa, S. Channel modeling and capacity analysis of large MIMO in real propagation environments. In Proceedings of the IEEE International Conference on Communications, London, UK, 8–12 June 2015; pp. 1447–1452.
5. Li, J.; Jiang, D.; Zhang, X. DOA estimation based on combined unitary ESPRIT for coprime MIMO radar. *IEEE Commun. Lett.* **2017**, *21*, 96–99. [CrossRef]
6. Fan, W.; Carton, I.; Kyosti, P.; Karstensen, A.; Jamsa, T.; Gustafsson, M.; Pedersen, G. A step toward 5G in 2020: Low-cost OTA performance evaluation of massive MIMO base stations. *IEEE Antennas Propag. Mag.* **2017**, *59*, 38–47. [CrossRef]
7. N5106A PXB Channel Emulator. Available online: https://www.keysight.com (accessed on 13 June 2020).
8. Azimuth ACE 400WB Channel Emulator. Available online: https://cdn.thomasnet.com (accessed on 13 June 2020).

9. Ghiaasi, G.; Ashury, M.; Vlastaras, D.; Hofer, M.; Xu, Z.; Zemen, T. Real-time vehicular channel emulator for future conformance tests of wireless ITS modems. In Proceedings of the 2016 10th European Conference on Antennas and Propagation (EuCAP), Davos, Switzerland, 10–15 April 2016; pp. 1–5.
10. Fard, S.F.; Alimohammad, A.; Cockburn, B.F. Single-field programmable gate array simulator for geometric multiple-input multiple-output fading channel models. *IET Commun.* **2011**, *5*, 1246–1254. [CrossRef]
11. Huang, P.; Du, Y.; Li, Y. Stability analysis and hardware resource optimization in channel emulator design. *IEEE Trans. Circ. Syst.* **2016**, *63*, 1089–1100. [CrossRef]
12. Alimohammad, A.; Fard, S.F. A compact architecture for simulation of spatial-temporally correlated MIMO fading channels. *IEEE Trans. Circ. Syst.* **2014**, *61*, 1089–1100.
13. Dahech, W.; Patzold, M.; Gutierrez, C.A.; Youssef, N. A nonstationary mobile-to-mobile channel model allowing for velocity and trajectory variations of the mobile stations. *IEEE Trans. Wirel. Commun.* **2017**, *16*, 1987–2000. [CrossRef]
14. Liang, X.; Zhao, X.; Li, Y.; Wang, Q. A non-stationary geometry-based street scattering model for vehicle-to-vehicle wideband MIMO channels. *IEEE Wirel. Commun.* **2016**, *90*, 325–338. [CrossRef]
15. Zhu, Q.; Li, W.; Yang, Y.; Xu, D. A general 3D nonstationary vehicle-to-vehicle channel model allowing 3D arbitrary trajectory and 3D-shaped antenna array. *Int. J. Antennas Propag.* **2019**, *2019*, 8708762. [CrossRef]
16. Ghazal, A.; Wang, C.X.; Ai, B.; Yuan, D.F.; Hass, H. A nonstationary wideband MIMO channel model for high-mobility intelligent transportation systems. *IEEE Trans. Intell. Transp. Syst.* **2015**, *16*, 885–897. [CrossRef]
17. Ghazal, A.; Yuan, Y.; Wang, C.X. A non-stationary IMT-advanced MIMO channel model for high-mobility wireless communication systems. *IEEE Trans. Wirel. Commun.* **2016**, *16*, 2057–2068. [CrossRef]
18. Zhu, Q.; Wang, Y.; Jiang, K.; Chen, X.; Zhong, W.; Ahmed, N. 3D non-stationary geometry-based MIMO channel model for UAV-Ground communication systems. *IET Microw. Antennas Propag.* **2019**, *13*, 1104–1112. [CrossRef]
19. Ren, F.; Zheng, Y.R. A novel emulator for discrete-time MIMO triply selective fading channels. *IEEE Trans. Circ. Syst.* **2010**, *57*, 2542–2551. [CrossRef]
20. Habib, B.; Zaharia, G.; Zein, G.E. Hardware Simulator for MIMO Propagation Channels: Time Domain Versus Frequency Domain Architectures. *Sci. J. Circ. Syst. Signal Proc.* **2013**, *2*, 37–55. [CrossRef]
21. Habib, B.; Zaharia, G.; Zein, G.E. MIMO hardware simulator design for outdoor time-varying heterogeneous channels. In Proceedings of the International Symposium on Signals, Circuits and Systems, Lasi, Romania, 11–12 July 2013; pp. 1–4.
22. Zhang, N.; Yang, G.; Zhai, J. Design and implementation of flexible $4M \times 4N$ MIMO channel emulator. In Proceedings of the IEEE Antennas and Propagation Society International Symposium, Memphis, TN, USA, 6–11 July 2014; pp. 713–714.
23. Tien, T.V.; Tien, T.M.; Khai, L.D. Hardware Implementation of a MIMO Channel Emulator for high speed WLAN 802.11 ac. In Proceedings of the 2018 5th NAFOSTED Conference on Information and Computer Science, Ho Chi Minh City, Vietnam, 23–24 November 2018; pp. 183–188.
24. Habib, B.; Baz, B. Digital architecture of 8×8 MIMO Hardware channel simulator for time-varying heterogeneous systems with LTE-A, 802.11 ac and VLC signals. In Proceedings of the 2016 3rd International Conference on Advances in Computational Tools for Engineering Applications (ACTEA), Beirut, Lebanon, 13–15 July 2016; pp. 195–200.
25. Hofer, M.; Xu, Z.; Vlastaras, D.; Schrenk, B.; Tufvesson, F.; Zemen, T. Real-time geometry-based wireless channel emulation. *IEEE Trans. Veh. Technol.* **2019**, *68*, 1631–1645. [CrossRef]
26. Ruiz, A.E.; Gutierrez, C.A.; Castillo, J.V.; Cortez, J. SDR-Based channel emulator for vehicular communications. In Proceedings of the 2019 IEEE Colombian Conference on Communication and Computing (COLCOM), Barranquilla, Colombia, 5–7 June 2019; pp. 1–6.
27. Zhu, Q.; Li, H.; Fu, Y.; Wang, C.X.; Tan, Y.; Chen, X. A novel 3D non-stationary wireless MIMO channel simulator and hardware emulator. *IEEE Trans. Commun.* **2018**, *66*, 3865–3878. [CrossRef]
28. Zhu, Q.; Liu, X.; Yin, X.; Chen, X.; Xue, C. A novel simulator of nonstationary random MIMO channels in Rayleigh fading scenarios. *Int. J. Antennas Propag.* **2016**, *2016*, 3492591. [CrossRef]
29. Nguyen, T.T.; Lanante, L.; Nagao, Y.; Kurosaki, M.; Ochi, H. MU-MIMO channel emulator with automatic channel sounding feedback for IEEE 802.11 ac. In Proceedings of the Wireless Communications and Networking Conference, Doha, Qatar, 3–6 April 2016; pp. 1–6.

30. Hua, J.; Yang, J.; Lu, W.; Meng, L.; Yu, X. Design of universal wireless channel generator accounting for the 3-D scatter distribution and hardware output. *IEEE Trans. Instrum. Meas.* **2014**, *64*, 2–13. [CrossRef]
31. Qiu, Z.; Chu, X.; Calvo-Ramirezand, C.; Briso-Rodriguez, C.; Yin, X. Low Altitude UAV Air-to-Ground Channel Measurement and Modeling in Semiurban Environments. *Wirel. Commun. Mob. Comput.* **2017**, *2017*, 1587412. [CrossRef]
32. Jiang, K.; Chen, X.; Zhu, Q.; Chen, L.; Xu, D.; Chen, B. A Novel Simulation Model for Nonstationary Rice Fading Channels. *Wirel. Commun. Mob. Comput.* **2018**, *2018*, 8086073. [CrossRef]
33. Zhu, Q.; Yang, Y.; Wang, C.X.; Tan, Y.; Sun, J.; Chen, X.; Zhong, W. Spatial correlations of a 3-D nonstationary MIMO channel model with 3-D antenna arrays and 3-D arbitrary trajectories. *IEEE Wirel. Commun. Lett.* **2019**, *8*, 512–515. [CrossRef]

applied sciences

MDPI

Article

Secure D2D Communication for 5G IoT Network Based on Lightweight Cryptography

Byoungjin Seok [1], Jose Costa Sapalo Sicato [1], Tcydenova Erzhena [1], Canshou Xuan [1], Yi Pan [2] and Jong Hyuk Park [1,*]

[1] Department of Computer Science and Engineering, Seoul National University of Science and Technology, Gongneung-ro, Nowon-gu, Seoul 01811, Korea; sbj7534@seoultech.ac.kr (B.S.); josecostasicato@seoultech.ac.kr (J.C.S.S.); etcydenova@seoultech.ac.kr (T.E.); shou19@seoultech.ac.kr (C.X.)

[2] Department of Computer Science, Georgia State University, Atlanta, GA 30302-5060, USA; yipan@gsu.edu

* Correspondence: jhpark1@seoultech.ac.kr; Tel.: +82-02-970-6702

Received: 12 November; Accepted: 23 December 2019; Published: 27 December 2019

Abstract: Device-to-device (D2D) communication is a direct means of communication between devices without an intermediate node, and it helps to expand cell coverage and to increase radio frequency reuse in a 5G network. Moreover, D2D communication is a core technology of 5G vehicle-to-everything (V2X) communication, which is an essential technology for autonomous driving. However, typical D2D communication in an 4G network which is typical telecommunication network has various security challenges including impersonation, eavesdropping, privacy sniffing, free-riding attack, etc. Moreover, when IoT technology emerges with 5G networks in massive machine type communication (mMTC) and ultra-reliable low latency communication (URLLC) application scenarios, these security challenges are more crucial and harder to mitigate because of the resource-constrained nature of IoT devices. To solve the security challenges in a 5G IoT environment, we need a lightweight and secure D2D communication system that can provide secure authentication, data confidentiality/integrity and anonymity. In this paper, we survey and analyze existing results about secure D2D communication systems in terms of their security considerations and limitations. Then, we lastly propose a secure D2D communication system to address the aforementioned security challenges and the limitations of the existing results. The proposed secure D2D communication was designed based on elliptic curve cryptography (ECC) and lightweight authenticated encryption with associated data (AEAD) ciphers to cover resource-constrained IoT devices.

Keywords: D2D communication; 5G IoT network; lightweight cryptography; authentication

1. Introduction

D2D communication is a peer-to-peer communication mechanism between devices without an intermediate node [1,2]. D2D communication has many advantages in mobile networks [3]. First, it can expand coverage of each cell in a cellular network as a communication bridge for transmitting data to the node located outside of cell coverage. Second, D2D communication helps to reduce the energy consumption of the base station by transmitting data directly between devices. Lastly, the efficiency of reusing the same radio frequency is increased. In D2D communication, the distance between devices is quite shorter than the distance between a device and a base station. This means the interference of radio frequency decrease in D2D communication scenario, and it helps to transmit the multiple data using the same radio frequency. Moreover, D2D communication is a core technology of V2X communication [4]. Due to these advantages, the 5G network also includes D2D communication technology such as the LTE-advanced (4G) network.

However, typical D2D communication in a mobile network has some security challenges [5]. The D2D communication mechanism consists of three procedures, device discovery, link setup and data transmission [6]. In this process, there is no authentication process for validating device identity. When a device sends a request for a setup link to transmit data, another node replies by sending an acknowledgement message. Moreover, D2D communication does not use encryption for confidentiality and message authentication for integrity in the communication process. This means the attacker can conduct attacks such as impersonation, eavesdropping, privacy sniffing, free-riding and location spoofing. Besides, IoT technology is combined with the 5G network to address their service demands [7], and it corresponds to mMTC and URLLC, which are the use-cases of the 5G network [5]. However, IoT applications deal with many sensitive data, and IoT devices have limited resources [8] in terms of performance, memory and power consumption. These features of IoT make the aforementioned security challenges more critical and harder to address because typical security solutions cannot be implemented or processed properly. To overcome the security challenges of D2D communication in the 5G IoT network, we need a secure D2D communication system that contains a proper authentication process between devices. Moreover, considering the resource-constrained environment, it has to be made light.

Lightweight cryptography can be a proper solution for covering resource-constrained devices. Elliptic curve cryptography (ECC), which is most representative of lightweight asymmetric-key algorithms, can provide 128-bit cryptographic security using a 256-bit key, which is significantly smaller than the 3072-bit key of the most widely used public-key encryption algorithm RSA [9]. ECC has been applied to various cryptographic algorithms including elliptic curve Diffie–Hellman(ECDH) and the elliptic curve digital signature algorithm (ECDSA). ECDH and ECDSA are both cryptographic public-key algorithms but they have different purposes: ECDH is used for key exchange and ECDSA is a variation of the digital signature algorithm. ECDH is a variation of the Diffie–Hellman algorithm for elliptic curves, which is a cryptographic key agreement protocol that allows two parties with public/private key pairs on elliptic curves to obtain a shared secret key using an unprotected communication channel. ECDSA is a public key algorithm for creating a digital signature, similar in structure to a DSA, but defined, in contrast to it, not above the ring of integers, but in a group of points of an elliptic curve. In the case of the symmetric-key algorithms, many lightweight AEAD ciphers have been proposed recently to deal with a resource-constrained environment; moreover, the standardization project at the National Institute of Standards and Technology (NIST) is in process [10]. AEAD ciphers can provide not only data confidentiality but also data integrity and authentication using a message authentication code (MAC) with associated data during the encryption process. These lightweight cryptographic algorithms help to make D2D communication secure and able to process communication efficiently.

In this paper, we propose a secure D2D communication system for a 5G IoT network based on lightweight cryptography ECC and the AEAD cipher. First, we analyze typical security threats and present security considerations for D2D communication in a 5G IoT network. Moreover, we survey the existing research on secure D2D communication schemes and make a taxonomy of these results based on our security considerations. Finally, we propose a lightweight cryptography-based secure D2D communication system that can provide anonymity, user authentication, data confidentiality/integrity and efficiency. Because of its lightweight construction, the proposed D2D communication system can be applied efficiently on the 5G IoT network. Our main contributions can be summarized below:

- We analyze existing typical security threats and secure D2D communication. Then we present security considerations for secure D2D communication for a 5G IoT network.
- We survey and analyze existing research based on our security considerations, including authentication, data confidentiality/integrity, anonymity and efficiency.
- We propose a lightweight and secure D2D communication system. The proposed D2D communication system is designed based on lightweight cryptography. It can be implemented simply and can efficiently process resource-constrained 5G IoT devices.

The remainder of this paper is organized as follows. Section 2 introduces related works where we surveyed D2D communication and analyzed security considerations for secure D2D communication.

In Section 3, we propose a secure D2D communication system for a 5G IoT network. In Section 4, we show the simulation results of our proposed D2D communication system. In Section 5, we analyze our proposed D2D communication system based on our security considerations and finally conclude in Section 6.

2. Related Work

2.1. Typical Security Threats of D2D Communication

D2D communication involves three steps, device discovery, link setup and data transmission, to make a direct connection between devices. In the device discovery step, the device searches for nearby devices. Then devices that are discovered in the previous step make a connection for transmitting data in the link setup step. After a connection is established, the data is transmitted through a direct link in the data transmission step. However, if there are no proper security measures, the data can be vulnerable to some security threats by attackers. Typical security threats of D2D communication introduced in [5] are as follows:

- Impersonation attack. In this attack the attacker acts like a legitimate user by using an identity such as an international mobile subscriber identity (IMSI). To prevent this attack proper authentication of users has to be considered.
- Eavesdropping. This is a type of attack where the attacker passively listens to communication between users and thereby the attacker can capture the transmitted data and also can fabricate the data. To prevent this attack, data confidentiality and integrity have to be considered.
- Privacy sniffing. D2D communication has to broadcast request messages to search for nearby devices. However, the attacker uses this feature to find and track the victim device. To mitigate this security threat, the devices have to use an anonymous identity, and it has to be authenticated.
- Free riding attack. Selfish devices receive the desired data from other devices but do not share their resources because of energy consumption and because of this they reduce system availability. To mitigate this attack, the user identity has to be authenticated and managed by a base station.
- Location spoofing. In this attack a malicious device may broadcast a request message with wrong or artificial location information to disrupt D2D communication in the device discovery step. To mitigate this attack, the request message has to be processed only from validated devices in D2D communication.

2.2. Security Considerations for a 5G IoT Network

In a 5G network, IoT applications correspond to mMTC and URLLC scenarios. For the security of D2D communication against threats, the D2D communication system has to provide security functions including authentication, data confidentiality/integrity and anonymity. However, IoT devices have limited resources in terms of performance, memory and power consumption. Therefore, the security functions must also provide efficiency, meaning that each security function has to be implemented lightly and run faster. The detailed description of security considerations are as follows:

- Authentication. Authentication is a key requirement for securing D2D communication in the 5G IoT network. For most types of attacks, proper user authentication is the most basic and appropriate solution. Every network should be able to verify the identity of users in order to guarantee the security of the network.
- Data confidentiality and Integrity. The data transmitted in the IoT network contains sensitive information, and due to a variety of attacks that can eavesdrop on or modify that information, confidentiality and integrity are a big concern. For providing this, we have to encrypt the transmitted data and use hash functions or message authentication algorithms.
- Anonymity. Anonymity refers to hiding the identity of origin and sensitive information such as location. Anonymity is a necessary security function to prevent attackers from targeting specific users for their purpose. In such cases, when anonymity is not provided, the attacker can choose a

specific target for the attack. If you take the example of autonomous vehicles, the attacker may decide to attack a specific car. Therefore, anonymity should be considered extensively.

- Efficiency. Efficiency is the communication system's ability to be implemented and to operate economically. This consideration is about availability, which means that authorized users can access the information at any time they request it. This consideration is especially critical when it comes to the IoT network because IoT devices have limited resources.

2.3. Existing Research

Mingsheng Cao et al. [11] proposed a secure lightweight D2D communication system with multiple sensors. Their proposed communication system is designed based on lightweight key generation and a distribution scheme by leveraging an acceleration sensor and secure near field authentication by using a device's microphone and speaker as sensors and for data transmission, which includes encryption/decryption by audio and RF channels. Adeel Abro et al. [12] proposed a lightweight authentication scheme based on elliptic ElGamal encryption, which is public key algorithm based on elliptic curve discrete logarithm problem (ECDLP). This paper presents an authentication scheme based on public key infrastructure (PKI) and uses a combination of ECC to select key pair and ElGamal encryption to exchange the secret key. Yasir Javed et al. [13] also proposed a lightweight security scheme based on ECC and ElGamal encryption over public key infrastructure. This paper uses ECC to create keys and ElGamal for encryption and decryption. Atefeh Mohseni-Ejiyeh et al. [14] proposed an incentive-aware lightweight secure data sharing scheme for D2D communication in 5G networks. In their proposed scheme, users obtain digital signatures to prove successful data sharing and, in the sharing process, the symmetric encryption algorithm and MAC are used. Haowen Tan et al. [15] proposed a D2D authenticating mechanism employing smartphone sensor behaviour analysis. Their authentication scheme is designed based on certificateless cryptography for group authentication and user's behavior analysis extracted from smartphone sensors is employed for continuous authentication. Sheeba Backia, Mary Baskaran et al. [16] proposed a lightweight key exchange mechanism for LTE-A assisted D2D communication that can be applied in 5G networks. Their mechanism is designed by using ECC-based symmetric keys. Yunqing Sun et al. [17] proposed privacy protection device discovery and an authentication mechanism for D2D using the identity-based prefix encryption and ECDH key agreement protocol. All of these studies can provide authentication and data confidentiality/integrity and most of them use ECC based cryptographic algorithms. However, they have some limitations in that some of the results cannot provide anonymity or the researches did not deeply consider the data transmission step of D2D communication. Moreover, most of the existing schemes use only lightweight public key algorithms not lightweight symmetric encryption algorithms. Table 1 shows a taxonomy of strategies of existing research in terms of the security functions provided (confidentiality/integrity, authentication, anonymity) and the steps considered (device discovery, link setup, data transmission). Since our proposed system uses ECC and lightweight AEAD cipher for covering our security considerations and all of the steps in D2D communication, it can improve the efficiency and security of D2D communication.

Table 1. Taxonomy of strategies of existing secure device-to-device (D2D) communication.

Ref.	Security Function			Considered D2D Step		
	Conf. / Int.	Auth.	Anon.	D.D.	L.S.	D.T.
Mingsheng Cao et al. [11]	✓	✓		✓	✓	✓
Adeel Abro et al. [12]	✓	✓		✓	✓	✓
Yasir Javed et al. [13]	✓	✓		✓	✓	✓
Atefeh Mohseni-Ejiyeh et al. [14]	✓	✓		✓	✓	✓
Haowen Tan et al. [15]	✓	✓	✓	✓	✓	
Sheeba Backia Mary Baskaran et al. [16]	✓	✓	✓	✓	✓	
Yunqing Sun et al. [17]	✓	✓	✓	✓	✓	

3. Secure D2D Communication

3.1. Proposed D2D System Model

In this section, we propose a secure D2D communication mechanism for a 5G IoT network based on lightweight AEAD ciphers. The proposed secure D2D communication model is shown in Figure 1. Objects participating in D2D communication consist of 5G network components including user equipment (UE), general node-B (gNB), access and mobility management function (AMF)/security anchor function (SEAF) and user data management (UDM). UE is a device that is a mobile entity in a 5G network, and UE is an actual device that communicates with other devices directly in our system. gNB is a base station responsible for connecting UE to mobile networks. In our system, gNBs share their public key with other gNBs in advance and use their private key to generate D2D tokens ($D2DTK_{gNB_x}$) via ECDSA. Moreover, AMF is responsible for the management of a mobile entity. SEAF is a middle entity of authentication between UE and a 5G network and is co-located with AMF. UDM stores information about mobile entities in a 5G network. A 5G network provides the authentication framework using 5G-AKA to verify the identity of the UE. 5G-AKA is used to authenticate the UE's validity before generating a D2D token for use in communication in the proposed secure D2D communication. It is corresponded to step 0 in the proposed D2D system, and this process is performed only once for each UE.

After generating a D2D token, the D2D communication process has three steps similar to a typical D2D communication system: Device discovery, link setup and secure data transmission. However, in each process, there are features for security such as anonymity, authentication and confidentiality/integrity. We will discuss the details of these features in Section 3.2. The brief descriptions of each process are as follows:

- Device discovery is a process that searches for nearby nodes. In this step, nodes in a network broadcast a request message to discover other nodes. If a node receives a request message, it sends a response message to another node. The broadcast or response message in this process includes each UE's encrypted identity SUCI and the issued D2D token.
- Link setup is a process for making a peer-to-peer connection between two nodes. During this process, each node sends a verification request to its base station, gNB, with the SUCI and D2D token of the target UE being received in the device discovery phase. After verification, ECDH is used to exchange secret keys for secure data transfer.
- Secure data transmission is a process where data is transmitted. The main feature of this step is that the data is encrypted using a lightweight AEAD cipher before transmission. In the encryption process, the sender node uses its D2D token identity and context sequence, thereby the confidentiality and integrity of the data are ensured. Moreover, authentication is processed in every transmission.

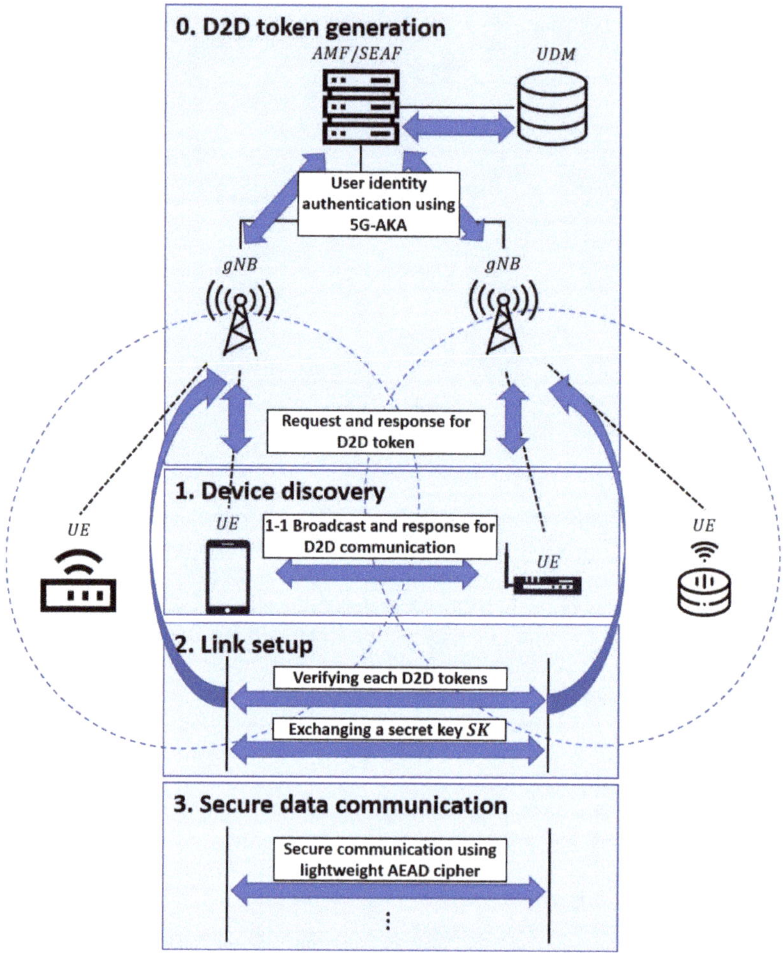

Figure 1. Secure D2D communication system model for a 5G Internet of Things (IoT) network.

3.2. Details of Communication Mechanism

This section deals with the detailed process for the proposed D2D communication system. As described in the system model, the proposed D2D communication system has four steps in total. These four steps may be classified into one pre-processing step performed before D2D communication and the remaining three steps in which actual D2D communication is performed. The pre-processing step is the D2D token generation step (corresponding to step 0), and the steps in which D2D communication is performed are device discovery, link setup and secure data transmission (corresponding to steps 1–3, respectively).

First, in the D2D token generation step, each UE sends a request to the gNB to generate a D2D token for later use in D2D communication. The gNB that receives the D2D token generation request first verifies the identity of the UE that sent the request. At this time, the identity of the UE verifies the SUCI, which is an encrypted identity that emerges for user privacy in a 5G network. Unlike IMSI, the identity of the UE used in existing 4G networks, the SUCI can provide anonymity for the UE as a result of encrypting the IMSI using a public key (PUK). The verification for SUCI is performed using 5G-AKA, an authentication framework for performing primary authentication of UE registration in

5G networks. The subject that performs the actual verification is AMF/SEAF, and the verification is performed by comparing the credentials obtained by decrypting SUCI with the user credentials stored in the UDM. When the SUCI verification is completed, the result is transmitted to the gNB, and accordingly the gNB generates a D2D token and transmits it to the requesting UE. The generation of the D2D token uses the digital signature value calculated by the ECDSA of the UE's SUCI using the gNB's private key (PRK). The issued D2D token may also give anonymity to the UE with a value generated through a cryptographic algorithm by using the identity of the UE like SUCI. The issued D2D token can be verified if the SUCI of the UE and the public key of the gNB are known (note that each gNB shares the public key we mentioned in the previous section). The D2D token generation procedure is shown in Figure 2.

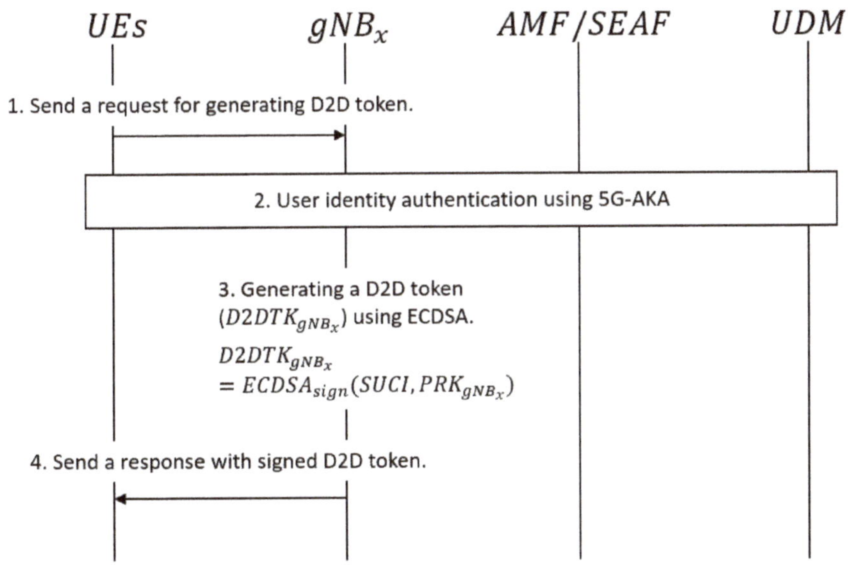

Figure 2. D2D token generation procedure.

From now on, actual D2D communication performing steps will be described. The whole process of proposed D2D communication is shown in Figure 3. Step 1 is device discovery, which is a process of searching for a nearby device with which to perform D2D communication. Here, each UE desiring D2D communication broadcasts a message requesting to perform D2D communication, and UEs in a state capable of D2D communication transmit a response message to the received D2D request message. Here, the broadcast message or response message includes the D2D token issued in step 0 and its SUCI. If a response message to the broadcasted request message is received, the process proceeds to the next step.

The second stage of D2D communication is the link setup to establish a communication session. In this step, prior to establishing a communication session, verification is performed on the D2D token exchanged through device discovery. The verification of the D2D token performed here is similar to the UE identity verification performed in the D2D token generation, but the authentication is performed in the gNB without connecting to the core network. The D2D token can be verified using the public key and SUCI of each gNB, which authenticates that the D2D token has been issued from the gNB by request by a pre-authenticated UE. When the verification of the D2D token is completed, the secret key exchange used in the encryption process of the data transmission step is performed according to the result. The exchanged secret key is a secret key derived from the secret keys of both UEs using ECDH.

Therefore, even if the attacker taps the data transmitted in the middle of the key exchange, the secret key cannot be derived.

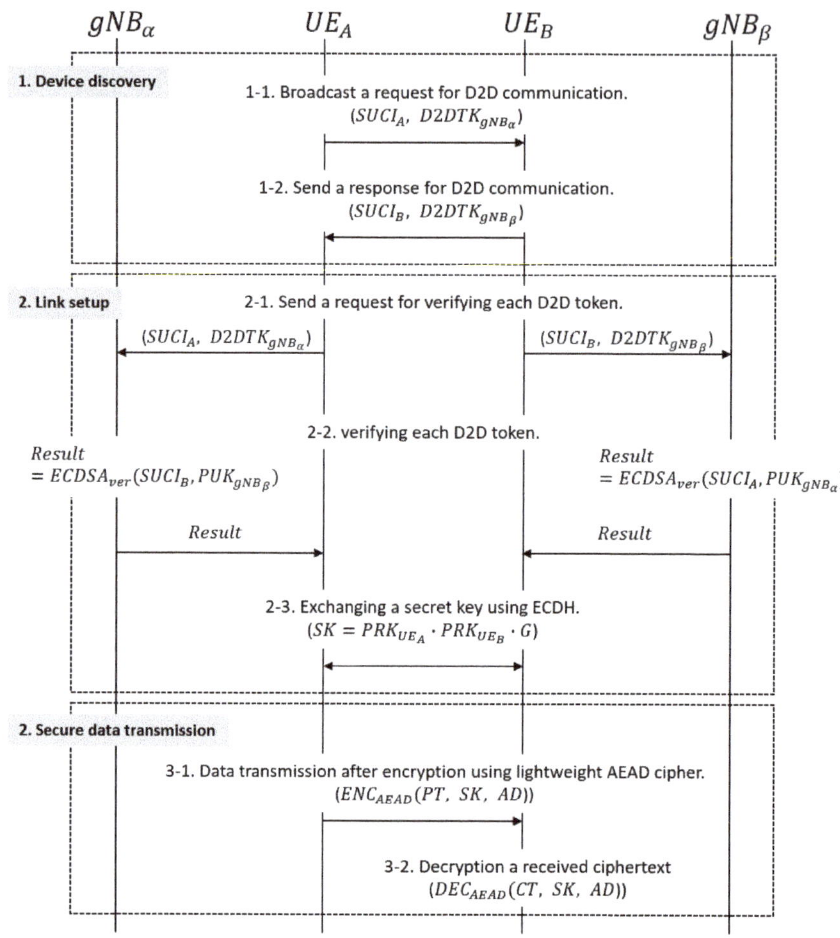

Figure 3. The whole process of secure D2D communication.

Secure data transmission, the last step of the proposed D2D communication, performs data encryption communication. At this time, encryption uses a lightweight AEAD cipher. The lightweight AEAD cipher is a cryptographic algorithm that provides not only confidentiality but also integrity and authenticity. It encrypts the data to be transmitted and creates a MAC for authenticating data integrity. Moreover, in the encryption process, the AEAD cipher uses additional information about a communication session and the other party, called the associated data (AD), thereby the AEAD cipher provides authenticity, which means the message is transmitted from the right party at the right time. In the proposed D2D communication system, the AD consists of the D2D token and context sequence information and manages the sequence for each transmission. Upon receiving the cipher text using the AD configured as described above, the UE may check whether the other UE performing D2D communication has received data corresponding to the current situation along with authentication. The data format used in secure data transmission is shown in Figure 4. In this step, any lightweight AEAD cipher can be applied according to available resources. Table 2 shows available lightweight

AEAD ciphers which are candidate cipher from NIST lightweight cryptography standardization (Round 2) [10].

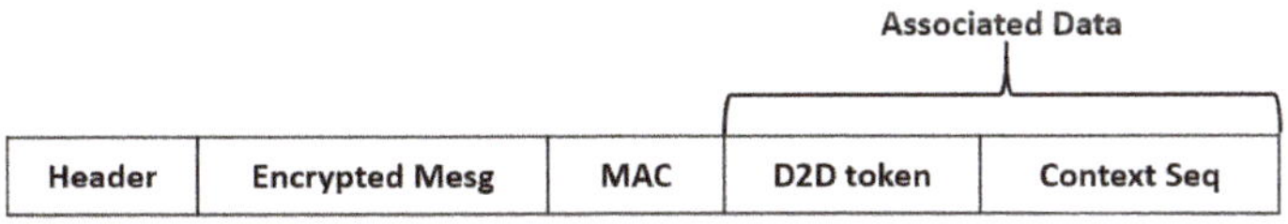

Figure 4. Data format for data communication.

Table 2. A list of lightweight authenticated encryption with associated data (AEAD) ciphers (National Institute of Standards and Technology (NIST) lightweight cryptography standardization (Round 2)).

Cadidates	Type	Functionality
ACE	Permutation based	AEAD and Hashing
ASCON	Permutation based	AEAD and Hashing
COMET	Block cipher based	AEAD only
DryGASCON	Permutation based	AEAD and Hashing
Elephant	Permutation based	AEAD only
ESTATE	Tweakable block cipher based	AEAD only
ForkAE	Tweakable block cipher based	AEAD only
GIFT-COFB	Block cipher based	AEAD only
Gimli	Permutation based	AEAD and Hashing
Grain-128AEAD	Stream cipher based	AEAD only
HYENA	Block cipher based	AEAD only
ISAP	Permutation based	AEAD only
KNOT	Permutation based	AEAD and Hashing
LOTUS-AEAD/LOCUS-AEAD	Tweakable block cipher based	AEAD only
mixFeed	Block cipher based	AEAD only
ORANGE	Permutation based	AEAD and Hashing
Oribatida	Permutation based	AEAD only
PHOTON-Beetle	Permutation based	AEAD and Hashing
Pyjamask	Block cipher based	AEAD only
Romulus	Tweakable block cipher based	AEAD only
SAEAES	Block cipher based	AEAD only
Saturnin	Block cipher based	AEAD and Hashing
SKINNY-AEAD/SKINNY-HASH	Tweakable block cipher based	AEAD and Hashing
SPARKLE	Permutation based	AEAD and Hashing
SPIX	Permutation based	AEAD only
SpoC	Permutation based	AEAD only
Spook	Tweakable block cipher based	AEAD only
Subterranean 2.0	Permutation based	AEAD and Hashing
SUNDAE-GIFT	Block cipher based	AEAD only
TinyJambu	Block cipher based	AEAD only
WAGE	Permutation based	AEAD only
Xoodyak	Permutation based	AEAD and Hashing

4. Simulation Results

In this section, we conduct a simulation to evaluate the proposed D2D communication system in terms of performance and efficiency. The performance in this section shows the whole processing time of the proposed D2D communication process. Moreover, for evaluating the efficiency of the proposed D2D communication, we perform analysis of implementation cost of lightweight AEAD ciphers, and simulate energy consumption according to AEAD ciphers.

The proposed D2D communication includes cryptographic algorithms for providing our security considerations (authentication, data confidentiality/integrity, anonymity). In detail, the applied cryptographic algorithms are the digital signature, the Diffie–Hellman key exchange algorithm and the AEAD cipher. We suppose that the processing time of each cryptographic algorithm is as follows.

The processing time for signing of a digital signature $t_{DS_{sign}}$, the processing time for verification of a digital signature $t_{DS_{ver}}$, the processing time for key exchange t_{DH} and the processing time for the AEAD cipher t_{AEAD}. Then we suppose the transmission latency in D2D communication is l_{tr}. Finally, we can estimate the total length of the D2D communication processing time t_{D2D} through Equation (1).

$$t_{D2D} = \sum l_{tr} + \sum t_{DS_{sign}} + \sum t_{DS_{ver}} + \sum t_{DH} + \sum t_{AEAD} \tag{1}$$

For calculating the summation of each processing time, we analyze the proposed D2D communication in terms of the number of transmissions and the usage count of the cryptographic algorithm at each step. In D2D token generation (step 0), there are two transmissions, request and response, for a D2D token; this step also includes 5G-AKA for user identity authentication. The 5G-AKA have 10 transmissions between UEs, gNB, AMF and UDM. In terms of the usage of the cryptographic algorithm, the D2D token generation step uses ECDSA-signing to process token generation. Moreover, we assume that the 5G-AKA consists of ECDSA-signing and ECDSA verification because the 5G-AKA is based on the ECC certificate. In device discovery (step 1), the requested UE broadcasts the request message; this means that the number of transmissions for a request message equal the number of devices (m), which are located near the sender UE. Moreover, in this step, there is a transmission to response. In link setup (step 2), when two devices set the connection, there are transmissions, including two for token verification, two for response of verification and two for key exchange, and there are the usages of the cryptographic algorithm, including two for ECDSA verification and one for ECDH. Lastly, the secure data transmission (step 3) has transmissions according to the amount of data (n bytes), and we assumed that data are transmitted in packets and in units of 1460 bytes, which is a general maximum transmission unit (MTU) size. Moreover, the AEAD cipher is used twice (encryption/decryption) in this step. Then we can finally calculate the summation of processing time by multiplying each processing time by the number of transmissions or the usage count of the cryptographic algorithm. Table 3 shows the summary of processing time at each step of proposed D2D communication.

Table 3. The summary of processing time of proposed D2D communication.

Step	Transmission Latency	Processing Time of Cryptographic Algorithm			
		ECDSA-Sign	ECDSA-Verify	ECDH	AEAD
Step 0	$(2+10) * l_{tr}$	$(1+1) * t_{DS_{sign}}$	$1 * t_{DS_{ver}}$	-	-
Step 1	$(m+1) * l_{tr}$	-	-	-	-
Step 2	$(2+2+2) * l_{tr}$	-	$2 * t_{DS_{ver}}$	$1 * t_{DH}$	-
Step 3	$(n/1460) * l_{tr}$	-	-	-	$2 * t_{AEAD}$
Total ($\sum l$ or $\sum t$)	$(19+m+n/1460) * l_{tr}$	$2 * t_{DS_{sign}}$	$3 * t_{DS_{ver}}$	$1 * t_{DH}$	$2 * t_{AEAD}$

When we simulate Equation (1) using processing time in Table 3, we set each time parameter based on 5G network requirements and existing implementation results of the cryptographic algorithm. The 5G network requires a transmission latency of 1 ms [18]; accordingly, we set l_{tr} as 0.001. Moreover, we set the processing time of the ECC-based algorithm based on the performance presented in [19] ($t_{DS_{sign}} = 0.122, t_{DS_{ver}} = 0.458, t_{DS_{DH}} = 0.1672$). In the case of t_{AEAD}, we can calculate processing time by multiplying the throughput (Mbps) of the algorithm by the amount of data (n' (Mb) $= n$ (MB) $* 8/10^6$). For simulating various AEAD ciphers, we set the parameter following five AEAD ciphers (AES-GCM, ASCON, SpoC, Spook and GIFT-COFB) based on the performance results presented in [20]. Each case of t_{AEAD} is as follows (power measured: 50 MHz): $t_{AES-GCM} = n'$ (Mb)/31.2 (Mbps), $t_{ASCON} = n'$ (Mb)/39.0 (Mbps), $t_{SpoC} = n'$ (Mb)/28.8 (Mbps), $t_{Spook} = n'$ (Mb)/88.3 (Mbps), $t_{GIFT-COFB} = n'$ (Mb)/120.8 (Mbps). Figure 5 shows the simulation result of the proposed D2D communication. The AEAD ciphers used in the simulation consist of one general-purpose

AEAD cipher (AES-AEAD) and four lightweight AEAD ciphers. Simulation results show that three lightweight AEAD ciphers (ASCON, Spook and GIFT-COFB) are faster than AES-GCM (optimized). In particular, GIFT-COFB shows about 18.71% faster performance than AES-GCM when transmitting 10 KB data.

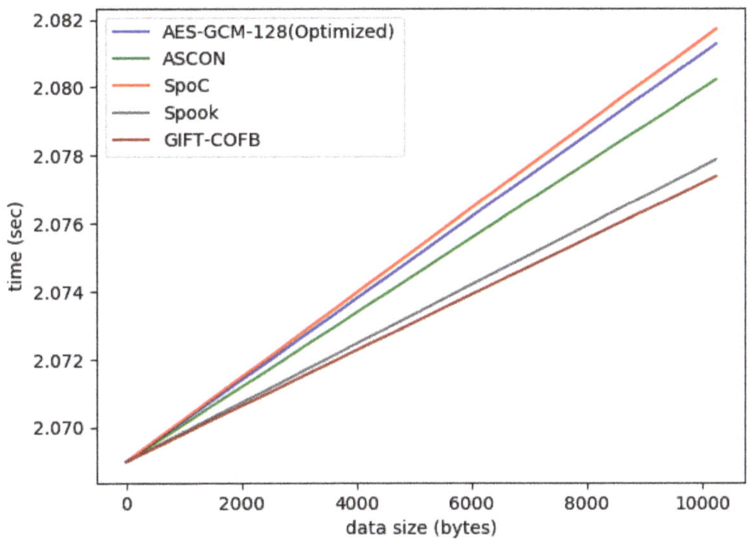

Figure 5. The processing time of the proposed D2D communication system.

However, because 5G IoT networks have limited resources, good performance of cryptographic algorithms may not cover all of the 5G IoT devices. This means the cryptographic algorithm has to be implemented lightly and must consume a small amount of power. Table 4 shows the hardware implementation results of AEAD block ciphers [19]. Even though Spook is faster than AES-GCM (optimized), Spook has the highest implementation cost, as in the mentioned area for implementing a look-up tables (LUTs).

Table 4. The hardware implementation result of AEAD ciphers.

AEAD Cipher	Area (LUTs)	Power (mW)	Throughput (Mbps)	Energy (nJ/bit)
AES-GCM (Optimized)	1532	35.9	31.2	1.15
ASCON	1808	33.6	39.0	0.86
SpoC	1344	34.7	28.8	1.20
Spook	7082	125.9	88.3	1.43
GIFT-COFB	2695	36.6	120.8	0.30

Figure 6 shows energy consumption by amount of data based on energy efficiency in Table 4. In terms of energy consumption, GIFT-COFB and ASCON consume less energy than AES-GCM (optimized), but SpoC and Spook consume more energy. Considering that both GIFT-COFB and ASCON show better performance than AES-GCM (optimized) in the performance simulation, when GIFT-COFB or ASCON is applied to the proposed D2D communication, both speed and energy efficiency of the proposed D2D communication are better than for AES-GCM (optimized)-based D2D communication.

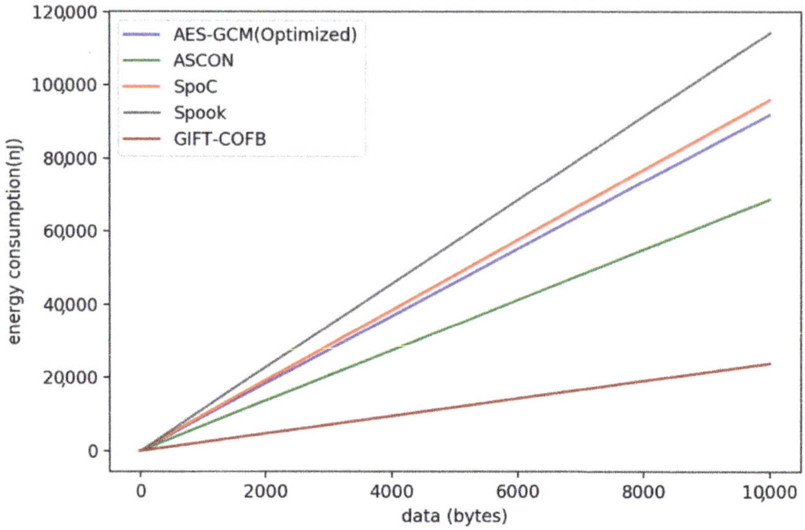

Figure 6. The energy consumption of AEAD ciphers.

5. Security Analysis

In this section, we perform security analysis of the proposed secure D2D communication system. As we mentioned before, secure D2D communication requires authentication, data confidentiality/integrity and anonymity. In addition, considering the resource-constrained nature of a 5G IoT network, it must be implemented lightly and must perform efficiently. We first discuss the proposed D2D communication system based on our security considerations. Moreover, we discuss security against typical threats of D2D communication.

5.1. Analysis Based on Security Considerations

- Authentication: The proposed D2D communication system performs primary authentication using 5G-AKA, which is an authentication framework provided by 5G, to perform authentication for a UE before issuing a D2D token. Moreover, the issued token can perform secondary authentication through verification of the process of creating a link of D2D communication through the gNB's public key and SUCI. Finally, in the data transmission step, the token is used as an AD to authenticate the other party for each transmission of data. In this way, authentication of the UE is performed in all processes of data communication before issuing a token for D2D communication so that more secure communication can be performed.
- Data confidentiality and integrity: The proposed D2D communication system generates D2D communication using SUCI, which is the encrypted UE identity, and the secret key of the gNB during the D2D generation process. In this process, there is no case where the identity of the unencrypted UE is transmitted. In addition, in the step of actual data transmition after creating a D2D link, encryption is performed using a lightweight AEAD cipher. AEAD cryptography can provide integrity and authentication as well as data confidentiality. Therefore, the proposed D2D communication system can guarantee the confidentiality/integrity of the identity and communication data of the UE.
- Anonymity: In 5G networks, SUCI is an encrypted identity for UE anonymity, which provides anonymity for the UE itself. Moreover, the D2D token used in the proposed D2D communication is a value obtained by signing SUCI with the private key of the gNB, which also provides anonymity by not being able to recognize the identity of the UE directly.

- Efficiency: Both the authentication process and the data encryption process used in the proposed D2D communication system are based on lightweight cryptography. The lightweight ciphers used in this paper are the ECC-based public key cryptosystem and the lightweight AEAD cipher. The ECC-based public key cryptosystem uses a 256-bit key and operates faster than RSA, which uses a 1024-bit key. Moreover, the lightweight AEAD cipher is designed to be efficiently implemented in a resource-constrained environment such as in IoT and provides data confidentiality/integrity and authentication.

5.2. Security against Typical Threats

- Impersonation attack. In D2D the token generation step, each UE is issued a D2D token, which is signed by the gNB. When gNBs generate the token, they authenticate the validity of the UE by comparing the identity of the UE in UDM. After this authentication process, gNBs complete the generation of D2D tokens by using their private key. Because of this procedure, the attacker cannot impersonate other UE.
- Eavesdropping. In a secure data transmission step, every instance of data transmission is protected by the lightweight AEAD cipher. In the AEAD encryption process, UE uses its D2D token and context sequence as associated data. Using this associated data, MAC is generated, and thereby UE can check the integrity of the message and the validity of the sender UE. For these reasons, the attacker cannot eavesdrop and cannot fabricate a message.
- Privacy sniffing. The proposed D2D communication system uses the D2D token, which is generated based on SUCI of UE and digital signature of gNB using ECDSA. The D2D token can provide anonymity as a cryptographic identity. For this reason, the attacker cannot recognize the original identity of the UE.
- Free riding attack and location spoofing. When the D2D token is generated by gNB, the validity of the UE is authenticated. This means that each instance of validating a UE is managed by gNB. The D2D token is authenticated in the link setup step in the proposed D2D communication system, and the data transmission is protected by AEAD encryption using a D2D token. Therefore, if a free-riding attack or location spoofing occurs in D2D communication, gNB can handle these situations by eliminating malicious UE.

6. Conclusions

In this paper, we propose a secure D2D communication system in a 5G IoT environment. The proposed D2D communication is designed based on an ECC-based public key cryptosystem and a lightweight AEAD cipher for efficiency in 5G use cases corresponding to IoT scenarios, mMTC and URLLC. Before the D2D communication is performed, the UE identity is verified based on the 5G-AKA provided by the 5G network, and then a token is used as the ECDSA for the D2D communication. The generated token could authenticate the legitimacy of the corresponding UE identity in the link setup process after device discovery. This can be done without connecting to the core network. In addition, by performing the encrypted communication through the lightweight AEAD cipher using the token as the associated data in the secure data communication step, the confidentiality/integrity of the data and authentication of the UE can be performed in each data transmission step. This approach can provide higher performance and energy efficiency than a general-purpose AEAD cipher-based communication system, and can also provide security against security threats such as impersonation, eavesdropping, privacy sniffing, free-riding and location spoofing.

Author Contributions: Conceptualization, B.S., J.H.P.; methodology, B.S. and J.C.S.S.; investigation, J.C.S.S., T.E. and C.X.; writing—original draft preparation, B.S.; project administration, J.H.P.; funding acquisition, J.H.P. and Y.P. All authors have read and agreed to the published version of the manuscript.

Acknowledgments: This study was supported by the Advanced Research Project funded by the SeoulTech (Seoul National University of Science and Technology).

Conflicts of Interest: The authors declare no conflict of interest.

References

1. Tehrani, M.N.; Uysal, M.; Yanikomeroglu, H. Device-to-device communication in 5G cellular networks: Challenges, solutions, and future directions. *IEEE Commun. Mag.* **2014**, *52*, 86–92. [CrossRef]
2. Jeong, M.; Ahn, S. A network coding-aware routing mechanism for time-sensitive data delivery in multi-hop wireless networks. *J. Inf. Process. Syst.* **2017**, *13*, 1544–1553.
3. Doppler, K.; Rinne, M.P.; Janis, P.; Ribeiro, C.; Hugl, K. Device-to-device communications; functional prospects for LTE-advanced networks. In Proceedings of the 2009 IEEE International Conference on Communications Workshops, Dresden, Germany, 14–18 June 2009; pp. 1–6.
4. Chen, S.; Hu, J.; Shi, Y.; Peng, Y.; Fang, J.; Zhao, R.; Zhao, L. Vehicle-to-everything (V2X) services supported by LTE-based systems and 5G. *IEEE Commun. Stand. Mag.* **2017**, *1*, 70–76. [CrossRef]
5. Zhang, S.; Wang, Y.; Zhou, W. Towards secure 5G networks: A Survey. *Comput. Netw.* **2019**, *162*, 106871. [CrossRef]
6. Lin, Z.; Du, L.; Gao, Z.; Huang, L.; Du, X. Efficient device-to-device discovery and access procedure for 5G cellular network. *Wirel. Commun. Mob. Comput.* **2016**, *16*, 1282–1289. [CrossRef]
7. Li, S.; Da Xu, L.; Zhao, S 5G Internet of Things: A survey. *J. Ind. Inf. Integr.* **2018**, *10*, 1–9. [CrossRef]
8. Daoud, W.B.; Obaidat, M.S.; Meddeb-Makhlouf, A.; Zarai, F.; Hsiao, K.F. TACRM: Trust access control and resource management mechanism in fog computing. *Hum.-Centric Comput. Inf. Sci.* **2019**, *9*, 28. [CrossRef]
9. Stallings, W. *Cryptography and Network Security: Principles and Practice*; Pearson: Upper Saddle River, NJ, USA, 2017.
10. NIST Computer Security Resource Center. Lightweight Cryptography Project. Available online: https://csrc.nist.gov/projects/lightweight-cryptography (accessed on 8 December 2019).
11. Cao, M.; Wang, L.; Xu, H.; Chen, D.; Lou, C.; Zhang, N.; Zhu, Y.; Qin, Z. Sec-D2D: A Secure and Lightweight D2D Communication System With Multiple Sensors. *IEEE Access* **2019**, *7*, 33759–33770. [CrossRef]
12. Abro, A.; Deng, Z.; Memon, K.A. A Lightweight Elliptic-Elgamal-Based Authentication Scheme for Secure Device-to-Device Communication. *Future Internet* **2019**, *11*, 108. [CrossRef]
13. Javed, Y.; Khan, A.S.; Qahar, A.; Abdullah, J. EEoP: A lightweight security scheme over PKI in D2D cellular networks. *J. Telecommun. Electron. Comput. Eng.* **2017**, *9*, 99–105.
14. Mohseni-Ejiyeh, A.; Ashouri-Talouki, M.; Mahdavi, M. An Incentive-Aware Lightweight Secure Data Sharing Scheme for D2D Communication in 5G Cellular Networks. *ISeCure* **2018**, *10*, 15–27.
15. Tan, H.; Song, Y.; Xuan, S.; Pan, S.; Chung, I. Secure D2D group authentication employing smartphone sensor behavior analysis. *Symmetry* **2019**, *11*, 969. [CrossRef]
16. Baskaran, S.B.M.; Raja, G. A Lightweight Incognito Key Exchange Mechanism for LTE-A Assisted D2D Communication. In Proceedings of the 2017 Ninth International Conference on Advanced Computing (ICoAC), Chennai, India, 14–16 December 2017; pp. 301–307.
17. Sun, Y.; Cao, J.; Ma, M.; Li, H.; Niu, B.; Li, F. Privacy-Preserving Device Discovery and Authentication Scheme for D2D Communication in 3GPP 5G HetNet. In Proceedings of the 2019 International Conference on Computing, Networking and Communications (ICNC), Honolulu, HI, USA, 18–21 February 2019; pp. 425–431.
18. Carugi, M. Key features and requirements of 5G/IMT-2020 networks. In Proceedings of the ITU Arab Forum on Emerging Technologies, Algiers, Algeria, 14–15 Feburary 2018.
19. Tschofenig, H.; Pegourie-Gonnard, M.; Unit, I.B. Performance of State-of-the-Art Cryptography on ARM-based Microprocessors. In Proceedings of the NIST Lightweight Cryptography Workshop 2015 Session VII: Implementations & Performance, Gaithersburg, MD, USA, 20–21 July 2015.
20. Rezvani, B.; Diehl, W. Hardware Implementations of NIST Lightweight Cryptographic Candidates: A First Look. In Proceedings of the NIST Lightweight Cryptography Workshop 2019, Gaithersburg, MD, USA, 4–6 November 2019.

applied sciences

MDPI

Article

Interference Avoidance Using TDMA-Beamforming in Location Aware Small Cell Systems

Khalid S. Mohamed *, Mohamad Y. Alias and Mardeni Roslee

Centre for Wireless Technology (CWT), Faculty of Engineering, Multimedia University, Cyberjaya 63100, Selangor, Malaysia; yusoff@mmu.edu.my (M.Y.A.); mardeni.roslee@mmu.edu.my (M.R.)
* Correspondence: khalidkaradh@hotmail.com

Received: 15 October 2019; Accepted: 12 November 2019; Published: 20 November 2019

Abstract: It is ascertained that the current communication systems will not be able to support the future network demands due to the increasing traffic, limitation of frequency resources, and high level of interference. Recently, beamforming techniques have been introduced to reduce the interference by redirecting the transmission towards the desired users only. While such beamforming enables better interference mitigation and improved network performance, the concerns on its effectiveness in dense deployment environments are arising. In this paper, the prospect of interference avoidance in location aware small cell environments using time division multiple access (TDMA) and beamforming is studied. The interference is reduced by identifying the aggressor small cell and transmit the beams towards the desired users at different times. Simulation results show that the proposed scheme is able to enhance the signal to interference plus noise ratio (SINR) by approximately 18 dB, enhance the user throughput by about 10 Mbps in comparison to small cell on/off control scheme with a discovery signal (SCon/off-DS), and improve the fairness index to about 95% in comparison to the baseline scheme. It is believed that the presented results promote the proposed scheme as an efficient interference management paradigm for the fifth generation (5G) communication systems.

Keywords: beamforming; interference management; throughput; fairness index, small cell networks

1. Introduction

The future communication systems that operate in higher frequency bands promise to exploit wider bandwidths to cater higher data rates for the end users [1]. For instance, the highest bandwidth range that is obtainable in millimetre wave (mmWave) systems is approximately 10 Gigahertz (GHz) over a frequency range of 30–300 GHz which achieves about 100 bits/sec/Hz spectral efficiency, and obtains terabyte data rate at the physical layer [2,3]. This is triggered by the limitation of the current wireless communication systems in terms of spectrum limitation as well as fairness issues. In wireless systems, different channel conditions and/or improper resource allocation and scheduling are experienced by signals. Therefore, fairness is one of the issues worth to look at in such systems. In [4], the downlink throughput of OFDMA small cells is maximised subject to short-term and long-term fairness indices. This is achieved using a fixed transmission power. The proposed scheme utilises cumulative fairness constraint to replace the long-term fairness and further improve the system's fairness. Similarly, the authors of [5] used non-orthogonal multiple access (NOMA) to maximise the fairness of drone small cells. More works on fairness can be found in [6–8].

Similarly, extensive researches focused on investigating bandwidths of up to 10 Terahertz (THz) because it offers extremely broader spectrum leading to huge capacity enhancements. Nevertheless, THz signals are very fragile in longer distances due to the high frequency. In other words, THz communication performs best in indoor capacities due to the proximity between transmitters and receivers [9,10]. On the other hand, multiple input multiple output (MIMO) technique is exploited to achieve enhanced beamsteering gains in which multiple antennas are tightly bundled together over a small spatial area. Subsequently, the technical and hardware complexities are few of many constraints that researchers had to realise in approaching the technique. The authors of [11] investigated the performance and the complications of signals phase-uncertainty when deploying low-complexity indoor wireless units that utilise hybrid beamforming techniques. The study concluded that the number and the size of antennas are proportional to the performance degradation.

The authors of [12] proposed a hybrid beamforming that is able to map a beamsteering codebook based on the channel state information (CSI). It consists of a digital beamformer that eliminates interference by dealing with regulated channel inversion (RCI) to reduce the overall complexity. It is meaningful to mention that the proposed algorithm is designed to eliminate inter-band-interference that is caused by orthogonal frequency division multiplexing (OFDM) carrier offsets (CFO) in [13].

The authors of [14] presented an experimental analysis of the in-band full-duplex communication (IBFD) end-fire arrangement array transceiver and then developed a MIMO testbed that communicates at 12.9 GHz carrier frequency. However, the scheme suppressed the self generated interference by using the OFDM techniques and did not show evaluations in terms of data rates and bit error rates. In view of that, the consideration of transmission and reception, proper channel modelling, and multipath and fading effects lead to the smart antennas techniques. In [15], smart antennas were categorised into adaptive arrays and switched beam antennas. While the latter describes the antennas that have multiple beams and able to switch to any beam at any given time, the former estimates the arrival direction and in-lines the beams with that direction, thereby resulting less interference.

Subsequently, the switched beam antennas do design the multi-user constellation depending on how interference is perceived. Interference is mostly eliminated or suppressed through minimum mean square error (MMSE) [16], zero-forcing (ZF) [17,18], or time division multiple access (TDMA) techniques [19]. TDMA techniques usually utilise time division multiplexing (TDM) training sequences to assist at the demodulating terminal. In that regard, the authors of [20] focused on achieving better synchronisation to realise the adaptive beamforming algorithms in real time co-channel communication systems.

The small cell technology on the other hand have expanded the capacity of the current communication systems by allowing more users to utilise more network resources with enhanced signals quality. Given the fact that most of future data exchange will originate from indoor environments along with the fragility of higher frequency signals; small cells are the most attractive solution that will solve for all the highlighted issues. With regards to that, service areas are often divided into equal set of squares whereby each square is served by single small cell to suppress interference power [21]. This is achieved by forcing other small cells within the service area to switch off to save power. However, interference caused by edge users transmitting with the maximum power may lead to grave performance deterioration. In [22] adaptive power control mechanism was proposed to reduce the interference by adaptively adjusting the transmission power to enhance the signal to interference plus noise ratio (SINR) and minimise the outage probability. The adaptation was done by either increasing or reducing the transmission power by a certain value and concurrently measure the SINR to achieve an optimal value. Additional works on interference management using power suppression techniques can be found in [23–25].

The authors of [26] presented a smart virtual antenna array that is deployable in 5G-IOT systems to avoid the interference by precisely directing the beams towards specific users. While transmitter signals are assumed to be in the form of generalised frequency division multiplexing (GFDM), the user signals are assumed to be OFDM signals. However, the authors of this paper understand that current wireless systems should be investigated extensively especially in views of small cells, beamforming, and TDMA technologies due to the existence of limited realistic channel models for future 5G communication systems and applications.

To the best of the authors knowledge, all works on beamforming techniques in long term evolutions (LTE) communication systems utilise location aware principles in approaching the objectives of the work. Without loss of generality, the authors have found no works that consider exploiting beamforming and TDMA techniques together to suppress the interference in dense small cell systems. Therefore, this paper proposes an interference avoidance algorithm that exploits beamforming and TDMA techniques in location aware small cell communication systems. This is achieved by redirecting the beams towards the users at different times in which the locations of the users are assumed to be known. The main contributions of this paper are summarised below:

1. Realistic interference estimation and avoidance based on line-of-sight conditions. Therefore, favourable settings for propagation conditions are noted.
2. Effective beamforming and TDMA algorithm for interference avoidance in dense indoor environments.
3. The proposed scheme improves the network data rate in broader bandwidth transmissions.
4. Realistic network model that is suitable for future communication systems.
5. Motivation for further research directions, such as: interference avoidance in ultra-high frequencies and beamsteering techniques.

The remainder of this paper is divided as follows; Section 2 describes the system modelling and problem formulation, the proposed model is described in Section 3, Section 4 presents the evaluations of the numerical results, and Section 5 concludes the paper.

2. System Modelling and Problem Formulation

Crucial challenges and limitations in terms of the practicality of any simulation-based wireless communication system, need to be addressed especially channel and network representations. Here, the realisation of the channel and the annotations used in this simulation are introduced. As highlighted earlier, indoor facilities are expected to embrace the greatest portion of data exchange since higher frequencies are assumed to be effective within indoor capacities. Therefore, all simulation parameters including channel modelling parameters are tailored with respect to line-of-sight indoor environments.

In this paper, the set of small cells is denoted by $\mathcal{S}$ whereby each small cell is assigned $\mathcal{K}$ subchannels. The set of users associated to each small cell is denoted by $\mathcal{U}$. Moreover, $\mathcal{P}^-$ and $\mathcal{P}^+$ describe the maximum transmission power and the received power of each small cell, respectively. Without loss of generality, the necessary assumptions used in this paper are described below:

- Small scale fading (Rayleigh fading) is represented by the instantaneous received signal strength, which varies frequently in response to the separation between the transmitter and receiver. This is theoretically modelled using zeros-mean exponential distribution with unity deviation. However, for more practical analysis, Jake's model [27] can be taken into account because it considers multipath, subchannel frequency, and the user speed.
- The service area (single indoor facility) is virtually controlled by small cell management system (ScMS) that is later connected to the LTE core network through the packets gateway (P-GW) [28].
- The network bandwidth $\mathcal{B}$ = 20 MHz.
- The locations of small cells and users are entirely random (random distribution) in the simulation. However, once the locations of the small cells are generated, they remain unchanged until the iteration cycle completes. It later changes when the number of users is changed. Moreover, all locations and angles are assumed to be known to the ScMS.

Other simulation parameters are described in Table 1.

Table 1. Simulation settings.

Parameter	Value
Service area	100 m × 100 m
$\mathcal{P}^{-}$ [29]	23 to 33 dBm
Frequency	3.5 GHz
SINR threshold [30]	2.2 dB
Antenna elements	6
Elements type	Patch
Inter-element spacing	$\frac{\lambda}{2}$

Signal propagation in the given scenario is dominated by line-of-sight path whereby any other reflected, scattered, or refracted signals are entirely neglected. Therefore, the channel gain g_{sku} between small cell and a user that are separated by a distance d_{su} is obtained by the following [31]:

$$g_{sku} = -[\mathcal{P}^{\mathcal{L}}_{sku} + \mathcal{L}^{\mathcal{H}}_{sku} + \mathcal{L}^{F}_{sku}] \tag{1}$$

whereby $\mathcal{P}^{\mathcal{L}}_{sku}$ is the pathloss that is calculated by $127 + 30log_{10}d_{su}$ [31], $\mathcal{P}^{\mathcal{H}}_{sku}$ is the shadowing loss, and $\mathcal{P}^{\mathcal{F}}_{sku}$ is the fading loss. It is also important to state that noise is Gaussian [31] whereby it equals to:

$$\mathcal{N} = \mathcal{N}_0 \times \mathcal{N}_d \times \mathcal{B}^{-} \tag{2}$$

$\mathcal{N}_0$ is the noise factor $\approx$2.5, $\mathcal{N}_d$ is the noise spectral density ≈ -174 dBm/Hz, and $\mathcal{B}^{-}$ is the subchannel bandwidth $\approx$180 kHz.

Additionally, the simulation of the array antenna with respect to the parameters of Table 1 gives a beamwidth of 40° in which the main lobe is directed towards the desired user at a specific angle. The antenna polar beam pattern is presented in Figure 1.

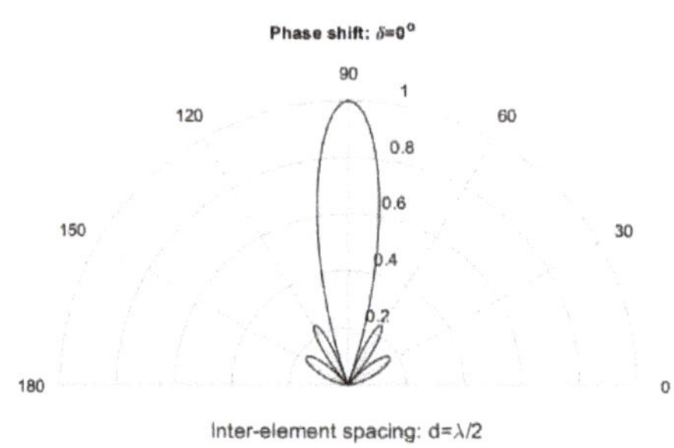

Figure 1. Polar beam pattern of an array antenna that has 6 elements, inter-element spacing = $\frac{\lambda}{2}$, and $\delta = 0°$.

2.1. Problem Formulation

Channel and interference in dense environments are unpredictable especially in random deployment scenarios where cells are distributed uniformly, which increases the complexity of channel estimation and interference cancellation processes. Omni-directional antennas *(wider beamwidths)* in conventional wireless communication systems have higher interference levels due to the reception from multiple angles. However, antennas in beamforming environments have narrow beamwidths and reception is limited to specific angles from the user point of view. Therefore, if the user has certain knowledge of the arrival angles of transmitted signals, it would curb down huge levels of interference.

Nevertheless, it increases the complexity of the system and addresses more challenges such as the synchronisation issue.

The ratio of the desired signals in comparison to other channel factors is familiarly described by the SINR metric that is obtained by the following [31]:

$$SINR_{su} = \sum_{\mathcal{k}\in\mathcal{K}} \frac{\mathcal{P}^{+}_{s\mathcal{k}u}\mathcal{g}_{s\mathcal{k}u}}{\sum\limits_{i\in\mathcal{S}/\{s\}} \mathcal{P}^{+}_{i\mathcal{k}u}\mathcal{g}_{i\mathcal{k}u} + \mathcal{N}} \times \alpha_{su} \tag{3}$$

whereby α is the user association matrix of $[s \times u]$ and that $\alpha_{su} = 1$ if the user u is associated to the small cell s, otherwise $\alpha_{su} = 0$.

Looking at Equation (3), it can be concluded that if the interference power *(denominator)* is reduced, SINR is enhanced. Therefore, the problem is formulated as follows:

$$\begin{aligned} &\arg\min \sum_{i\in\mathcal{S}/\{s\}} \sum_{\mathcal{k}\in\mathcal{K}} \mathcal{P}^{+}_{i\mathcal{k}u} \\ &s.t. \\ &\quad \sum_{i\in\mathcal{S}/\{s\}} \alpha_{ij} \neq 0 \\ &\quad ,\forall j \in \mathcal{U}/\{u\} \end{aligned} \tag{4}$$

2.2. Evaluation Metrics

In this section, the parameters used to assess the performance of the proposed algorithm are introduced.

1. Throughput [31]: this describes the competence of the network to deliver data. It also describes the effects of interference on the signals quality. Higher throughput means less interference, high SINR, and efficient resource allocation scheme. Throughput is obtained by the following:

$$\mathcal{T}_s = \sum_{u\in\mathcal{U}} \mathcal{T}_u = \sum_{u\in\mathcal{U}} [\mathcal{B}^{-} \times log_2(1 + SINR_{su})] \tag{5}$$

   It is noteworthy to mention that Equation (5) is valid for the maximum obtainable throughput. However, the amount of successful packets delivery in real-time scenarios represents the network throughput.

2. Jain's Fairness model [32]: described by Equation (6); it is used to analyse how fairly the users can access the required throughput. Fairness index varies between 0 and 1 whereby the higher the index, the fairer the system.

$$\mathcal{R}_s = \frac{\left[\sum\limits_{u\in\mathcal{U}} \mathcal{T}_{su}\right]^2}{\left[\mathcal{U} \times \sum\limits_{u\in\mathcal{U}} \mathcal{T}^2_{su}\right]} \tag{6}$$

3. SINR: this compares the ratio of the received signal to interference plus noise.

3. Proposed Algorithm

As highlighted in Section 2, the synchronisation of both users and base stations increases the complexity of the beamforming system. To avoid this, we introduce the parameter β at the ScMS, which is a matrix of $[s \times u]$ that describes which user is receiving from which base station, β_{su} is described below:

$$\beta_{su} = \begin{cases} 1 & \theta_{su} - \frac{w_{su}}{2} \leq \theta_{su} \geq \theta_{su} + \frac{w_{su}}{2} \\ 0 & else \end{cases} \tag{7}$$

whereby w_{su} is the beamwidth, and θ_{su} is the angle at which the user is located. Algorithm 1 describes how β_{su} and $\beta_{iu} \forall i \in \mathcal{S}/\{s\}$ are derived from α_{su}. When user u is associated to small cell s, α_{su} is said to equal 1. Subsequently, if the angle of user u is located within angles of the beam of small cell s, β_{su} is set to 1. Similarly, β_{iu} is set to 1 if user u is located within angles of the beam of small cell $i, \forall i \in \mathcal{S}/\{s\}$

Algorithm 1 Initialise β_{su}

```
function ScMS
    for s = 1 : S do
        for u = 1 : U do
            Set β_su = 0
            if α_su = 1 then
                if θ_su − w_su/2 ≤ θ_su ≥ θ_su + w_su/2 then
                    Set β_su = 1
                end if
                for i = 1 : S do
                    if i ≠ s then
                        if d_si > 2 × radius_s then
                            Set β_iu = 0
                            if θ_iu − w_iu/2 ≤ θ_su ≥ θ_iu + w_iu/2 then
                                Set β_iu = 1
                            end if
                        end if
                    end if
                end for
            end if
        end for
    end for
end function
```

Additionally, we introduce the parameter τ, which is also a matrix of $[s \times u]$. It describes the transmission time slot for user u, and obtained by the following:

$$\tau_{su} = (u-1) \times [\mathcal{U} \times \mathbb{TTI}] + 1 \tag{8}$$

whereby $\mathbb{TTI}$ is the transmission time interval for a user in LTE.

In view of that, the ScMS forces s to transmit at the angle θ_{su} where the user is located. The maximum signal response will result at θ_{su} if $\gamma = -\frac{2\pi}{\lambda} l \sin\theta_{su}$, which is used in obtaining the summation of all antenna elements radiations i.e., array factor [33] as given by Equation (9).

$$\mathcal{F}(\theta_{su}) = \sum_{c \in C} \mathcal{X}_c e^{j(\frac{2\pi}{\lambda} cl \sin\theta_{su} + c\gamma)} \tag{9}$$

whereby $\mathcal{X}_c$ is the complex weight of the cth antenna element, λ is the wavelength, l is the inter-element spacing, and γ is the phase lead of the cth element.

At the ScMS, the base station transmission can now be described in terms of time and angle whereby the transmission of small cell s is set with respect to Table 2. Note that in Table 2 subchannels are considered in the transmission whereby the transmission is said to be in the downlink only i.e., only the small cells are instructed to adjust the transmission. For two interfering small cells s and i, and after τ_{su} is set according to Equation (8), β_{iu} is updated if user u is not receiving small cell s i.e., no interference. This is shown in Algorithm 2.

Table 2. ScMS transmission codebook for cell s.

	User u				
Subchannel k	1	2	3	…	$\mathcal{U}$
1	$\mathcal{F}(\theta_{s11}), \tau_{s11}$	$\mathcal{F}(\theta_{s12}), \tau_{s12}$	$\mathcal{F}(\theta_{s13}), \tau_{s13}$	…	$\mathcal{F}(\theta_{s1\mathcal{U}}), \tau_{s1\mathcal{U}}$
2	$\mathcal{F}(\theta_{s21}), \tau_{s21}$	$\mathcal{F}(\theta_{s22}), \tau_{s22}$	$\mathcal{F}(\theta_{s23}), \tau_{s23}$	…	$\mathcal{F}(\theta_{s2\mathcal{U}}), \tau_{s2\mathcal{U}}$
3	$\mathcal{F}(\theta_{s31}), \tau_{s31}$	$\mathcal{F}(\theta_{s32}), \tau_{s32}$	$\mathcal{F}(\theta_{s33}), \tau_{s33}$	…	$\mathcal{F}(\theta_{s3\mathcal{U}}), \tau_{s3\mathcal{U}}$
⋮	⋮	⋮	⋮	⋮	⋮
$\mathcal{K}$	$\mathcal{F}(\theta_{s\mathcal{K}1}), \tau_{s\mathcal{K}1}$	$\mathcal{F}(\theta_{s\mathcal{K}2}), \tau_{s\mathcal{K}2}$	$\mathcal{F}(\theta_{s\mathcal{K}3}), \tau_{s\mathcal{K}3}$	…	$\mathcal{F}(\theta_{s\mathcal{K}\mathcal{U}}), \tau_{s\mathcal{K}\mathcal{U}}$

Algorithm 2 Update β_{iu}

```
function ScMS
    for s = 1 : S do
        for u = 1 : U do
            if β_su = 1 then
                for i = 1 : S do
                    if i ≠ s then
                        if β_iu = 1 then
                            if τ_su ≠ τ_iu then
                                β_iu = 0
                            else
                                Break
                            end if
                        end if
                    end if
                end for
            end if
        end for
    end for
end function
```

It can now be claimed that the maximum $SINR_s$ is obtained if $\sum_{i \in \mathcal{S}/\{s\}} \beta_{iu} = 0$. However, enhancements can be obtained if β_{iu} is minimised. Therefore, the $SINR_s$ in Equation (3) can now be obtained by the following:

$$\begin{aligned} &\arg\max SINR_{\mathcal{S}} \\ &= \sum_{u \in \mathcal{U}} \sum_{k \in \mathcal{K}} \frac{\mathcal{P}^{+}_{sk_u} g_{sk_u} \times \beta_{su}}{\sum_{i \in \mathcal{S}/\{s\}} \mathcal{P}^{+}_{ik_u} g_{ik_u} \times \beta_{iu} + \mathcal{N}} \\ &s.t.: \\ &\sum_{s \in \mathcal{S}} \beta_{su} = 1 \end{aligned} \tag{10}$$

Obtaining the correct angles where the users are located makes it easier to redirect the beams towards the desired user. Therefore, the minimum β is obtained by Algorithm 2.

4. Results and Discussion

Here, the simulation results are presented to evaluate the relevance of the findings. The results of the algorithm presented in Section 3 are compared with the small cell on/off control scheme with a discovery signal (SCon/off-DS), location-aware self-optimisation (LASO) in [29]. When a nearby user requests specific amount of data rate from the small cell, the macrocell can provide the required coverage and mobility. Therefore, the SCon/off-DS scheme achieves downlink interference avoidance, and mitigation at the macrocell and the small cells, respectively. The LASO scheme improves the throughput by adjusting the power offset in downlink transmission by effectively locating the users. The authors have categorised the coverage area into two categories, namely: the non-dominantly interfered region (NDIR), and the interfered region (IR). The users in the NDIR experience high SINR, whereby the users in the IR experience low SINR. Additionally, beamforming scheme without TDMA technique is referred to as Pl-beamforming, the proposed algorithm is referred to as Sl-beamforming. Subsequently, the fairness of the beamforming schemes is later compared with baseline (BL) scheme in which interference is assumed to be zero i.e., the best performance, which is used to assess the feasibility and reliability of the proposed algorithm. Simulation results presented in this section are in-line with the assumptions of $\mathcal{U}$ = [10–100] users, 20 MHz bandwidth, $\mathcal{B}^{-}$ = 180 kHz, service area = 100 m × 100 m, and 100 iterations.

The user throughput performance is illustrated in Figure 2a. Although the throughput achieved by the SCon/off-DS and the LASO in NDIR is relatively high when there are 10 users, the number quickly degrades as the number of users increase. However, both the Pl-beamforming and the Sl-beamforming schemes maintain a throughput level of approximately 16 Mbps and 18 Mbps, respectively. This is because the beams are effectively redirected towards the desired users only. Thus, reduces the interference to other users.

This is evidenced by the SINR levels shown in Figure 2b whereby the superiority of the beamforming schemes is clear. This shows how significantly can the beamforming schemes reduce the interference in comparison to the other schemes. However, interference can still occur when the number of users increase due to the proximity of users.

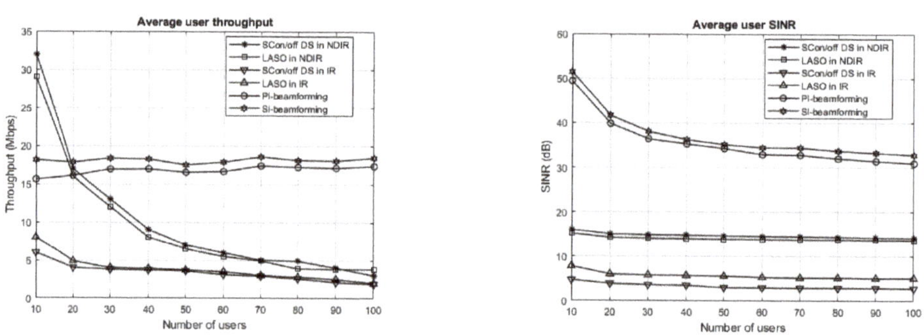

(**a**) Throughput (**b**) SINR

Figure 2. Throughput and SINR comparison when increasing the number of users in the service area for all schemes.

Figure 3 shows the fairness level obtained by the Sl-beamforming. It is noteworthy to mention that the fairness results are compared to the BL scheme because the fairness was not evaluated in [29]. Nevertheless, the Sl-beamforming is able to perform better than the Pl-beamforming because narrower beamwidth and TDMA leads to less interference. Therefore, the Sl-beamforming is able to achieve about 95% fairness levels for 100 users. The overall performance of the Sl-beamforming promotes it as an efficient interference mitigation paradigm for future communication systems where the number of base stations is expected to increase.

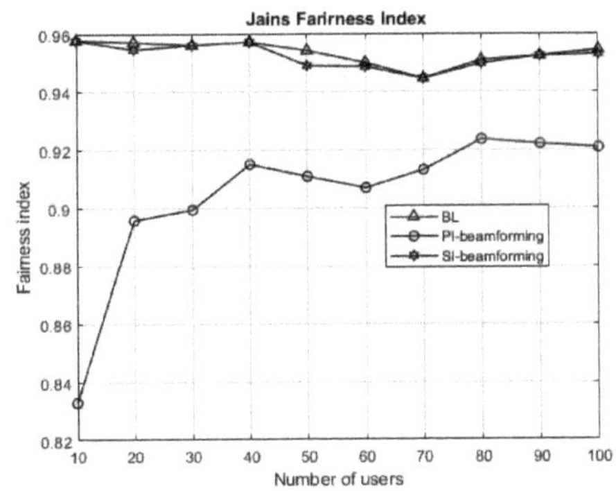

Figure 3. Network fairness in comparison to the baseline scheme.

5. Conclusions

In this paper, the applicability of TDMA beamforming for future communication systems is studied and assessed. The paper addressed the concerns of future wireless communication followed by the features and the operations of beamforming. Subsequently, the Sl-beamforming is proposed for line-of-sight small cell systems in LTE architecture. The modelling of the network is tailored to give insights on the technical concerns and the achievements of the proposed scheme. Empirical simulations are demonstrated to enforce the applicability of the Sl-beamforming scheme whereby it is shown that the proposed scheme enhances the network SINR by about 18 dB, and user average throughput by about 10 Mbps in comparison to the NDIR schemes. It also increases the network fairness to approximately 95%. Future research directions will demonstrate the performance over non-line-of-sight scenarios, uplink communication, and ultra-dense environments. Resource scheduling algorithms should be added to the design, such as round robin and proportional fairness for more factual simulation. Furthermore, investigations over 5G communication channels and frequencies are recommended.

Author Contributions: Methodology, investigation, data curation, writing–original draft preparation, and writing–review and editing, K.S.M.; conceptualization, validation, formal analysis, supervision, and project administration, M.Y.A. and M.R.

Funding: This research received no external funding.

Conflicts of Interest: The authors declare no conflict of interest.

Abbreviations

The following abbreviations are used in this manuscript:

5G	Fifth Generation
BL	Base Line
CFO	Carrier Offsets
CSI	Channel State Information
GFDM	Generalised Frequency Division Multiplexing
GHz	Giga Hertz
IBFD	In-Band Full-Duplex
IoT	Internet of Things
IR	Interfered Region
LASO	Location Aware Self-Optimisation
LTE	Long Term Evolution
MIMO	Multiple Input Multiple Output
MMSE	Minimum Mean Square Error
mmWave	Millimetre Wave

NDIR	Non-Dominantly Interfered Region
OFDM	Orthogonal Frequency Division Multiplexing
P-GW	Packets Gateway
RCI	Regulated Channel Inversion
ScMs	Small cell Management System
SCon/off-DS	Small Cell on/off control scheme with Discovery Signal
SINR	Signal to Interference plus Noise Ratio
TDM	Tim Division Duplexing
TDMA	Time Division Multiple Access
THz	Tera Hertz
ZF	Zero Forcing

References

1. Akyildiz, I.F.; Jornet, J.M.; Han, C. TeraNets: Ultra-broadband communication networks in the terahertz band. *IEEE Wirel. Commun.* **2014**, *21*, 130–135. [CrossRef]
2. Faisal, A.; Sarieddeen, H.; Dahrouj, H.; Al-Naffouri, T.Y.; Alouini, M.S. Ultra-massive MIMO systems at terahertz bands: Prospects and challenges. *arXiv* **2019**, arXiv:1902.11090.
3. Rangan, S.; Rappaport, T.S.; Erkip, E. Millimeter wave cellular wireless networks: Potentials and challenges. *arXiv* **2014**, arXiv:1401.2560.
4. Guo, C.; Sheng, M.; Wang, X.; Zhang, Y. Throughput maximization with short-term and long-term Jain's index constraints in downlink OFDMA systems. *IEEE Trans. Commun.* **2014**, *62*, 1503–1517. [CrossRef]
5. Sohail, M.F.; Leow, C.Y. Maximized fairness for NOMA based drone communication system. In Proceedings of the 2017 IEEE 13th Malaysia International Conference on Communications (MICC), Johor Bahru, Malaysia, 28–30 November 2017; pp. 119–123. [CrossRef]
6. Lee, Y.L.; Loo, J.; Chuah, T.C.; El-Saleh, A.A. Fair resource allocation with interference mitigation and resource reuse for LTE/LTE-A femtocell networks. *IEEE Trans. Veh. Technol.* **2016**, *65*, 8203–8217. [CrossRef]
7. Ertürk, M.C.; Güvenç, I.; Mukherjee, S.; Arslan, H. Fair and QoS-oriented resource management in heterogeneous networks. *EURASIP J. Wirel. Commun. Netw.* **2013**, *2013*, 121. [CrossRef]
8. Lu, Z.; Bansal, T.; Sinha, P. Achieving user-level fairness in open-access femtocell-based architecture. *IEEE Trans. Mob. Comput.* **2012**, *12*, 1943–1954. [CrossRef]
9. Lin, C.; Li, G.Y.L. Terahertz communications: An array-of-subarrays solution. *IEEE Commun. Mag.* **2016**, *54*, 124–131. [CrossRef]
10. Federici, J.; Moeller, L. Review of terahertz and subterahertz wireless communications. *J. Appl. Phys.* **2010**, *107*, 6. [CrossRef]
11. Lin, C.; Li, G.Y. Indoor terahertz communications: How many antenna arrays are needed? *IEEE Trans. Wirel. Commun.* **2015**, *14*, 3097–3107. [CrossRef]
12. Yuan, H.; Yang, N.; Yang, K.; Han, C.; An, J. Hybrid Beamforming for MIMO-OFDM Terahertz Wireless Systems over Frequency Selective Channels. In Proceedings of the 2018 IEEE Global Communications Conference (GLOBECOM), Abu Dhabi, UAE, 9–13 December 2018; pp. 1–6. [CrossRef]
13. Sathananthan, K.; Tellambura, C. Probability of error calculation of OFDM systems with frequency offset. *IEEE Trans. Commun.* **2001**, *49*, 1884–1888. [CrossRef]
14. Honma, N.; Heianna, S.; Kawagoe, A.; Tada, S.; Yamamoto, Y.; Yuan, Q.; Chen, Q. Enabling Full-Duplex MIMO Communication Exploiting Array Antenna Arrangement. In Proceedings of the 2018 International Symposium on Antennas and Propagation (ISAP), Busan, Korea, 23–26 October 2018; pp. 1–2.
15. Misra, G.; Agarwal, K.; Agarwal, A.; Ghosh, K.; Agarwal, S. Smart Antenna for Wireless Cellular Communication-A Technological Analysis on Architecture, Working Mechanism, Drawbacks and Future Scope. In Proceedings of the 2018 2nd International Conference on (I-SMAC), Palladam, India, 30–31 August 2018; pp. 37–41. [CrossRef]
16. Patcharamaneepakorn, P.; Armour, S.; Doufexi, A. On the equivalence between SLNR and MMSE precoding schemes with single-antenna receivers. *IEEE Commun. Lett.* **2012**, *16*, 1034–1037. [CrossRef]

17. Peel, C.B.; Hochwald, B.M.; Swindlehurst, A.L. A vector-perturbation technique for near-capacity multiantenna multiuser communication-part I: channel inversion and regularization. *IEEE Trans. Commun.* **2005**, *53*, 195–202. [CrossRef]
18. Wiesel, A.; Eldar, Y.C.; Shamai, S. Zero-Forcing Precoding and Generalized Inverses. *IEEE Trans. Signal Process.* **2008**, *56*, 4409–4418. [CrossRef]
19. Khang, Y.Y.; Zhangt, J.K.; Yu, H.Y. Finite-alphabet beamformed NOMA for multiuser MISO broadcast visible light communications. In Proceedings of the 2018 IEEE 10th Sensor Array and Multichannel Signal Processing Workshop (SAM), Sheffield, UK, 8–11 July 2018; pp. 563–567. [CrossRef]
20. Leary, J.; Gooch, R. Adaptive beamforming for TDMA signals. In Proceedings of the Twenty-Ninth Asilomar Conference on Signals, Systems and Computers, Pacific Grove, CA, USA, 30 October–1 November 1995; Volume 2, pp. 1378–1382. [CrossRef]
21. Busson, A.; Zitoune, L.; Vèque, V.; Jabbari, B. Outage analysis of integrated mesh LTE femtocell networks. In Proceedings of the IEEE Global Communications Conference (GLOBECOM), Austin, TX, USA, 8–12 December 2014; pp. 187–192. [CrossRef]
22. Saad, S.A.; Ismail, M.; Nordin, R.; Ahmed, A.U. A fractional path-loss compensation based power control technique for interference mitigation in LTE-A femtocell networks. *Phys. Commun.* **2016**, *21*, 1–9. [CrossRef]
23. Lin, S.; Ni, W.; Tian, H.; Liu, R.P. An evolutionary game theoretic framework for femtocell radio resource management. *IEEE Trans. Wirel. Commun.* **2015**, *14*, 6365–6376. [CrossRef]
24. Nam, C.; Joo, C.; Bahk, S. Joint subcarrier assignment and power allocation in full-duplex OFDMA networks. *IEEE Trans. Wirel. Commun.* **2015**, *14*, 3108–3119. [CrossRef]
25. LeAnh, T.; Tran, N.H.; Saad, W.; Le, L.B.; Niyato, D.; Ho, T.M.; Hong, C.S. Matching theory for distributed user association and resource allocation in cognitive femtocell networks. *IEEE Trans. Veh. Technol.* **2017**, *66*, 8413–8428. [CrossRef]
26. Datta, J.; Lin, H.P. Interference Avoidance using Spatial Modulation based Location Aware Beamforming in Cognitive Radio IOT Systems. *Adv. Sci. Technol. Eng. Syst. J.* **2018**, *3*, 49–57. [CrossRef]
27. Jakes, W. *Microwave Mobile Communications*; IEEE and John Wiley & Sons: Hoboken, NJ, USA, 1974.
28. Lee, Y.L.; Loo, J.; Chuah, T.C. Dynamic resource management for lte-based hybrid access femtocell systems. *IEEE Syst. J.* **2016**, *12*, 959–970. [CrossRef]
29. Choi, J.H.; Shin, D.J. Location-Aware Self-Optimization for Interference Management in Ultra-Dense Small Cell Networks. *IEEE Commun. Lett.* **2018**, *22*, 2555–2558. [CrossRef]
30. Ibrahim, L.F.; Salman, H.A.; Sery, S.Y.; Taha, Z. Using clustering techniques to plan indoor Femtocell base stations layout in multi-floors. *Comput. J.* **2019**, *62*, 919–930. [CrossRef]
31. Loong, L.Y. Radio Resource Management for Long-term Evolution Based Heterogeneous Cellular Networks. Ph.D. Thesis, Multimedia University, Cyberjaya, Malaysia, 2016.
32. Jain, R.; Durresi, A.; Babic, G. *Throughput Fairness Index: An Explanation*; ATM Forum Contribution, 99, 1999. Available online: https://www.cse.wustl.edu/~jain/atmf/ftp/af_fair.pdf (accessed on 1 October 2019).
33. Litva, J.; Lo, T.K. *Digital Beamforming in Wireless Communications*; Artech House, Inc.: Norwood, MA, USA, 1996.

applied sciences

MDPI

Article

A Novel Self-Interference Cancellation Method Using an Optimized LMS Algorithm in CCFD Systems for a 5G Communication Network

Zeng-You Sun and Yu-Jie Zhao *

School of Electrical Engineering, Northeast Electric Power University, NO.169 ChangChun Road, ChuanYing District, Jilin 132012, China

* Correspondence: neepu_zhao@163.com; Tel.: +86-188-4322-7633

Received: 4 July 2019; Accepted: 8 August 2019; Published: 12 August 2019

Abstract: The Co-frequency Co-time Full Duplex (CCFD) is a key concept in 5G wireless communication networks. The biggest challenge for CCFD wireless communication is the strong self-interference (SI) from near-end transceivers. Aiming at cancelling the SI of near-end transceivers in CCFD systems in the radio frequency (RF) domain, a novel time-varying Least Mean Square (LMS) adaptive filtering algorithm which is based on step-size parameters gradually decrease with time varying called the DTV-LMS algorithm is proposed in this paper. The proposed DTV-LMS algorithm in this paper establishes the non-linear relationship between step factor and the evolved arct-angent function, and using the relationship between the time parameter and error signal correlation value to coordinately control the step factor to be updated. This algorithm maintains a low computational complexity. Simultaneously, the DTV-LMS algorithm can also attain the ideal characteristics, including the interference cancellation ratio (ICR), convergence speed, and channel tracking, so that the SI signal in the RF domain of a full duplex system can be effectively cancelled. The analysis and simulation results show that the ICR in the RF domain of the proposed algorithm is higher than that in the compared algorithms and have a faster convergence speed. At the same time, the channel tracking capability has also been significantly enhanced in CCFD systems.

Keywords: Co-time Co-frequency Full Duplex (CCFD); self-interference (SI) signal cancellation; RF domain; optimize Least Mean Square (LMS) algorithm; interference cancellation ratio (ICR)

---

1. Introduction

The increasing lack of spectrum resources has become a bottleneck, restricting the development of the fifth generation (5G) of mobile communication technology [1–3]. Therefore, alleviating the shortage of wireless spectrum resources and improving the utilization of spectrum resources has become one of the important problems of current wireless communication research. Co-time Co-frequency Full Duplex (CCFD) transmission technology, one of the key 5G technologies, can alleviate this lack of spectrum resources to a certain extent [4]. Transceivers in CCFD systems can occupy the same frequency band to transmit and receive data from the uplink and downlink at the same time. Compared to traditional half-duplex systems, such as a frequency division duplex (FDD) system and a frequency division duplex (TDD) system, the spectrum efficiency of the CCFD systems can be double. This increase has a significant advantage on throughput by reducing congestion [5,6]. However, the data from the uplink and downlink are transmitted while in the same frequency band, and the near-end receiver of the full duplex system will be subjected to high-power interference from the local transmitter. We called this interference self-interference (SI) [7]. Suffering from the limited quantization dynamic range of analog-to-digital converters (ADCs), the interference to signal ratio (ISR) of the ADC input cannot be too large or it will cover the desired signal with additional quantization noise [8]. Therefore, in order to ensure that the ADC is not blocked, the receivers of the CCFD system need to suppress SI signals in the radio frequency (RF) domain before quantization of the ADC [9,10]. Thus, the purpose of this paper is to suppress strong SI signals in the RF domain of the CCFD system.

At present, most of the studies in the field of RF SI cancellation in the CCFD system make use of direct radio frequency coupling cancellation (DRFCC) structures [11]. The basic idea of SI using DRFCC is to directly couple a part signal of the RF channel to the transmitter as a reference signal. By adjusting the phase, amplitude, and delay of the reference signal, the SI signal can be reconstructed, so that both the linear and non-linear SI can be effectively suppressed [12–14]. Compared with indirect radio frequency coupling cancellation (IRFCC), which needs an additional separate transmitting channel to reconstruct the interference signal, the DRFCC does not require extra runtime for the separate channel, thereby reducing the quantity of memory data that have to be transmitted at run-time, so that the runtime overhead of DRFCC can reduce greatly. The Least Mean Square (LMS) algorithm is an optimization extension of the Wiener filtering theory using the fast descent method [15]. It has the advantages of a simple principle, low computational complexity, and easy implementation [16,17]. However, there is an obvious disadvantage of the LMS algorithm: the convergence speed and steady-state error (SSE) are mutually constrained [18]. Reference [19] proposed a Variable Step Size LMS (VSSLMS) algorithm to overcome the drawbacks of the LMS algorithm. VSSLMS uses a prediction error signal to control the step size factor. However, the practicability of the algorithm is poor, and in the process of convergence, a steady-state misalignment (SSM) easily occurs, which does not guarantee precision. Reference [20] improved the VSSLMS algorithm based on the sigmoid function (SVSLMS). The algorithm has the characteristics of a fast convergence speed and small SSE, but the algorithm has high computational complexity and a large computational load, and the step size of the error will be mutated when the value approaches near zero, which is bad for the stability of the algorithm. Reference [21] proposed an modified variable step size LMS (MVSSLMS) algorithm, which assumes that the input power remains unchanged and makes use of the average estimation of the autocorrelation of the error signal function to the control iteration of the step factor, but MVSSLMS is incompatible with a full-duplex system with variable input power. Reference [22] proposed a scheme based on optimal dynamic power allocation, with the objective of maximizing the rate of convergence, but the capability of channel tracking was not well considered. In reference [23], a relationship between the step and the mean square instantaneous error was built. This algorithm has strong tracking ability and fast convergence speed, but a weak anti-jamming capability. An algorithm for tuning the parameters of a multiple-tap analog SI canceller by channel estimation is presented in reference [24], but a performance analysis is not provided. In [25,26], a variable step size algorithm is proposed, which has a larger value in the initial stage of the algorithm, to improve the convergence speed, and a smaller value near the convergence time to reduce the SSM error. However, this method is easily affected by the related noise and other factors. Reference [27] adopts a self-mixing RF SI cancellation structure to achieve better SI cancellation for a gaussian minimum shift keying(GMSK) narrowband full duplex signal at a lower hardware complexity, but its application scope is limited, and the improvement of RF cancellation abilities based on this structure is not large.

In order to solve the above problems of the existing SI cancellation methods in the RF domain, and to suppress SI signals more effectively, this paper proposes an optimized LMS algorithm based on the step-size attenuation of convergence parameters. This algorithm is called DTV-LMS. In this paper, Interference cancellation in wireless networks has been addressed through exact mathematical optimization methods [28] within optimal wireless network design [29]. Compared to the RF domain SI cancellation reference algorithms mentioned above, the proposed method in this paper makes use of an evolved arctangent function, with the help of the relationship between the time parameter and error signal correlation value to coordinately control the step-size. Thus, this algorithm can maintain ideal characteristics, such as fast convergence and a good interference cancellation ratio (ICR) at low computational complexity.

This paper is structured as follows. Section 2 presents the system model and the mathematical model of the CCFD system separately. The DTV-LMS algorithm and an analysis of its related abilities as they apply to SI cancellation in the RF domain are introduced in Section 3. Simulation results and analyses are given in Section 4. Finally, Section 5 summarizes this whole paper and proposes the prospect of future work.

2. System and Mathematical Models

2.1. System Model

In a double-node CCFD system, the near-end and far-end nodes can be defined so that both sides of the communication can simultaneously transmit and receive data in the same frequency band [30]. The transceiver model of the system is shown in Figure 1. An orthogonal dual-tap direct RFCC structure is adopted and uses a near-end transceiver as an example. At the transmitting end, the transmitted signal, through digital modulation, transforms to *s(n)*, and then the RF output signal *s(t)* is obtained through digital-to-analog conversion (DAC), up-conversion, and power amplification (PA), and fed into the transmitting antenna. At the receiving end, the received signal $r_0(t)$ includes not only the desired signal $r_u(t)$ from the far-end nodes, but also the high-power SI signal $r_{SI}(t)$ from the near-end nodes and the additive white Gaussian noise (AWGN) n(t). Therefore, the full-duplex receiver needs to suppress the SI signal in the RF domain after obtaining the received signal.

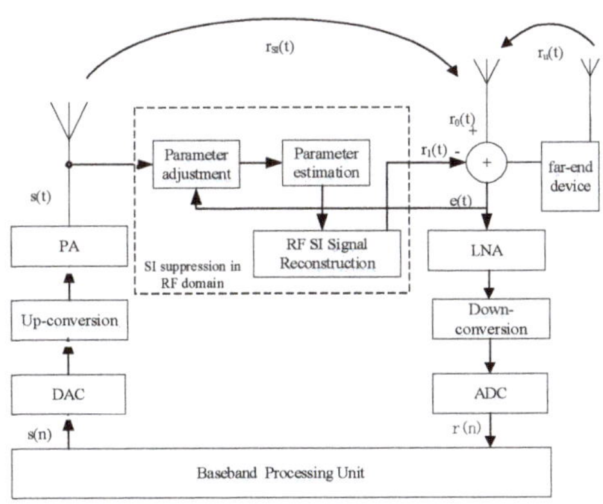

Figure 1. Near-end transceiver model for the Co-frequency Co-time Full Duplex (CCFD) system.

For the transmitter of the near-end transceiver in the full duplex system, the output s(t) can be expressed as follows:

$$s(t) = \sqrt{2P_n}s(n)\cos(2\pi f_c t + \varphi_n) \quad (1)$$

where P_n is the near-end transmitted power, f_c is carrier frequency, and φ_n is the initial phase. According to the transceiver model of the system, the signal received by the full duplex communication node entering the near-end is obtained as

$$r_0(\mathrm{t}) = r_{SI}(\mathrm{t}) + r_u(\mathrm{t}) + \mathrm{n}(\mathrm{t}). \quad (2)$$

The near-end SI signal can be specifically expressed as follows:

$$r_{SI}(t) = h_{SI}\ (t) * s(t) \quad (3)$$

$$h_{SI}\ (t) = k_n e^{-2jp_n f_c t_n} \quad (4)$$

where $h_{SI}(t)$ is the impulse response of the SI channel, k_n and t_n are the amplitude attenuation factor and time delay of the SI channel, separately. Therefore, the SI signal can be represented as

$$r_{SI}(t) = h_{SI}(t) * s(t) = k_n s(t - t_n). \tag{5}$$

In this paper, the SI signal in the RF domain of the near-end receiver of the full-duplex system is estimated using the DTV-LMS algorithm. The implementation process of this algorithm is described in detail in Section 3. The residual SI signal after SI cancellation in the RF domain (i.e., the error signal) can be expressed as

$$e(t) = r_0(t) - r_1(t) = r_{SI}(t) + r_u(t) + n(t) - r_1(t) \tag{6}$$

where $r_1(t)$ is the estimated SI signal (i.e., the reconstructed SI signal).

2.2. Mathematical Model

The main mathematical principle of RF cancellation in a CCFD system is the composition of vectors [31]. The following notation is utilized in developing a mathematical model for the problem of SI cancellation in the RF domain addressed in this paper: $s(t)$ is the near-end transmitting signal, $s_i(t)$ is the same phase component as $s(t)$, $s_q(t)$ is the orthogonal component obtained by $s_i(t)$ through a 90 degree phase shifter, $r_1(t)$ is the estimated SI signal, $r(t)$ is the near-end receiving signal, and $e(t)$ is the error signal. Assuming that the SI signal in the RF domain is a vector in the space rectangular coordinate system, the signal can be cancelled by synthesizing another vector with the same information characteristics as the vector. As shown in Figure 2, the near-end transmitting signal $s(t)$ is divided into two branches, $s_i(t)$ and $s_q(t)$, after group delay, t. The delay block is the group delay of the estimate interference signal.

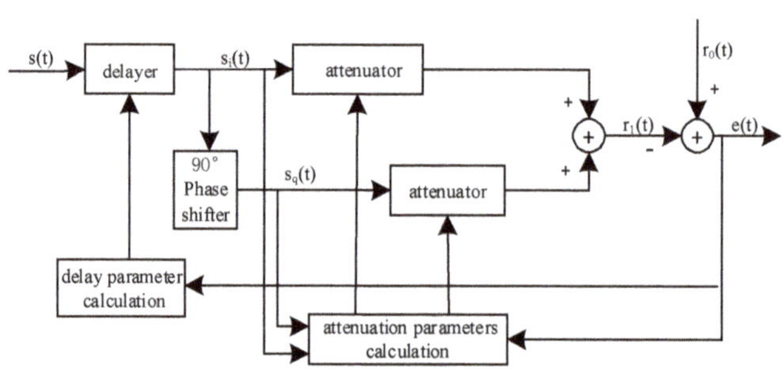

Figure 2. Principle diagram of self-interference (SI) cancellation in the radio frequency (RF) domain.

Let $S(t) = \left[s_i(t)\ s_q(t)\right]^T$, where T is the transpose of the matrix, and the weight vector is $W(t) = \left[\omega_i(t)\ \omega_q(t)\right]^T$, that is the amplitude attenuation controlled by in-phase and orthogonal branch attenuator, which is used to adjust the amplitude and phase of $r_1(t)$. Therefore, the basic process of cancellation mentioned above can be expressed by

$$r_1(t) = [W(t)]^T S(t) = \left[\omega_i(t)\ \omega_q(t)\right]\begin{bmatrix} s_i(t) \\ s_q(t) \end{bmatrix} = \omega_i(t)s_i(t) + \omega_q(t)s_q(t) \tag{7}$$

$$e(t) = [r_{SI}(t) - r_1(t)] + r_u(t) + n(t) = \left\{r_{SI}(t) - [\omega_i(t)s_i(t) + \omega_q(t)s_q(t)\right\} + r_u(t) + n(t) \tag{8}$$

$$W(t+t) = \left[\omega_i(t+\Delta t)\ \omega_q(t+\Delta t)\right]^T = \left[\omega_i(t)\ \omega_q(t)\right]^T + \left[\omega_{i,\Delta t}\ \omega_{q,\Delta t}\right]^T = W(t) + \mu\, e(t)S(t). \tag{9}$$

Equation (9) is the iteration formula for updating the weight vector of the LMS algorithm, where μ is a fixed step factor and Δt is the step interval of feedback control. The above mentioned cancellation process, the amplitude of the in-phase component, and the orthogonal component can be changed by adjusting the value of $W(t)$, and then the SI signal can be reconstructed. Therefore, the weight vector is the key factor affecting the performance of SI cancellation. The adjustment of the weight vector based on the minimum mean square error (MMSE) criterion is expressed as

$$\min\left\{E\left(|e(t)|^2\right)\right\} = min\left\{E\left(|r_0(t) - r_1(t)|^2\right)\right\}. \tag{10}$$

3. DTV-LMS Algorithm and Analysis

3.1. DTV-LMS Algorithm

As shown in Equation (9) above, in a classical LMS algorithm, μ is a fixed constant value whose convergence range is between $[0, 1/\lambda_{max}]$, and λ_{max} is the maximum eigenvalue of the autocorrelation matrix of the input signal. The convergence speed of the LMS algorithm became faster with an increase of μ, but at the same time, it produces a problem by increasing SSM. An optimized LMS algorithm called the DTV-LMS algorithm is proposed in this paper. This algorithm makes the value of the step factor change over time (from large to small) to reduce SSM. Based on the iterative formula in the reference [20,21], we make further improvements the weight and step updating formulas of the DTV-LMS algorithm, as given in Equations (11) and (12), respectively:

$$\mu(t) = \alpha tan^{-1}\left(-\beta|e(t)e(t-\Delta t)|\right) + 0.4\mu(t-1) \tag{11}$$

$$W(t+\Delta t) = W(t) + \left[\frac{\mu(t)}{\rho + S^T(t)S(t)}\right]e(t)S(t) \tag{12}$$

where α, β are constants, which are used to control the step-size in order to accelerate the convergence speed while keeping a low steady-state error; ρ is also a constant and can solve the numerical problem that a minimal value exists when $S(t)$ is too small.

The autocorrelation of the error signal is used from Formula (11) to reduce the impact of burst impulse noise interference, thereby improving ICR and the cancellation performance. The step updating equation uses the arctangent function to establish a new non-linear relationship with the time parameters. Compared to the sigmoid function and hyperbolic tangent function used in references, the simulation results are shown in Figure 3.

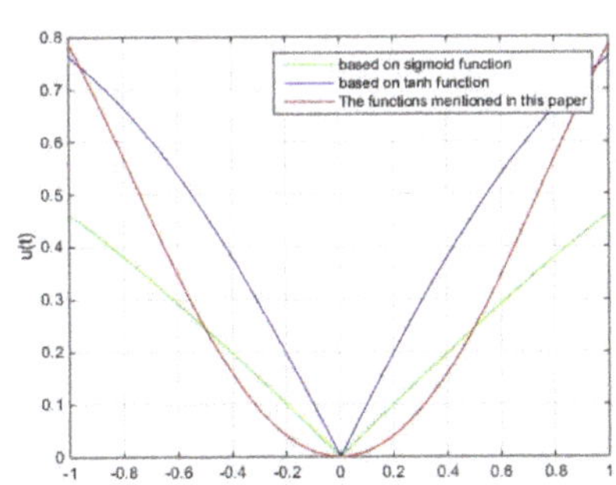

Figure 3. The curves of the step-size characteristics.

From the curve shown in Figure 3, it is obvious that when the error factor is close to zero, the sigmoid function and tanh function will change dramatically, and the small error changes will result in large step size changes. The optimized algorithm proposed in this paper overcomes this shortcoming very well, and $\mu(t)$ changes slowly when $e(t)$ approaches zero, so that the adaptive weight vector value of the algorithm is closer to the actual value when it reaches a steady state.

From Equation (11), we can see that the updating of $\mu(t)$ is influenced by the statistical characteristics of the near-end transmitting signal, and the adjustment of the weight vector is controlled by the step factor. That is to say, the algorithm uses the statistical characteristics of the transmitting signal to realize an effective adjustment of the weight.

In addition, we adjust the coefficients α, β to change the convergence performance. The curves of the relationship between $\mu(t)$ and $e(t)$ are shown in Figures 4 and 5, with a change in the values of α and β, respectively. From the two figures, we can see when the slope gets larger, the step causes a mutation just as the error approaches 0, so 'α' should not be too large. Similarly, 'β' also needs to be selected reasonably.

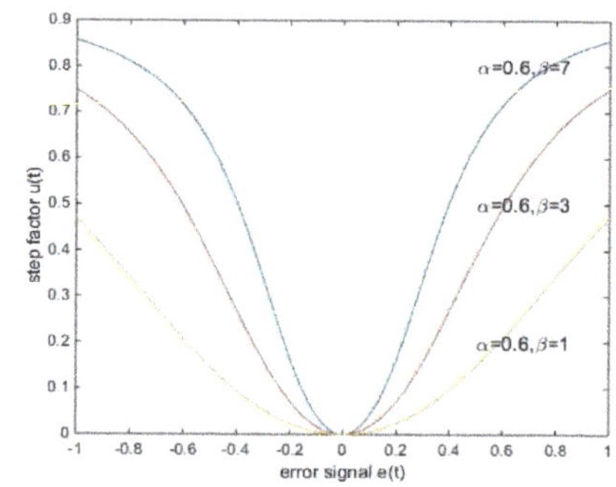

Figure 4. The variation of curves; when α = 0.6, β changes.

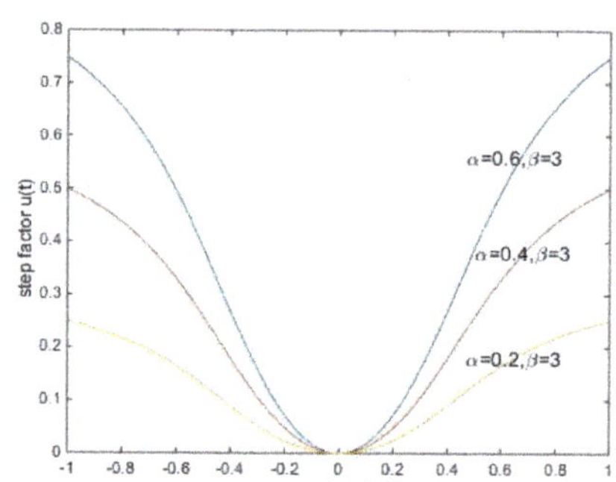

Figure 5. The variation of curves; when β = 0.3, α changes.

According to the model of the CCFD system and the structure of RF cancellation mentioned in Section 2, the detailed steps of the SI cancellation method in the RF domain are given as follows:

Step 1. The group delay time of the transmitted and received signals is estimated by the training sequence.

Step 2. The optimal values of parameters α, β are determined according to various parameters of the CCFD system model.

Step 3. Give the initial value to the weight vector $W(0)$, assuming $e(t) = r_0(t)$, when $t < 0$.

Step 4. The $\mu(t)$ is calculated at intervals of time Δt, and the weights are updated while adjusting the gains of the two in-phase and orthogonal branches.

Step 5. Calculate the real-time error signal $e(t)$.

Step 6. Return to (4).

3.2. Performance Analysis

In a full duplex system, the SI signals at the near-end transceivers and the useful signals at the far-end are independent of each other. In order to make the analysis more convenient, assuming that the error of time delay is 0, the coefficients of the in-phase component and the orthogonal component corresponding to the SI signal $r_{SI}(\mathrm{t})$ are, respectively, $\omega_i{}^*, \omega_q{}^*$. The in-phase and orthogonal components are wide stationary random signals with a mean of 0. Assuming that their power is P_s and the power of the useful signal $r_u(t)$ is P_f at the far end, the mean square of the error signal during Δt can be represented by

$$\begin{aligned} \mathrm{E}\left[\mathrm{e}^2(\mathrm{t})\right] &= \mathrm{E}\left\{\left[(r_{SI}(\mathrm{t}) - r_1(t)) + r_u(t) + n(t)^2\right]\right\} \\ &= E\left\{\left\{\left[(\omega_i{}^* - \omega_i)\left(\omega_q{}^* - \omega_q\right)\right]S(t)\right\}^2\right\} + \mathrm{E}\left\{[r_u(t) + n(t)]^2\right\} \\ &+2E[(r_{SI}(\mathrm{t}) - r_1(t))(r_u(t) + n(t))] \\ &= P_s\left[(\omega_i{}^* - \omega_i)^2 + \left(\omega_q{}^* - \omega_q\right)^2\right] + P_f + \sigma_n^2 \end{aligned} \tag{13}$$

where σ_n^2 is the mean power of AWGN, and the weight vector formula is further simplified as

$$\begin{aligned} W(t+\Delta t) &= W(t) + \left[\frac{\mu(t)}{\rho + S^T(t)S(t)}\right]e(t)S(t) \\ &= W(t) + \left[\frac{\mu(t)}{\rho+P_s}\right]\left[W^T(t) - W^{*T}\right]P_s + \left[\frac{\mu(t)}{\rho+P_s}\right][ru(t) + n(t)]S(t) \\ &= \left[I - \frac{\mu(t)P_s}{\rho+P_s}\right](W(t) - W^*) + \left[\frac{\mu(t)}{\rho+P_s}\right][ru(t) + n(t)]S(t) \end{aligned} \tag{14}$$

where W^* is the weight of the real SI signal, and I is the identity matrix. Combining Formula (13) and (14), (15) can be obtained as follows:

$$\begin{aligned} \mathrm{E}\left[\mathrm{e}^2(\mathrm{t})\right] = P_s E\{[W(t-\Delta t) - W^*]^T\left[I - \frac{\mu(t)P_s}{\rho+P_s}\right]^2[W(t-\Delta t) - W^*]\} \\ +\left[\frac{\mu(t)P_s}{\rho+P_s}\right]^2 P_s{}^2(P_f + \sigma_n^2) + P_f + \sigma_n^2. \end{aligned} \tag{15}$$

The initial value to the weight vector $\mathrm{W}(0) = [00]^T$. Assuming $\mathrm{t} = \Delta \mathrm{nt}$, according to Equation (15), the relationship between the mean square error (MSE) of the nth iteration and the initial MSE can be obtained as follows:

$$\begin{aligned} \mathrm{E}\left[\mathrm{e}^2(\mathrm{t})\right] &= \left[(\omega_i{}^*)^2 + \left(\omega_q{}^*\right)^2\right]P_s \prod_{k=0}^{n-1}\left[1 - \frac{\mu(\mathrm{k}\Delta\mathrm{t})P_s}{\rho+P_s}\right]^2 + (P_f \\ &+\sigma_n^2)\left[1 + 2\left(\frac{\mu(\mathrm{t}-\Delta\mathrm{t})P_s}{\rho+P_s}\right)^2\right] + 2(P_f \\ &+\sigma_n^2)\left\{\sum_{k=0}^{n-2} P_n^2\left[\frac{\mu(\mathrm{k}\Delta\mathrm{t})P_s}{\rho+P_s}\right]^2 \prod_{i=k+1}^{n-1}\left[1 - \frac{\mu(\mathrm{i}\Delta\mathrm{t})P_s}{\rho+P_s}\right]^2\right\}. \end{aligned} \tag{16}$$

It can be seen that the convergence conditions of the above equation are $\left|1 - \frac{\mu(\mathrm{k}\Delta\mathrm{t})P_s}{\rho+P_s}\right| < 1$, $\mathrm{k} \in \mathrm{N}^*$. Due to ρ generally being very small, the condition for the convergence can be simplified as $\left|1 - \mu(\mathrm{k}\Delta\mathrm{t})\right| < 1$, and then we can express $\left|\mu\right| < 1$,which is consistent with the convergence conditions of the classical LMS algorithm given above. It can be seen from Equation (16) that the convergence of the algorithm is determined by the trend of the cumulative product value in the error convergence function.

Let the convergence factors of the LMS algorithm and the DTVLMS algorithm be $u_{LMS}(\Delta\mathrm{nt})$. According to Equation (16), we can obtain $u_{DTVLMS}(\mathrm{n}\Delta\mathrm{t})$ as

$$u_{LMS}(\mathrm{n}\Delta\mathrm{t}) = 1 - \mu P_s{}^{2n} \tag{17}$$

$$u_{DTVLMS}(\mathrm{n}\Delta \mathrm{t}) \begin{aligned} &= \prod_{k=0}^{n-1}\left[1-\frac{\mu(\mathrm{k}\Delta \mathrm{t})P_s}{\rho+P_s}\right]^2 \\ &= \prod_{k=0}^{n-1}\left\{1-\frac{\left[\alpha tan^{-1}\left(-\beta\left|e(\mathrm{k}\Delta \mathrm{t})e(\mathrm{k}\Delta \mathrm{t}-\Delta t)\right|+0.4\mu(t-1)\right)\right]P_s}{\rho+P_s}\right\}^2 \\ &\approx \prod_{k=0}^{n-1}\left\{1-\left[\alpha tan^{-1}\left(-\beta\left|e(\mathrm{k}\Delta \mathrm{t})e(\mathrm{k}\Delta \mathrm{t}-\Delta t)\right|+0.4\mu(t-1)\right)\right]\right\}^2. \end{aligned} \tag{18}$$

It can be seen from the above two formulas, compared with the convergence factors of the traditional fixed-step LMS algorithm, that the convergence speed of the DTV-LMS algorithm is almost independent of the power of the near-end transmitted signal in the RF domain, which can effectively avoid the interference of the gradient noise amplification. In addition, by using the correlation value of the error signal to adjust the step size, the algorithm can take into account the performance of the convergence speed and error. At the same time, the problem of intermittent communication or burst noise interference by relying only on the time parameter can also be effectively solved.

According to reference [24], the equation for ICR is $\mathrm{ICR} = 10\mathrm{lg}\left(P_{before}/P_{after}\right)$, where P_{before} and P_{after} is the power of the near-end SI signal before and after the SI cancellation in RF domain, respectively. Let $\mathrm{K} \in N^*$. When $\mathrm{t} \geq \mathrm{K}\Delta \mathrm{t}$, the MSE of the DTV-LMS algorithm tends to be stable, and Formula (16) can be simplify as

$$\begin{aligned} \mathrm{E}\left[\mathrm{e}^2(\mathrm{t})\right] \approx\ & 2\left[(\omega_i{}^*)^2+\left(\omega_q{}^*\right)^2\right]P_s\left[1-\frac{\mu(\mathrm{t}\geq \mathrm{K}\Delta \mathrm{t})P_s}{\rho+P_s}\right]^{2(n-K)}+2(P_f \\ &+\sigma_n^2)\left[\frac{\mu(\mathrm{t}\geq \mathrm{K}\Delta \mathrm{t})P_s}{\rho+P_s}\right]^2\frac{1-\left[1-\frac{\mu(\mathrm{t}\geq \mathrm{K}\Delta \mathrm{t})P_s}{\rho+P_s}\right]^{2(n-K)}}{1-\left[1-\frac{\mu(\mathrm{t}\geq \mathrm{K}\Delta \mathrm{t})P_s}{\rho+P_s}\right]^2}+(P_f+\sigma_n^2). \end{aligned} \tag{19}$$

Because $\left|1-\frac{\mu(\mathrm{k}\Delta \mathrm{t})P_s}{\rho+P_s}\right|<1$, when $\mathrm{K}\rightarrow\infty$, $\lim\limits_{K\rightarrow\infty}\left[1-\frac{\mu(\mathrm{t}\geq \mathrm{K}\Delta \mathrm{t})P_s}{\rho+P_s}\right]^{2(n-K)}\rightarrow 0$, and we can get the formula of MSE as

$$\begin{aligned} \mathrm{E}\left[e^2(\infty)\right] &\approx (P_f+\sigma_n^2)+2(P_f+\sigma_n^2)\left[\frac{\mu(\mathrm{t}\geq \mathrm{K}\Delta \mathrm{t})P_s}{\rho+P_s}/\left(2-\frac{\mu(\mathrm{t}\geq \mathrm{K}\Delta \mathrm{t})P_s}{\rho+P_s}\right)\right] \\ &= (P_f+\sigma_n^2)\left[\left(2+\frac{\mu(\mathrm{t}\geq \mathrm{K}\Delta \mathrm{t})P_s}{\rho+P_s}\right)/\left(2-\frac{\mu(\mathrm{t}\geq \mathrm{K}\Delta \mathrm{t})P_s}{\rho+P_s}\right)\right]. \end{aligned} \tag{20}$$

Thus, the ICR of the DTV-LMS algorithm can be given as

$$\begin{aligned} \mathrm{ICR} &= 10\mathrm{lg}\left\{\frac{P_{r_0}-(P_f+\sigma_n^2)}{\mathrm{E}[e^2(\infty)]-(P_f+\sigma_n^2)}\right\} \\ &\approx -10lg\left\{2\frac{\mu(\mathrm{t}\geq \mathrm{K}\Delta \mathrm{t})P_s}{\rho+P_s}(P_f+\sigma_n^2)/P_{SI}\left[2-\frac{\mu(\mathrm{t}\geq \mathrm{K}\Delta \mathrm{t})P_s}{\rho+P_s}\right]\right\}. \end{aligned} \tag{21}$$

According to the above equation, ICR will be applied to the simulation experiment in the next section, and its performance will be verified from the simulation results.

4. Simulation and Result Analysis

4.1. Simulation Parameter Setting

In this section, we make use of a MATLAB simulation to test the performance of the convergence speed, ICR, tracking speed, and so on. The conditions of the computer simulation parameters are:

1. Simulating modulation using quadrature phase shift keying (QPSK) with a transmission rate of 10 Mbps and a carrier frequency of 2.1 GHz, regardless of the nonlinear and ADC quantization noise effects;
2. Setting the noise limit of the received channel to −95 dBm by referring to reference [25]; the power of desired signal is −85 dBm, the SI signal is −10 dBm (i.e., the modulated channel ISR = 75 dB), and the SNR = 10 dB;

3. The step interval is $\Delta t = \Delta 0.03$ms. Let t = nt, and change the phase of the SI signal at the 500th (i.e., t = 15 ms) iteration;
4. Extract 1000 sampling points and perform 200 statistically independent simulation experiments.

The improved LMS algorithm proposed in this paper is compared with the existing RF cancellation algorithms of SVSLMS [20], VSSLMS [19], and MVSSLMS [21]. Refer to the parameter setting principle of reference in [19–21,29], as the parameters of these algorithms are selected to produce a comparable level of mis-adjustments, assuming that the delay estimation error is 0, and the power value of the reference signal is 0 dBm [30], and give the specific parameters of the four algorithms as shown in Table 1.

Table 1. The corresponding parameter value of each algorithm.

Algorithm	$\mu(t)$	Parameter 1	Parameter 2
SVSSLMS	$\mu(t) = \beta(1/\left(1 + e^{-\alpha\|t\|}\right) - 0.5)$	$\alpha = 1.0$	$\alpha = 0.83$
		$\beta = 0.1$	$\beta = 0.09$
VSSLMS	$\mu(t) = \beta\left(1 - e^{-\alpha\|e^2(t)\|}\right)$	$\alpha = 0.7$	$\alpha = 0.12$
		$\beta = 0.7$	$\beta = 0.1$
MVSSLMS	$\mu(t) = \alpha\mu(t-1) + \beta p^2(t)$ $p(t) = \gamma p(t-1) + (1-\gamma)e(t)e(t-1)$	$\alpha = 0.97$	$\alpha = 0.76$
		$\beta = 0.03$	$\beta = 0.03$
		$\gamma = 0.98$	$\gamma = 0.91$
DTV-LMS	$\mu(t) = \alpha tan^{-1}\left(-\beta\left\|e(t)e(t-\Delta t)\right\|\right)$	$\alpha = 0.075$	$\alpha = 0.071$
		$\beta = 3$	$\beta = 8$

4.2. Results Analysis

From the simulation curves of Figures 6 and 7, we can see that in terms of convergence speed under the condition of parameter 1, The DTV-LMS algorithm can reach a convergence state approximately during its 25th iteration. The MVSSLMS algorithm can reach a convergence state at about the 36th iteration. The VSSLMS algorithm and SVSLMS algorithm can reach convergence states at about the 70th iteration and the 120th iteration, respectively. Similarly, under the conditions of parameter 2, although their convergence speeds differ slightly, the convergence speed of the DTV-LMS algorithm proposed in this paper is obviously faster than the other three compared algorithms. When the phase change of the signal (i.e., the 500th iteration, the tracking, and the SSM performance of the DTV-LMS and the SVSLMS algorithms) is better than that of the other two algorithms, they can quickly converge again. However, the computational complexity of the DTV-LMS algorithm is lower than that of the SVSLMS algorithm. Thus, the algorithm proposed in this paper is slightly better.

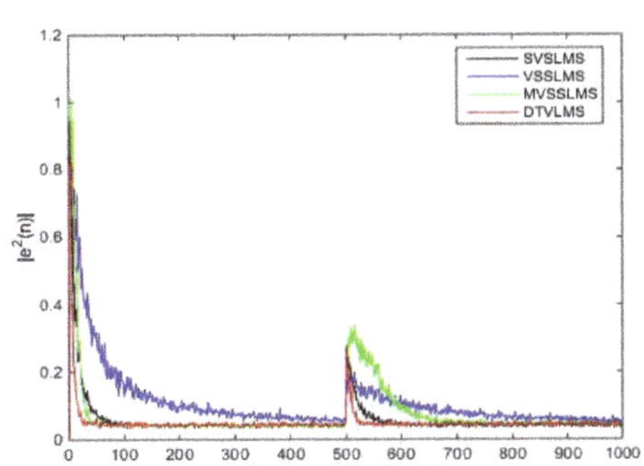

Figure 6. The curves of the algorithm convergence in parameter 1.

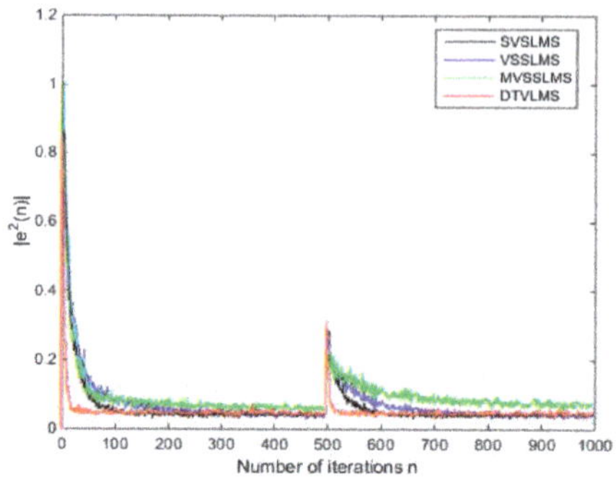

Figure 7. The curves of the algorithm convergence in parameter 2.

It can be seen from the curve of the ICR shown in Figures 8 and 9 that under the condition of parameter 1, the final ICR values of the SVSLMS, MVSSLMS, and VSSLMS algorithms are 74.98 dB, 74.76 dB, and 74.15 dB, respectively; their ICR are close to each other. The final ICR value of the DTV-LMS algorithm proposed in this paper is 78.21 dB, which increased by at least 5 dB compare to the other three algorithms. Under the condition of parameter 2, the ICR values of DTV-LMS, SVSLMS, MVSSLMS, and VSSLMS are 79.96 dB, 77.03 dB, 78.12 dB, and 75.65 dB, respectively, and the ICR value of the proposed algorithm is still the highest. When the phase suddenly changes its signal (i.e., when the 15 ms signal from the graph of the curve changes), we can see that the DTV-LMS algorithm clearly achieves its final stable ICR value faster than the other three compared algorithms—that is to say, the algorithm proposed in this paper can track and adapt to channel changes in a short time, and return to the stable state before channel mutation. This algorithm is verified to have a good adaptive channel tracking capability.

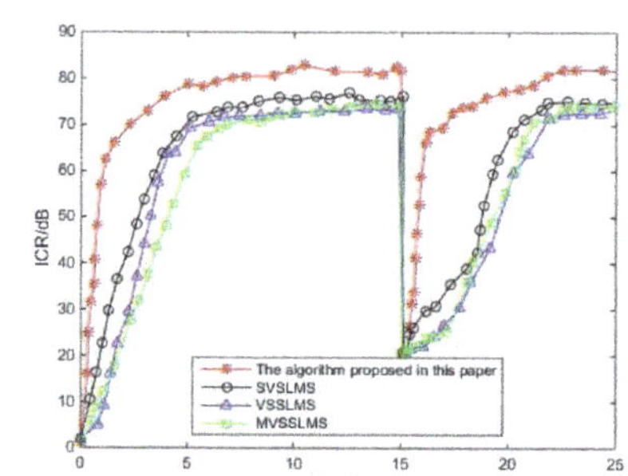

Figure 8. The curves of the algorithm's interference cancellation ration (ICR) in parameter 1.

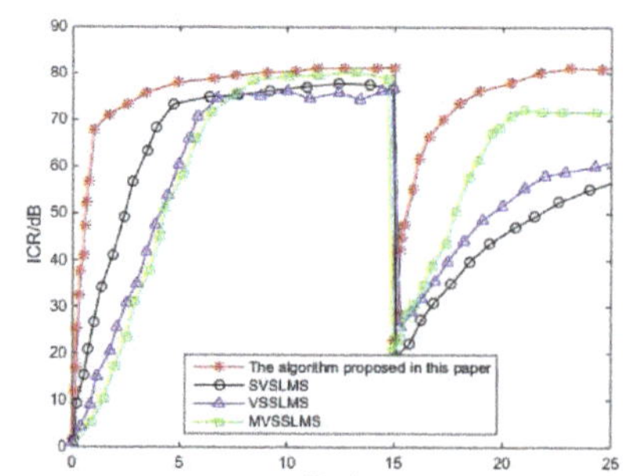

Figure 9. The curves of the algorithm's ICR in parameter 2.

It can be seen from the curve of the ICR shown in Figures 8 and 9 that under the condition of parameter 1, the final ICR values of the SVSLMS, MVSSLMS, and VSSLMS algorithms are 74.98 dB, 74.76 dB, and 74.15 dB, respectively; their ICR are close to each other. The final ICR value of the DTV-LMS algorithm proposed in this paper is 78.21 dB, which increased by at least 5 dB compare to the other three algorithms. Under the condition of parameter 2, the ICR values of DTV-LMS, SVSLMS, MVSSLMS, and VSSLMS are 79.96 dB, 77.03 dB, 78.12 dB, and 75.65 dB, respectively, and the ICR value of the proposed algorithm is still the highest. When the phase suddenly changes its signal (i.e., when the 15 ms signal from the graph of the curve changes), we can see that the DTV-LMS algorithm clearly achieves its final stable ICR value faster than the other three compared algorithms—that is to say, the algorithm proposed in this paper can track and adapt to channel changes in a short time, and return to the stable state before channel mutation. This algorithm is verified to have a good adaptive channel tracking capability.

In order to further explore the influence of intervals of time Δt on SI cancellation in the RF domain of the DTV-LMS algorithm, we changed the step interval, and other parameters remain unchanged under the condition of parameter 1. As shown in Figure 10, we can see that the convergence speed of the DTV-LMS algorithm slows down when the time interval for the feedback control in the system is larger, but it has no effect on the final ICR value of the DTV-LMS algorithm.

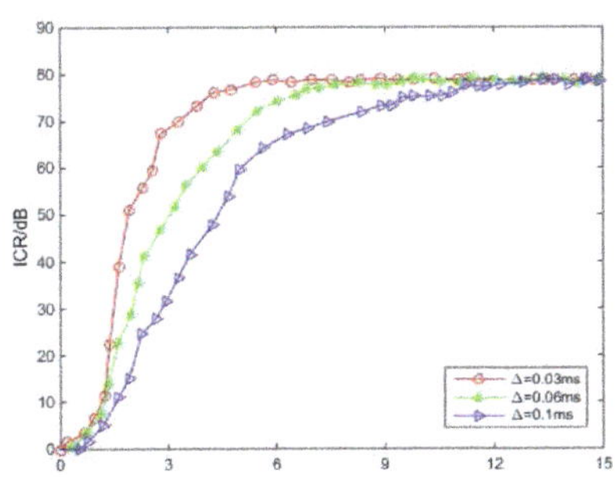

Figure 10. Impact of the time interval on ICR.

5. Conclusions

In this paper, an optimized LMS algorithm based on the attenuation of the step of the convergence parameters is proposed, and we called this novel algorithm the DTV-LMS algorithm. The DTV-LMS algorithm can solve the problem of the strong SI of RF domain in the CCFD system by cooperatively controlling the autocorrelation values between the time factor and the error signal to update the step size of the algorithm and provide a better quality of service for 5G communication networks. The proposed DTV-LMS algorithm is compared with the SVSLMS, MVSSLMS, VSSLMS algorithms and used to verify the superiority of the proposed algorithm by analyzing the convergence speed, the ICR, and the channel tracking capability among these algorithms. The final ICR value of the DTV-LMS algorithm proposed in this paper is 78.21 dB approximately, and the final ICR values of the SVSLMS, MVSSLMS, and VSSLMS algorithms are about 74.98 dB, 74.76 dB, and 74.15 dB. Meanwhile, the convergence speed of the algorithm is also obviously improved compared to the other reference algorithms mentioned in this paper, according to the experimental results. The theoretical analysis and simulation results show that the proposed algorithm achieves a good balance between convergence speed and SSE, which can maintain a small value of SSE at a faster convergence speed, and it also has a good channel tracking capability when dealing with burst noise interference. Therefore, it can be better applied to a 5G wireless communication environment.

However, we experimented with simulations under ideal conditions in this study—for example, the effects of nonlinear interference are not considered, so it has certain limitations. We are committed to researching and solving possible problems in real-life scenarios in future work. In addition, our test cases also have limitations, and we should enrich our test cases in the next work. Since some residual SI signals are still not eliminated after SI cancellation at the RF domain, we will also research how to eliminate residual SI completely in the digital domain in our next work. All in all, we still have many things to improve, and we will work hard to find the perfect way to solve them in our future work.

Author Contributions: Z.-Y.S. conceptualized the study and provided resources. Y.-J.Z. designed the experiments, conducted data analyses, wrote the original draft of the manuscript. Z.-Y.S. and Y.-J.Z. reviewed and improved the manuscript.

Funding: This research received no external funding.

Conflicts of Interest: The authors declare that there is no conflict of interests regarding the publication of this article.

References

1. Hong, S.; Brand, J.; Choi, J.I.; Jain, M.; Mehlman, J. Applications of Self-Interference Cancellation in 5G and Beyond. *IEEE Commun. Mag.* **2014**, *52*, 114–121. [CrossRef]
2. Aly, R.M.; Zaki, A.; Badawi, W.K.; Aly, M.H. Time Coding OTDM MIMO System Based on Singular Value Decomposition for 5G Applications. *Appl. Sci.* **2019**, *9*, 2691. [CrossRef]
3. Panwar, N.; Sharma, S.; Singh, A.K. A Survey on 5G: The Next Generation of Mobile Communication. *Phys. Commun.* **2016**, *18*, 64–84. [CrossRef]
4. Akyildiz, I.F.; Nie, S.; Lin, S.C.; Chandrasekaran, M. 5G Roadmap: 10 Key Enabling Technologies. *Comput. Netw.* **2016**, *106*, 17–48. [CrossRef]
5. Liu, G.; Yu, F.R.; Ji, H.; Leung, V.C.M.; Li, X. In-Band Full-Duplex Relaying for 5G Cellular Networks with Wireless Virtualization. *IEEE Netw.* **2015**, *29*, 54–61. [CrossRef]
6. Li, G.; Sun, C.; Zhang, J.; Jorswieck, E.; Xiao, B.; Hu, A. Physical Layer Key Generation in 5G and Beyond Wireless Communications: Challenges and Opportunities. *Entropy* **2019**, *21*, 497. [CrossRef]
7. Zhao, H.; Wang, J.; Tang, Y. Performance Analysis of RF Self-Interference Cancellation in Broad Band Full Duplex Systems. In Proceedings of the 2016 IEEE International Conference on Communications Workshops (2016 ICC), Kuala Lumpur, Malaysia, 23–27 May 2016.
8. Zhang, Z.; Long, K.; Uasilakos, A.V. Full-duplex wireless communications: Challenges, solutions, and future research diretions. *Proc. IEEE.* **2016**, *104*, 1369–1409. [CrossRef]
9. Atapattu, S.; He, Y.; Dharmawansa, P.; Evans, J. Impact of residual self-interference and direct-link interference on full-duplex relays. In Proceedings of the IEEE International Conference on Industrial & Information Systems, Peradeniya, Sri Lanka, 15–16 December 2017.
10. Debaillie, B.; Broek, D.J.; Lavin, C. Analog/RF Solutions Enabling Compact Full-Duplex Radios. *IEEE J. Sel. Areas Commun.* **2014**, *32*, 1662–1673. [CrossRef]
11. Do, D.; Van Nguyen, M.; Hoang, T.; Lee, B.M. Exploiting Joint Base Station Equipped Multiple Antenna and Full-Duplex D2D Users in Power Domain Division Based Multiple Access Networks. *Sensors.* **2019**, *19*, 2475. [CrossRef]
12. Ro, J.; Park, S.; Song, H. An Interference Cancellation Scheme for High Reliability Based on MIMO Systems. *Appl. Sci.* **2018**, *8*, 466.
13. Xia, W.; Xia, X.; Li, H.; Liu, W.; Hu, J.; He, Z. A noise-constrained distributed adaptive direct position determination algorithm. *Signal Process.* **2017**, *135*, 9–16. [CrossRef]
14. Sakai, M.; Lin, H.; Yamashita, K. Self-Interference Cancellation in Full-Duplex Wireless with IQ Imbalance. *Phys. Commun.* **2016**, *18*, 2–14. [CrossRef]
15. Kim, B.; Yoon, J. Modified LMS Strategies Using Internal Model Control for Active Noise and Vibration Control Systems. *Appl. Sci.* **2018**, *8*, 1007.
16. Vázquez, Á.A.; Maya, X.; Avalos, J.G.; Sánchez, G.; Sánchez, J.C.; Pérez, H.M.; Sánchez, G. A Time-Efficient Method for Determining an Optimal Scaling Factor and the Encoder Resolution in the Multichannel FXECAP-L Algorithm with Evolving Order for Active Noise Control. *Appl. Sci.* **2019**, *9*, 560. [CrossRef]
17. Le-Ngoc, T.; Masmoudi, A. *Full-Duplex Wireless Communications Systems: Self-Interference Cancellation*; Springer: Cham, Switzerland, 2017; Volume 54, pp. 1001–1006.

18. Wang, J.; Zhao, H.; Tang, Y. A RF Adaptive Least Mean Square Algorithm for Self-interference Cancellation in Co-frequency Co-time Full Duplex Systems. In Proceedings of the 2014 IEEE International Conference on Communications (ICC), Sydney, NSW, Australia, 10–14 June 2014.
19. Lee, H.S.; Kim, S.E.; Lee, J.W. A Variable Step-Size Diffusion LMS Algorithm for Distributed Estimation. *IEEE Trans. Signal Process.* **2015**, *63*, 1808–1820. [CrossRef]
20. Zou, Y.Z.; Xin, X.; Zhang, J.Y. Improved Time-Varying Step-Size NLMS Algorithm for RF Self-Interference Cancellation in Full-Duplex. *J. Signal Process.* **2017**, *32*, 787–791.
21. Ao, W.; Xiang, X.Q.; Zhang, Y.P. A New Variable Step Size LMS Adaptive Filtering Algorithm. In Proceedings of the International Conference on Computer Science & Electronics Engineering, Hangzhou, China, 23–25 March 2012.
22. Cheng, W.; Zhang, X.; Zhang, H. Optimal Dynamic Power Control for Full-Duplex Bidirectional-Channel Based Wireless Networks. In Proceedings of the IEEE INFOCOM, Turin, Italy, 14–19 April 2013.
23. Vaňuš, J.; Styskala, V. Application of optimal settings of the LMS adaptive filter for speech signal processing. In Proceedings of the International Multiconference on Computer Science & Information Technology, Wisla, Poland, 18–20 October 2010.
24. Masmoudi, A.; Le-Ngoc, T. Channel Estimation and Self-Interference Cancellation in Full-Duplex Communication Systems. *IEEE Trans.* **2017**, *66*, 321–334.
25. Dinesh, B.; Emily, M.; Sachin, K. Full Duplex Radios. *ACM SIGCOMM Comput. Commun. Rev.* **2013**, *43*, 375–386.
26. Melissa, D. Full-Duplex Wireless: Design, Implementation and Characterization. Ph.D. Thesis, Rice University, Houstin, TX, USA, 2012.
27. Lu, H.T.; Shao, S.H.; Tang, Y.X. Self-Mixed Self-Interference Analog Cancellation in Full-Duplex Communication. *Sci. China Inf. Sci.* **2016**, *59*, 042303. [CrossRef]
28. Angelakis, V.; Chen, L.; Yuan, D. Optimal and Collaborative Rate Selection for Interference Cancellation in Wireless Networks. *IEEE Commun. Lett.* **2011**, *15*, 819–821. [CrossRef]
29. D'Andreagiovanni, F.; Mannino, C.; Sassano, A. GUB Covers and Power-Indexed Formulations for Wireless Network Design. *Manag. Sci.* **2013**, *59*, 142–156. [CrossRef]
30. Al-Zahrani, A.Y. A Game Theoretic Interference Management Scheme in Full Duplex Cellular Systems under Infeasible QoS Requirements. *Future Internet.* **2019**, *11*, 156. [CrossRef]
31. Qiao, G.; Gan, S.; Liu, S.; Ma, L.; Sun, Z. Digital Self-Interference Cancellation for Asynchronous In-Band Full-Duplex Underwater Acoustic Communication. *Sensors* **2018**, *18*, 1700. [CrossRef]

applied sciences

MDPI

Article

Time Coding OTDM MIMO System Based on Singular Value Decomposition for 5G Applications

Rana M. Aly, Amira Zaki, Waleed K. Badawi and Moustafa H. Aly *

College of Engineering and Technology, Arab Academy for Science, Technology and Maritime Transport, Alexandria 1029, Egypt

* Correspondence: mosaly@aast.edu; Tel.: +20-100-663-9473

Received: 14 May 2019; Accepted: 17 June 2019; Published: 2 July 2019

Abstract: For 5G and beyond cellular communication systems, new coding and modulation techniques are suggested to reach the requirements of high data rate and quality of service. In this paper, a new space-time coded orthogonal transform division multiplexing (STC OTDM) technique is proposed for 5G applications. The proposed system is used to enhance the data rate and performance of the orthogonal transform division multiplexing (OTDM) technique. The proposed system is based on using space-time coding (STC) with OTDM to increase the system diversity and consequently the system performance. The OTDM technique is based on transmitting data on orthogonal basis functions obtained from the Singular Value Decomposition (SVD) of the channel impulse response of the desired user. Various modulation techniques like QPSK, 64-QAM, and 256-QAM are investigated using different subcarriers and channel models. The simulation results show that the proposed system achieved a better performance when compared to classical and recent multicarrier techniques. The proposed technique increases the diversity gain resulting in a decrease in the fading effect of the multipath channel and an enhancement in the bit error rate (BER) performance. The proposed technique also provides a secure data transmission to the desired user as his data is sent on the basis functions extracted from his own channel impulse response that cannot be decoded by other users.

Keywords: bit error rate (BER); space-time coding (STC); Singular Value Decomposition (SVD); massive MIMO; orthogonal transform division multiplexing (OTDM); Toeplitz matrix

1. Introduction

Wireless systems suffer from a rapid increasing demand for high data rates and quality of service. Therefore, it is substantial to design a practical physical layer technique to provide high data rates with high system performance [1]. This technique provides reliable advanced communication system like 5G and beyond [2]. This paper focuses on the cellular communication system although the proposed technique can be extended to different advanced wireless communication networks. The development of several modulation multicarrier techniques that are combined with the massive Multiple-Input-Multiple-Output (MIMO) is essential, which is the key to all the advanced cellular wireless systems [3].

Boroujeny and Moradi showed in [4] that, in the past, orthogonal frequency division multiplexing (OFDM) was one of the most popular modulation technique in wired and wireless systems, due to its ability to combat the frequency selectivity of the transmission channels and achieving high rate without intersymbol interference (ISI) [5,6]. Now, new techniques are needed to enhance the bit error rate (BER) and increase diversity.

In [7], Bharti and Rawat showed a combination between the OFDM and Alamouti space time block code (STBC) [8] to increase the performance of BER in multipath fading channels of the wireless communication systems by using multiple transmit and receive antennas. Additionally, it extended the diversity order as it used $4 \times N$ Transmitters-Receivers.

Fernando et al. displayed in [9,10] that the self-heterodyne OFDM (self-het. OFDM) technique gives a better performance than the OFDM. It is used with STC in [9] to enhance the performance of MIMO millimeter wave communication.

In [11] Bariah et al. displayed a non-orthogonal multiple access (NOMA) technique due to its ability to improve the overall spectral efficiency of wireless systems. They show that the maximum possible order of diversity is proportional to the order of the user. The error probability expressions obtained are used to formulate an optimization problem that minimizes the overall bit error rate under power and error rate threshold constrains.

Ghaffari et al. presented in [12] a new non-orthogonal multiple access technique called Sparse Code Multiple Access (SCMA). It offers better performance and higher spectral efficiency than other comparable techniques. However, these improvements come at the expense of complex decoders.

The upcoming 5G systems are required to provide higher performance and quality of services compared to the currently deployed long-term evolution (LTE) systems. These requirements can be achieved by providing a significant increase in the system date rate, power consumption efficiency, spectral efficiency throughput, size of coverage area, number of devices connecting (users), and low latency. To fulfill the requirements of the coming generations, new techniques, like orthogonal time-frequency space (OTFS) modulation, was proposed by Hadani et al. in [13]. Moreover, the BER has been enhanced by the OTDM technique, besides the improvement of the security which was illustrated by Hamamreh and Arslan in [14].

In this paper, an STC OTDM technique is introduced to meet some of the challenges of the upcoming 5G system by increasing the data rate and the throughput of the system. Moreover, the STC OTDM increases the security level of transmission for the wireless system. A STC OTDM technique is proposed to reduce the BER using the STC, which also increases the diversity order. This is carried out by using Singular Value Decomposition (SVD) of the estimated channel, where every two symbols are modulated using orthonormal basis functions of SVD. Then, the modulated symbols are encoded using an Alamouti encoder for a 2×2 MIMO system and are transmitted over two consecutive symbol durations. This technique improves diversity and reduces the BER as a result of increasing the signal to noise ratio (SNR), in addition to the advantages of the OTDM mentioned in [15]. This technique is promising for the future high performance and secure 5G systems when compared to the conventional OFDM system. The proposed technique increases the diversity gain resulting in a decrease in the fading effect of the multipath channel, thus enhancing the bit error rate (BER) performance and throughput of the system. The proposed technique also provides a secure data transmission to the desired user as his data is sent on the basis functions extracted from his own channel impulse response that cannot be decoded by other users, thus the desired user can only receive his data correctly.

The remainder of the paper is organized as follows. Section 2 is devoted to a detailed description of the proposed system. System evaluation and simulation results are presented in Section 3. This is followed by the main conclusions showing the merits of the proposed system in Section 4.

2. System Model and Analysis

2.1. STC OTDM Transmitter

The proposed system transmitter is illustrated in Figure 1, which shows the sequence of the system transmitter architecture. At the transmitter side, the number of symbols per frame is N. The data symbols S are transmitted in a number of frames, each of length N symbols and each frame is divided into two sequences; even sequence, S_1 , and odd sequence S_2. Both S_1 and $S_2 \in \mathrm{C}\,[\frac{\mathrm{N}}{2} \times 1]$.

Thus,

$$S_1 = \left[\mathrm{s}(2(m - 1))\right]^T \text{ and} \tag{1}$$

$$S_2 = \left[\mathrm{s}(1 + 2(m - 1))\right]^T \tag{2}$$

where $m = 1, 2, \ldots, \frac{\mathrm{N}}{2}$

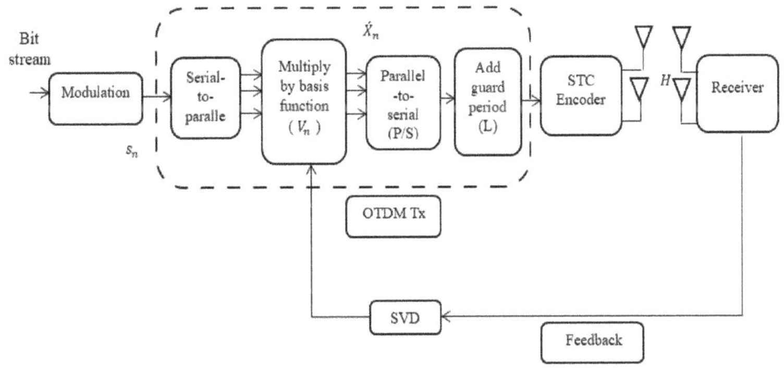

Figure 1. Space-time coded orthogonal transform division multiplexing (STC OTDM) transmitter technique.

After modulation with M subcarriers and passing through a serial-to-parallel (S/P) converter, each of the modulated complex value symbols is carried by a channel-based orthogonal transform basis. The mapping method is basically done by a direct multiplication operation between each data symbol and the orthogonal basis vector. These orthogonal basis vectors are extracted from the SVD decomposition of the Toeplitz matrix of the channel impulse response Hb_{ij} between the ith receiving antenna and the jth transmitting antenna, where $i = 1, 2$ and $j = 1, 2$. This can be represented by

$$Hb_{ij} = \begin{bmatrix} h_{ij}[0] & 0 & \cdots & \cdots & 0 \\ \vdots & h_{ij}[0] & 0 & \ddots & \vdots \\ \vdots & \vdots & h_{ij}[0] & \ddots & \vdots \\ h_{ij}[L-1] & \vdots & \vdots & \ddots & 0 \\ 0 & h_{ij}[L-1] & \vdots & \ddots & h_{ij}[0] \\ \vdots & 0 & h_{ij}[L-1] & \ddots & \vdots \\ \vdots & \vdots & 0 & \ddots & \vdots \\ 0 & 0 & \cdots & \cdots & h_{ij}[L-1] \end{bmatrix} \tag{3}$$

then,

$$Hb_j = \sum\nolimits_{i=1}^{2} Hb_{ij} \tag{4}$$

where $j = 1, 2$ and $Hb_j \in \mathrm{C}^{[(\frac{N}{2}+L-1)\times\frac{N}{2}]}$.

The receiver and the transmitter can extract the basic functions by decomposing the Toeplitz matrices using the SVD. Therefore, the Toeplitz matrices can be expressed in terms of the three new matrices as

$$Hb_j = U_n E_n V_n^H \tag{5}$$

where $n = 1, 2$ and U_n and V_n^H are orthogonal matrices that are $\in \mathrm{C}^{[(\frac{N}{2}+L)\times\frac{N}{2}]}$ and $\in \mathrm{C}^{[\frac{N}{2}\times\frac{N}{2}]}$, respectively, and E_n is a diagonal matrix with real entires $\in \mathrm{C}^{[\frac{N}{2}\times\frac{N}{2}]}$.

The transformation matrix V_n is given by

$$V_n = v_n(0)\ v_n(1) \ldots v_n(\frac{N}{2}-1) \tag{6}$$

where $v(i)$ is the orthogonal basis vector of length $(\frac{N}{2}-1)$.

The transmitter maps each symbol to its corresponding basis functions by simple multiplication. This leads to two blocks of samples X_1 and X_2 referred to two OTDM symbols. This process can mathematically be expressed as

$$X_n = \sum_{i=0}^{N-1} s_n[i] v_n[i] = V_n S_n \ldots\ldots\ldots \in \mathbb{C}^{[\frac{N}{2}\times 1]} \quad (7)$$

where $n = 1, 2$ and V_n is the Hermitian of V_n^H.

The transmitted data blocks X_n of length $\frac{N}{2}$ can be represented by simple multiplication of Toeplitz matrices $Hb_j \in \mathbb{C}^{[(\frac{N}{2}+L-1)\times\frac{N}{2}]}$ with the transmitted data blocks. The transmitted data blocks pass through a parallel-to-serial (P/S) converter before adding the guard period (L).

Then, Alamouti space-time coding is applied to X_1 and X_2 in the frequency domain to generate the following code word matrix:

$$X = \overset{\text{Space}\ \rightarrow}{\begin{bmatrix} X_1 & X_2 \\ -X_2^* & X_1^* \end{bmatrix}} \downarrow \text{Time} \quad (8)$$

So, the received OTDM signals R_{ij} can be represented as

$$R_{11} = (H_{11}\, X_1) + (H_{12}\, X_2) + z_1, \quad (9)$$

$$R_{21} = (H_{21}\, X_1) + (H_{22}\, X_2) + z_2, \quad (10)$$

$$R_{12} = (-H_{11}\, X_2^*) + (H_{12}\, X_1^*) + z_3, \text{ and} \quad (11)$$

$$R_{22} = (-H_{21}\, X_2^*) + (H_{22}\, X_1^*) + z_4, \quad (12)$$

where i is the number of receive antennas and j is the number of transmit antennas, and z_1, z_2, z_3, and z_4 are the additive white Gaussian noise (AWGN) with zero mean and variance σ^2. H_{ij} are the channel response between the ith receiving antennas and jth transmitting antennas. Then, these signals are exposed to the receiver sequence.

2.2. STC OTDM Receiver

At the receiver side, the received signals pass through STC decoder as shown in Figure 2.

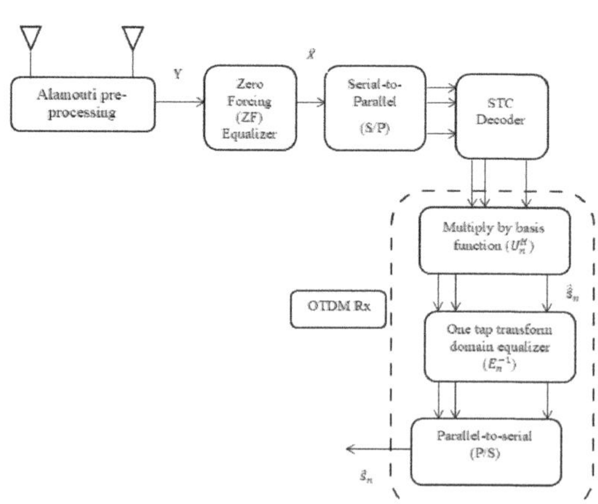

Figure 2. STC OTDM receiver technique.

The received symbol matrix in the MIMO system can be represented by [9]

$$\underbrace{\begin{bmatrix} Y_{11} \\ Y_{21} \\ Y_{12}^{*} \\ Y_{22}^{*} \end{bmatrix}}_{Y} = \underbrace{\begin{bmatrix} \alpha^{*}H_{11} & \alpha^{*}H_{12} \\ \beta^{*}H_{21} & \beta^{*}H_{22} \\ \alpha H_{12}^{*} & -\alpha H_{11}^{*} \\ \beta H_{22}^{*} & -\beta H_{21}^{*} \end{bmatrix}}_{Heq} \begin{bmatrix} X_1 \\ X_1 \end{bmatrix} + \underbrace{\begin{bmatrix} z_1 \\ z_2 \\ z_3 \\ z_3 \end{bmatrix}}_{Z} \tag{13}$$

where $\alpha = H_{11}+H_{12}$ and $\beta = H_{21}+H_{22}$.

The ZF equalizer is used to recover the data symbols by considering $\acute{H}_{eq}\, Y$, where $\acute{H}_{eq}$ is the pseudo-inverse of H_{eq} which can be represented as

$$\acute{H}_{eq} = \frac{1}{\Lambda} H_{eq}^{H} \tag{14}$$

where Λ is the diversity gain represented as [7]

$$\Lambda = |\alpha|^2(\Sigma_{m=1}^{2}|H_{1m}|^2) \ + |\beta|^2(\Sigma_{m=1}^{2}|H_{2m}|^2) \tag{15}$$

So, the estimated data symbols can be represented by

$$\hat{X} = \begin{bmatrix} \hat{X}_1 \\ \hat{X}_2 \end{bmatrix} = \begin{bmatrix} \alpha H_{11}^{*} & \beta H_{21}^{*} & \alpha^{*}H_{12} & \beta^{*}H_{22} \\ \alpha H_{12}^{*} & \beta H_{22}^{*} & -\alpha^{*}H_{11} & -\beta^{*}H_{21} \end{bmatrix} \begin{bmatrix} Y_{11} \\ Y_{21} \\ Y_{12}^{*} \\ Y_{22}^{*} \end{bmatrix} \text{and} \tag{16}$$

$$\hat{X}_n = X_n\, \Lambda + Z. \tag{17}$$

After that, $\hat{X}$ is passes through the S/P converter followed by the STC decoder. The receiver applies the SVD on the Toeplitz matrix of its channel response. The receiver uses the Hermitian U_n as inverse basis functions to extract the data symbols from the received symbols without interference where $n = 1, 2$. This can be enforced as

$$\hat{\hat{s}}_n = \Sigma_{i=0}^{N+L-1}\, \hat{x}_n[i] u_n^{*}[i] \ldots\ldots \in C^{[\frac{N}{2}]+L\times 1}. \tag{18}$$

Then, a one tap ZF equalization process for $\hat{\hat{s}}_n$ is performed by the receiver using E_n to get the equalized data symbol block $\hat{s}_n$. This process can mathematically be expressed as

$$\begin{aligned} \hat{s}_n = E_n^{-1}\hat{\hat{s}}_n &= E_n^{-1}\, U_n^{H}\hat{X}_n \\ &= E_n^{-1}\, U_n^{H}\, (X_n\, \Lambda\, +\, Z) \\ &= E_n^{-1}\, U_n^{H}\, (Hb_j\, V_n\, S_n\, \Lambda\, + Z) \\ &= E_n^{-1}\, U_n^{H}\, (U_n\, E_n\, V_n^{H}\, V_n\, S_n\, \Lambda\, + Z) \\ &= S_n\, \Lambda + E_n^{-1}\, U_n^{H}\, Z. \end{aligned} \tag{19}$$

This is clearly shown in the system receiver architecture in Figure 2.

2.3. BER Analysis

Finally, we need to calculate the power of the signal and the variance of AWGN with zero mean to be able to calculate the BER of the proposed system. So, the power of the signals can be derived as

$$P_S = \sum\nolimits_{n=1}^{2} E[|S_n\Lambda|^2] \tag{20}$$

and the noise variance yields to

$$\sigma^2 = \sum\nolimits_{n=1}^{2} E[|E_n^{-1}U_n^{H}Z|^2]. \tag{21}$$

Therefore, the total BER can be represented as [10]

$$\text{BER} = 2\,Q\left(\sqrt{\frac{P_s}{\sigma^2}}\right)\left(1 - \frac{1}{2}\,Q\left(\sqrt{\frac{P_s}{\sigma^2}}\right)\right) \tag{22}$$

where $Q(.)$ is the standard Q-function.

Finally, from Equation (22), we proved that the proposed system enhances the performance of the BER.

3. Simulation Results

In this section, the proposed STC OTDM technique is evaluated and compared to conventional OFDM [5] and recent multicarrier techniques like STC OFDM [7] and OTDM [14]. Set of experiments are conducted in terms of modulation techniques, number of subcarriers and channel taps. These experiments are discussed in the following subsections. The simulation results are displayed and discussed. An STC OTDM system is considered with N = 16, 64, and 128. The modulation schemes QPSK, 64-QAM, and 256-QAM are investigated with a number of channel taps equal to L = 9, 12 and 15. For the sake of comparison, a standard OFDM system, STC OFDM and standard OTDM are also considered with N = 16, 64, 128 sub-carriers. In this section, also, the effect of the modulation scheme on all the techniques is studied. The study includes the effect of increasing the number of subcarriers and the effect of increasing the number of channel taps. All the parameters used in simulations are collected in Table 1.

Table 1. Simulation parameters.

Type of Channel	Rayleigh Fading Channel
Number of channel taps, L	6, 12 and 15
Type of modulation	QPSK, 64-QAM and 256-QAM
Number of subcarrier, N	16, 64 and 128

3.1. Effect of Modulation Scheme

Figure 3 shows the BER of the proposed system, compared to the OFDM, STC OFDM and OTDM using QPSK as a modulation technique with a number of subcarriers N = 16. It is clear that the BER of the proposed system outperforms the other techniques. The difference in SNR between the proposed system and OTDM technique equals 6 dB at BER = 10^{-3}. If we compare the STC OTDM with STC OFDM, we find that at BER = 10^{-3}, the difference in SNR between them approximately equals 0.7 dB and the SNR between the proposed system and the OFDM technique equals 16.5 dB. Furthermore, the eavesdropper cannot decode the data correctly because the eavesdropper and the transmitter are not in the same channel or both have different basis functions as clearly shown in Figure 3.

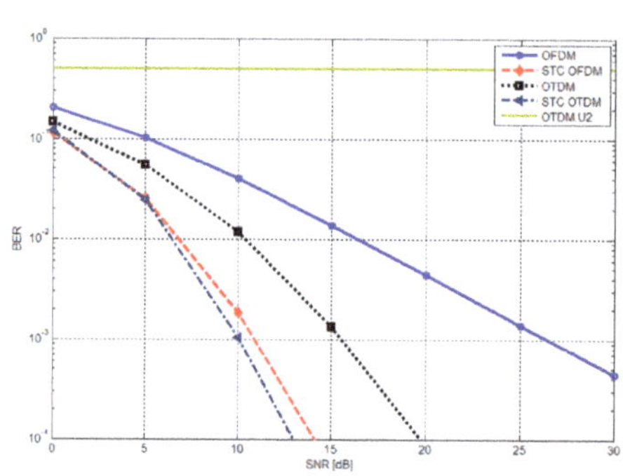

Figure 3. Bit error rate (BER) of the proposed technique (STC OTDM) in comparison with other multicarrier techniques at N = 16 and L = 9 using QPSK.

In Figure 4, 64-QAM is used with the same value of N. The effect of increasing the M-ary appears as the BER increases with the modulation order. This is because the constellation becomes closer to each other [16]. This appears at BER = 10^{-3}, where the difference in SNR between the proposed system and OTDM system is approximately equal to 5 dB, where the SNR of the proposed system is 22.5 dB and 10 dB at 64-QAM and QPSK, respectively, at BER = 10^{-3}. Thus, in 256-QAM, the performance of the proposed system degrades to reach 28 dB at the same BER as shown in Figure 5.

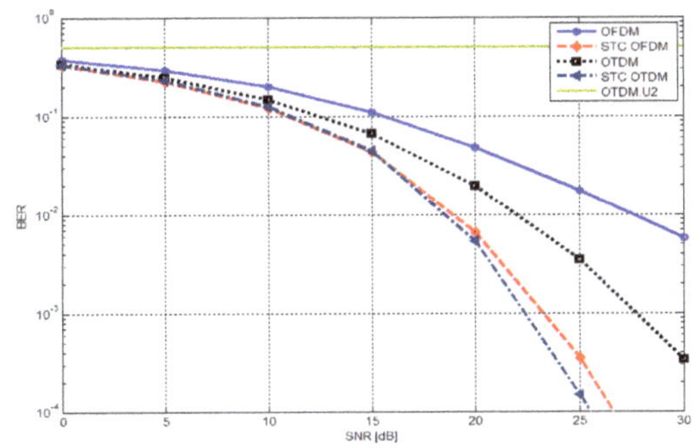

Figure 4. BER of the proposed technique (STC OTDM) in comparison with other multicarrier techniques at N = 16 and L = 9 using 64-QAM.

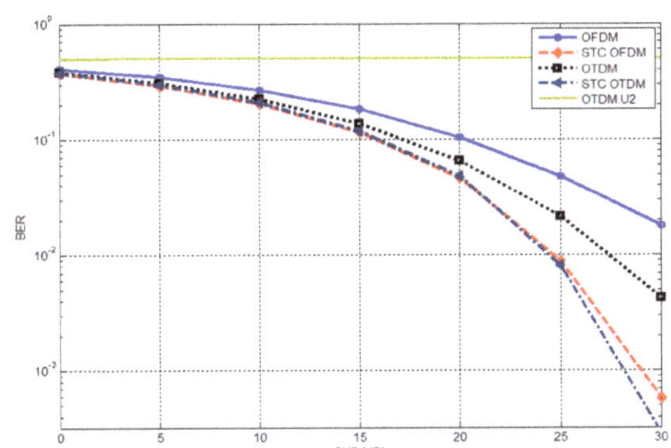

Figure 5. BER of the proposed technique (STC OTDM) in comparison with other multicarrier techniques at N = 16 and L = 9 using 256-QAM.

3.2. *Effect of Subcarriers*

Figure 6 shows the effect of increasing the number of subcarriers at all modulation techniques. The proposed system gives SNR = 10 dB at BER = 10^{-3} at QPSK modulation technique when N = 16. The performance of STC OTDM when N = 64 and 128 is nearly equal. So, this is great evidence to prove that STC OTDM gives high performance with less number of subcarriers at all the modulation techniques as it clear in the figure. The performance decreases at 64-QAM and 256-QAM to SNR = 22.5 dB and 28 dB, respectively, at BER = 10^{-3}. Thus, the required performance can be obtained with a lower number of subcarriers and then lower complexity. This is in addition to the fact that the proposed technique is based on linear equalization and an SVD algorithm which is widely used in different present wireless communication systems like in channels estimation and beamforming algorithms [15]. The complexity of the SVD computation is of order O (mn^2), where m and n are the matrix dimensions that will apply the SVD on it [17]. Thus, the proposed system is applicable and can practically be deployed.

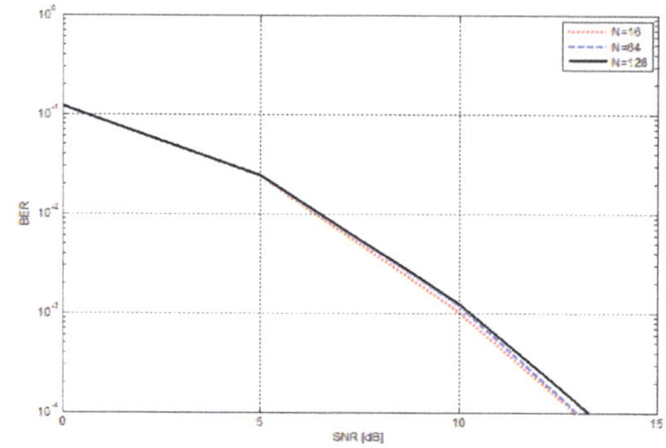

(**a**)

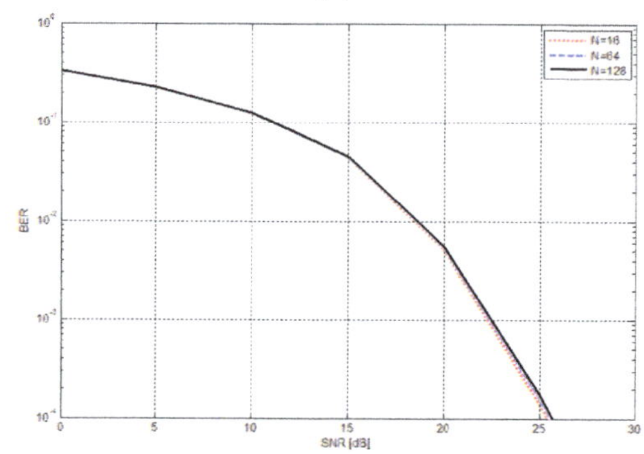

(**b**)

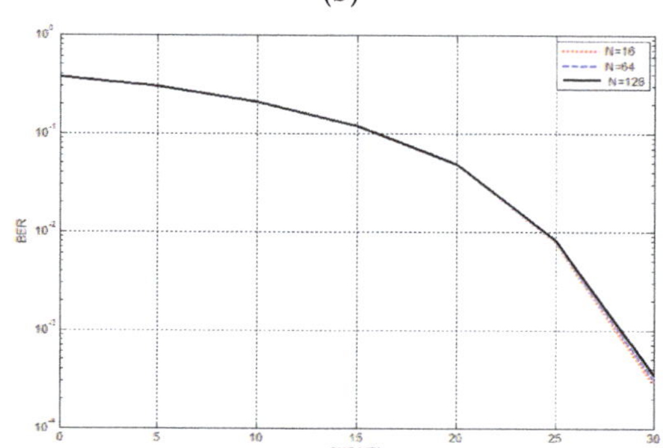

(**c**)

Figure 6. BER of STC OTDM at different subcarriers and L = 9 for: (**a**) QPSK; (**b**) 64-QAM; and (**c**) 256-QAM.

3.3. Effect of Channel Taps

The effect of increasing the taps of the channel on the proposed system appears in Figure 7. Increasing the channel taps from 9 to 12 leads to a little bit increase in the performance of the proposed system at SNR greater than 10 dB. This is in a fair agreement with the work of Hamamreh and Arslan in [14] as the OTDM is designed especially to work better with the frequency selective channels, which is the case in most broadband systems.

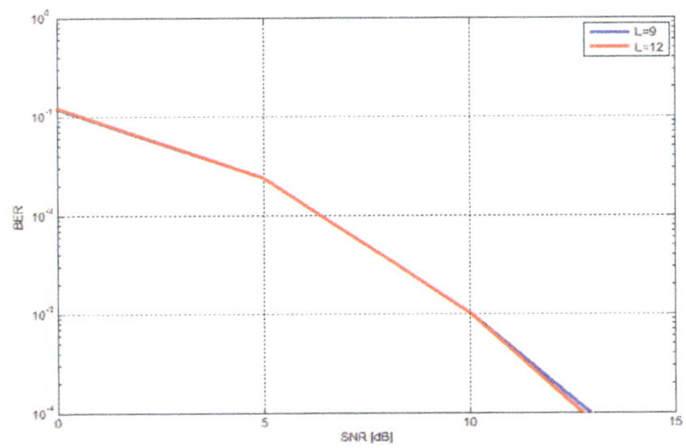

Figure 7. BER of STC OTDM at different channel taps L using QPSK.

The procedure is repeated for all modulation schemes at all values of N with channel taps of 9, 12, and 15. The obtained results at BER = 10^{-3} are summarized in Table 2.

Table 2. The SNR (dB) simulation results at BER = 10^{-3} for different modulation schemes, different subcarriers, and channels models.

	SNR (dB)								
L		**9**			**12**			**15**	
N	**16**	**64**	**128**	**16**	**64**	**128**	**16**	**64**	**128**
				QPSK					
OFDM	26.5	27	27	26.8	27	27	26.4	26.6	26.8
STC OFDM	10.7	11	11	10.7	10.9	10.9	10.9	11	11
OTDM	16	22	24	14	21	23.2	13.4	20	22.4
STC OTDM	10	10.1	10.1	10	10.01	10.01	10	10.01	10.01
				64-QAM					
OFDM	34	35	35	35	35.2	35.1	35.2	35.3	35.4
STC OFDM	23	23	23	23.2	23.15	23.1	23.2	23	23
OTDM	27.5	31	31.4	26.4	32.7	32.7	25.4	31.5	33
STC OTDM	22.3	22.4	22.4	22.2	22.4	22.4	22	22.3	22.4
				256-QAM					
OFDM	37.7	38.5	39	38.5	39	39.3	39.8	39.9	38.6
STC OFDM	28.3	28.5	28.7	28.2	28.4	28.29	29	28.89	28.96
OTDM	32.5	37	37.4	31	36.8	37	32	36	37
STC OTDM	28.2	28.3	28.32	28	28.2	28.28	28.1	28.12	28.2

4. Conclusions

An STC OTDM architecture is proposed for enhancing the performance of the BER and increasing the diversity order. We considered a system with N = 16, 64, and 128. The modulation schemes QPSK, 64-QAM, and 256-QAM were investigated with a number of channel taps equaling 9, 12, and 15. For the sake of comparison, we also considered a standard OFDM system, STC OFDM, and standard OTDM with the same active sub-carriers, a cyclic prefix (CP) of length L for the OFDM and STC OFDM techniques, and simple zero-forcing equalization. The proposed system shows its superiority over the OFDM, STC OFDM, and OTDM in enhancing the performance of BER. The simulation results show that the SNR of the proposed system using QPSK modulation technique is equal to 10 dB at BER =10^{-3} when 16 subcarriers are used in a 9-tap Rayleigh fading channel. Also, the results show that the BER performance of the proposed technique is nearly constant for a different number of subcarriers. For example, the proposed technique using 64-QAM achieved the same BER performance, 10^{-3} at

SNR = 22.5 dB, for the different subcarriers 64 and 128. Thus, the required system performance can be obtained with less number of subcarriers and thus less system complexity for the different modulation techniques. When using 64-QAM and 256-QAM the performance of BER is nearly the same and the proposed system achieves high performance at high data rates. Simulations also showed that the STC OTDM improves the security of the wireless communication system, where only the desired user can decode the transmitted data.

The proposed technique is basically composed of a linear equalizer and SVD algorithm which is widely used in different present wireless communication systems. The complexity of the SVD computation is of order $O(mn^2)$ where m and n are the matrix dimensions that will apply the SVD on it. Thus, the proposed system is applicable and can be practically deployed.

Author Contributions: Conceptualization, A.Z. and W.K.B.; Methodology, R.M.A., A.Z. and W.K.B.; Software, R.M.A., A.Z. and W.K.B.; Validation, A.Z., W.K.B. and M.H.A.; Formal Analysis, A.Z. and W.K.B.; Investigation, R.M.A.; Resources, R.M.A., A.Z. and W.K.B.; Data Curation, R.M.A.; Writing—Original Draft Preparation, R.M.A.; Writing—Review & Editing, A.Z., W.K.B. and M.H.A.; Visualization, R.M.A., A.Z. and W.K.B.; Supervision, A.Z., W.K.B. and M.H.A.; Project Administration, R.M.A., A.Z. and W.K.B.

Funding: This research has no external funding.

Conflicts of Interest: The authors declare no conflict of interest.

References

1. Vij, S.; Jain, A. 5G: Evolution of a secure mobile technology. In Proceedings of the 2016 3rd International Conference on Computing for Sustainable Global Development (INDIACom), New Delhi, India, 16–18 March 2016; pp. 2192–2196.
2. Yang, N.; Wang, L.; Geraci, G.; Elkashlan, M.; Yuan, J.; Renzo, M.D. Safeguarding 5G wireless communication networks using physical layer security. *IEEE Commun. Mag.* **2015**, *53*, 20–27. [CrossRef]
3. Gerzaguet, R.; Medjahdi, Y.; Demmer, D.; Zayani, R.; Dor, J.B.; Shaiek, H.; Roviras, D. Comparison of promising candidate waveforms for 5G: WLOA-OFDM versus BF-OFDM. In Proceedings of the 2017 International Symposium on Wireless Communication Systems (ISWCS), Bologna, Italy, 28–31 August 2017; pp. 355–359.
4. Boroujeny, B.F.; Moradi, H. OFDM inspired waveforms for 5G. *IEEE Commun. Surv. Tutor.* **2016**, *18*, 2474–2492. [CrossRef]
5. Goldsmith, A. *Wireless Communications*; Cambridge University Press: Cambridge, UK, 2005.
6. Li, Y.; Stüber, G.L. *Orthogonal Frequency Division Multiplexing for Wireless Communications*; Springer: New York, NY, USA, 2006.
7. Bharti, P.K.; Rawat, P. Alamouti-STBC based performance estimation of multi Tx and Rx antenna over MIMO-OFDM. In Proceedings of the 2018 2nd International Conference on Trends in Electronics and Informatics (ICOEI), Tirunelveli, India, 11–12 May 2018; pp. 1277–1281.
8. Alamouti, S. A simple transmit diversity technique for wireless communications. *IEEE J. Sel. Areas Commun.* **1998**, *16*, 1451–1458. [CrossRef]
9. Fernando, N.; Hong, Y.; Viterbo, E. MIMO self-heterodyne OFDM. *IEEE Trans. Veh. Technol.* **2015**, *65*, 1271–1280. [CrossRef]
10. Fernando, N.; Hong, Y.; Viterbo, E. Self-heterodyne OFDM transmission for frequency selective channels. *IEEE Trans. Commun.* **2013**, *61*, 1936–1946. [CrossRef]
11. Bariah, L.; Al-Dweik, A.; Muhaidat, S. On the performance of non-orthogonal multiple access systems with imperfect successive interference cancellation. In Proceedings of the 2018 IEEE International Conference on Communications Workshops (ICC Workshops), Kansas City, MO, USA, 20–24 May 2018; pp. 1–6.
12. Ghaffari, A.; Leonardon, M.; Cassagne, A.; Leroux, C.; Savaria, Y. Toward high-performance implementation of 5G SCMA algorithms. *IEEE J. Mag.* **2019**, *7*, 10402–10414. [CrossRef]
13. Hadani, R.; Rakib, S.; Molisch, A.F.; Ibars, C.; Monk, A.; Tsatsanis, M.; Delfeld, J.; Goldsmith, A.; Calderbank, R. Orthogonal time frequency space (OTFS) modulation for millimeter-wave communications systems. In Proceedings of the 2017 IEEE MTT-S International Microwave Symposium (IMS), Honololu, HI, USA, 4–9 June 2017; pp. 681–683.

14. Hamamreh, J.M.; Arslan, H. Secure orthogonal transform division multiplexing (OTDM) waveform for 5G and beyond. *IEEE Commun. Lett.* **2017**, *21*, 1191–1194. [CrossRef]
15. Alluhaibi, O.; Nair, M.; Hazzaa, A.; Mihbarey, A.; Wang, J. 3D beamforming for 5G millimeter wave systems using singular value decomposition and particle swarm optimization approaches. In Proceedings of the 2018 International Conference on Information and Communication Technology Convergence (ICTC), Jeju Island, Korea, 17–19 October 2018; pp. 15–19.
16. Sadinov, S.M. Simulation study of M-ARY QAM modulation techniques using Matlab/Simulink. In Proceedings of the 2017 40th International Convention on Information and Communication Technology, Electronics and Microelectronics (MIPRO), Opatija, Croatia, 22–26 May 2017; pp. 547–554.
17. Golub, G.H.; Van Loan, C.F. *Matrix Computations*; The Johns Hopkins University Press: Baltimore, MD, USA, 2013.

applied sciences

MDPI

Letter

Intrinsic Interference Use for FBMC-IOTA Systems

Lei Wen [1,2], Razieh Razavi [2] and Jing Lei [1,*]

[1] College of Electronics Science, National University of Defense Technology, Changsha 410000, Hunan, China
[2] 5GIC, University of Surrey, Guildford GU2 7XH, UK
* Correspondence: leijing@nudt.edu.cn

Received: 4 June 2019; Accepted: 2 August 2019; Published: 7 August 2019

Abstract: In this paper, the intrinsic interference of filter bank-based multicarrier systems (FBMC) systems with isotropic orthogonal transfer algorithm (IOTA) pulse-shaping is analyzed and used. Such intrinsic interference is treated as a parity symbol, and an iterative soft-in-soft-out (SISO) detector, which is based on message-passing algorithm (MPA), is proposed to exploit the useful information of the intrinsic interference. The performance of the intrinsic interference user (IIU) is investigated.

Keywords: filter bank multicarrier; isotropic orthogonal transfer algorithm; intrinsic interference

1. Introduction

To mitigate the effect of multipath fading, a cyclic prefix (CP) must be inserted to orthogonal frequency division multiplexing (OFDM) systems, resulting in a reduced spectral efficiency and an increased power consumption. By partitioning the wideband channel into a large number of parallel narrow band sub-channels, the task of high-data-rate transmission over a frequency selective channel has been transformed into number of parallel low-data-rate transmissions which need equalization techniques. Some improvements for OFDM have been reported to combat frequency dispersion sensitivity by exploiting ICI self-cancellation methods [1,2] or to explore space and time diversity in dispersive channels through fractional sampling [3,4]. And numerous research efforts have been spent on PAPR reduction techniques. The usage of CP, however, is retained to combat ISI in such techniques which aim to enhance OFDM [5,6]. In [7–10], with reference to the problem of joint equalization and narrowband interference (NBI) suppression in OFDM systems, synthesis and analysis of both unconstrained and constrained optimum equalizers are carried out, based on the maximum signal-to-noise-plus-interference (SINR) criterion. An alternative approach is to use pulse-shaping filters. With offset modulations, for example, OFDM/offset quadrature amplitude modulation (OQAM), the orthogonality can be maintained with proper pulse-shaping [11,12]. OFDM with offset QAM (OFDM/OQAM) which transmits real symbols with double lattice density has shown some advantages over OFDM, but faces difficulties of channel estimation and equalization [13]. It is shown that OFDM/OQAM with isotropic orthogonal transfer algorithm (IOTA) pulse-shaping, which is called FBMC-IOTA, has optimal localization property [14]. FBMC-IOTA has been studied in [15] and was shown to outperform conventional OFDM over different types of wireless propagation channels.

Despite all the advantages, the currently developed FBMC-IOTA system is not yet fully used to its best achievable performance. In the FBMC-IOTA receiver, there is intrinsic interference which contains rich information, and its effect on the spectral efficiency has been analyzed from information theoretical perspective in [16]. However, to the best of our knowledge, there does not exist any method to exploit the extra information offered by the intrinsic interference in the symbol detection process. This motivates us to develop an algorithm to use the intrinsic interference to materialize the performance gain predicted by the information theoretical study. In this contribution we show that the intrinsic interference can be determined by multiplying the neighboring symbols with a weighting matrix, the elements of which can be calculated using the ambiguity function of the pulse shape filter. This is essentially the principle of block coding and the intrinsic interference serves as non-binary parity symbol. Using this property and message-passing algorithm (MPA) [17], we propose a way which is termed as intrinsic interference user (IIU) to improve the system performance. Although FBMC-IOTA can mitigate ISI, if the receiver has estimated the channel well enough, then ISI can be further cleaned up using MPA on the demodulated reception. This translates to roughly 1 dB bit error rate (BER) improvement across SNR's for a particular channel model.

The rest of the paper is outlined as follows. Section 1 presents the maximum a posteriori (MAP) detection on weight matrix of FBMC-IOTA. In Section 2, MPA on the intrinsic interference of FBMC-IOTA is presented. Section 3 presents simulation results. Conclusions are drawn in Section 4.

2. MAP Detection on Weight Matrix

Consider a FBMC-IOTA system with N subcarriers and L symbols, the transmitted signal can be written as

$$s(t) = \sum_{n=0}^{L-1} \sum_{m=0}^{N-1} a_{m,n} g_{m,n}(t) \tag{1}$$

where $a_{m,n}$ and $g_{m,n}(t)$ represent the symbol modulated by the mth subcarrier at the nth symbol time and the synthesis basis obtained by the IOTA function $g(t)$, respectively.

After passing by a doubly dispersive channel, the demodulator output can be expressed as [15]

$$\begin{aligned} \hat{a}_{k,l} = {} & n_{k,l} + a_{k,l} H_{k,l} \\ & + \underbrace{\sum_{p,q \neq 0} a_{k+p,l+q} j^{p+q+p(q+2l)} A_g^*(q\tau_0, p\nu_0)\, H_{k,l}}_{I_{k,l}} \end{aligned} \tag{2}$$

where $H_{k,l}$ is the channel coefficient at the lth symbol and the kth subcarrier frequency, $A_g(\tau, \nu)$ is the auto-ambiguity function of $g(t)$, $n_{k,l}$ is the channel noise, $I_{k,l}$ is the intrinsic interference and can be written in the form of $I_{k,l} = j a_{k,l}^{(i)}$. Such intrinsic interference contains rich inherent information that has never been used. In fact, the intrinsic interference can be determined by multiplying the neighboring symbols with a weighting matrix, the elements of which can be calculated using the ambiguity function of the employed pulse shape filter [16]. The matrix that presents the weights corresponding to each neighboring symbol is shown below

$$W_{l_{\text{even}}} = \begin{bmatrix} 0 & 0 & -\eta & 0 & \eta & 0 & 0 \\ -\lambda & -\zeta & -\gamma & -\delta & -\gamma & -\zeta & -\lambda \\ -\theta & 0 & -\beta & 1 & \beta & 0 & \theta \\ -\lambda & \zeta & -\gamma & \delta & -\gamma & \zeta & -\lambda \\ 0 & 0 & -\eta & 0 & \eta & 0 & 0 \end{bmatrix} \tag{3}$$

$$W_{l_{\text{odd}}} = \begin{bmatrix} 0 & 0 & -\eta & 0 & \eta & 0 & 0 \\ \lambda & \zeta & \gamma & \delta & \gamma & \zeta & \lambda \\ -\theta & 0 & -\beta & 1 & \beta & 0 & \theta \\ \lambda & -\zeta & \gamma & -\delta & \gamma & -\zeta & \lambda \\ 0 & 0 & -\eta & 0 & \eta & 0 & 0 \end{bmatrix} \tag{4}$$

where l_{even} and l_{odd} respectively corresponds to the weight for real and imaginary symbols, and $\delta = 0.2486$, $\beta = 0.5756$, $\gamma = 0.1898$, $\eta = 0.0021$, $\theta = 0.0956$, $\lambda = 0.0473$, $\zeta = 0.0991$.

For each transmitted information symbol $a_{k,l}$, its associated intrinsic interference $a_{k,l}^{(i)}$ is determined by neighboring symbols and can be calculated as in (2) using the above weighting matrices. We find that such principle is essentially the same as block coding where each symbol $a_{k,l}$ is associated with some non-binary parity symbols $P_{k,l} = a_{k,l}^{(i)}$. In the following we will focus on the real branch of FBMC-IOTA receiver, and the analysis can be applied to the imaginary branch straightforwardly.

Given $\boldsymbol{p} = [P_{k,l} : k \in [0, \ldots, N-1]\ ,\ l \in [0, \ldots, L-1]]$ as the vector of observed parity symbols at real branch, the MAP detection will estimate $\hat{\boldsymbol{a}}$ that maximizes the joint a posteriori probability mass function (PMF) of the transmitted symbol vector $\boldsymbol{a} = [a_{m,n} : m \in [0, \ldots, N-1]\ ,\ n \in [0, \ldots, L-1]$, i.e.,

$$\hat{\boldsymbol{a}} = \arg\max_{\boldsymbol{a}} p(\boldsymbol{a}|\boldsymbol{p}). \tag{5}$$

The MAP detector can be implemented by two approaches: individual and joint optimum detection [18,19]. The joint approach maximizes the joint a posteriori PMF of the transmitted symbol vector as shown above, while the individual optimum MAP detection maximizes the a posteriori PMF, $p(a_{k,l}|\boldsymbol{p})$, of each individual symbol. Let $\mathbb{X}$ be the constellation alphabet, from which $a_{m,n}$ will take its value, the estimation of $a_{k,l}$ with individual MAP detection can be written as

$$\hat{a}_{k,l} = \arg\max_{a_{m,n} \in \mathbb{X}} p(a_{m,n}|\boldsymbol{p}) \tag{6}$$

A posteriori PMF for $a_{k,l}$ can be found by calculating the marginal of the joint a posteriori PMF, thus (6) can be written as

$$\hat{a}_{k,l} = \arg\max_{a_{m,n} \in \mathbb{X}} \sum_{\boldsymbol{a} \in \mathbb{X}^{N,L}} p(\boldsymbol{a}|\boldsymbol{p}) \tag{7}$$

Let $Pr(a)$ be the priori probability of symbol a. According to Bayes' rule, we have

$$p(\boldsymbol{a}|\boldsymbol{p}) \propto p(\boldsymbol{p}|\boldsymbol{a})Pr(\boldsymbol{a}) \tag{8}$$

where $Pr(\boldsymbol{a})$ is the joint a priori PMF of all symbols assuming that they are independent to each other. Therefore, the estimation function can be modified to

$$\hat{a}_{k,l} = \arg\max_{a_{m,n} \in \mathbb{X}} \sum_{\boldsymbol{a} \in \mathbb{X}^{N \times L}} p(\boldsymbol{p}|\boldsymbol{a}) \prod_{a \in \boldsymbol{a}} Pr(a) \tag{9}$$

As can be seen from (9) this computation includes a marginalization process which is NP-hard [20]. Furthermore, with the assumption that the noise vector is identically and independently distributed (i.i.d.) and is uncorrelated with the transmitted symbols, we can factorize

$$p(\boldsymbol{p}|\boldsymbol{a}) = \prod_{k',l'} p(P_{k',l'}|\boldsymbol{a}) \tag{10}$$

Please note that $P_{k,l}$ is the parity symbol related to the symbol to be detected. Therefore, in this equation a general form of $P_{k',l'}$ is applied. Since a limited number of symbols interfere on each parity symbol, the calculation of this function can be simplified as

$$p(\boldsymbol{p}|\boldsymbol{a}) = \prod_{k',l'} p(P_{k',l'}|\boldsymbol{a}^{[k',l']}) \tag{11}$$

where $\boldsymbol{a}^{[k',l']}$ is the vector of symbols that interfere on the parity symbol with indices k', l'. Substituting (11) into (9) yields

$$\hat{a}_{k,l} = \arg\max_{a_{m,n} \in \mathbb{X}} \sum_{\boldsymbol{a} \in \mathbb{X}^{N \times L}} Pr(\boldsymbol{a}) \prod_{k',l'} p(P_{k',l'}|\boldsymbol{a}^{[k',l']}) \tag{12}$$

From (12), we can see that the FBMC-IOTA structure allows us to translate the MAP detection problem into the marginalize product of functions (MPF) problem. The local observation at parity symbol $P_{k',l'}$ is given by

$$p(P_{k',l'}|\boldsymbol{a}^{[k',l']}) = \frac{1}{\sqrt{2\pi\sigma}} \exp\left(\frac{1}{2\pi\sigma^2}\left\|P_{k',l'} - \boldsymbol{a}^{[k',l']}\boldsymbol{s}^T\right\|^2\right) \tag{13}$$

where $\boldsymbol{s}$ is the vector of calculated weights according to (3) and (4).

3. MPA on Intrinsic Interference

Although the joint MAP detection problem has been translated into an MPF problem as shown in (12) which is much simpler to resolve, the brute-force solution for (11) requires exponential complexity. In order to reduce the complexity, we propose a novel iterative symbol detector IIU based on the generic MPA [21], where MPA is applied to the weight matrix to iteratively approximate the solution of the MPF problem.

Let $l_{a\to P}(a_{k,l}, P_{k',l'})$ and $l_{a\leftarrow P}(a_{k,l}, P_{k',l'})$ be the message in the form of log-likelihood ratio (LLR) delivered from $a_{k,l}$ and $P_{k',l'}$, respectively. We denote $\xi_{k,l}$ as the set of positions (k', l') that $a_{k,l}$ interferes on and $\zeta_{k',l'}$ as the set of positions (k, l) that interfere on $P_{k',l'}$, then

$$\begin{aligned} &l_{a\to P}(a_{k,l}, P_{k',l'}) \\ &= \log \frac{Pr_{\text{ext}}(a_{k,l}=1)}{Pr_{\text{ext}}(a_{k,l}=-1)} \\ &= \sum_{(n,m)\in\xi_{k,l}\setminus(k',l')} l_{a\leftarrow P}(a_{k,l}, P_{n,m}) \end{aligned} \tag{14}$$

By using the message formulated in (14), we have

$$Pr_{\text{ext},k',l'}(a_{k,l}) = \lambda_{k,l} \exp\left(\frac{a_{k,l}}{2} l_{a\to P}(a_{k,l}, P_{k',l'})\right) \tag{15}$$

where $\lambda_{k,l}$ is chosen such that $Pr_{\text{ext},k',l'}(a_{k,l}=+1) + Pr_{\text{ext},k',l'}(a_{k,l}=-1) = 1$, and the subscript "ext" denotes that only the extrinsic information is used.

The inference of parity symbol $P_{k',l'}$ to data symbol $a_{k,l}$ is updated as follows

$$\begin{aligned} &l_{a\leftarrow P}(a_{k,l}, P_{k',l'}) \\ &= \log \frac{p_{\text{ext},k',l'}(a_{k,l}=+1|P_{k',l'}, \boldsymbol{a}^{[k',l']}\setminus a_{k,l})}{p_{\text{ext},k',l'}(a_{k,l}=-1|P_{k',l'}, \boldsymbol{a}^{[k',l']}\setminus a_{k,l})} \\ &= \log \frac{p_{\text{ext},k',l'}(P_{k,l}|\boldsymbol{a}^{[k',l']}, a_{k,l}=+1) Pr(\boldsymbol{a}^{[k',l']}\setminus a_{k,l})}{p_{\text{ext},k',l'}(P_{k,l}|\boldsymbol{a}^{[k',l']}, a_{k,l}=-1) Pr(\boldsymbol{a}^{[k',l']}\setminus a_{k,l})} \end{aligned} \tag{16}$$

where the second equality holds following the Bayes rule of (8). Based on MPA, the a priori PMF of $a_{k,l}$ is not included in the computation of a posteriori PMF of $a_{k,l}$.

Substituting (13)–(15) into (16), we can derive $l_{a\leftarrow P}(a_{k,l}, P_{k',l'})$ in the expression of (17), where $\setminus$ in (16) represents exception.

$$l_{a\leftarrow P}(a_{k,l}, P_{k',l'}) = \log \frac{\sum_{\substack{\boldsymbol{a}^{[k',l']}\in\mathbb{X} \\ a_{k,l}=+1}} \exp\left(\sum_{(n,m)\in\zeta_{k',l'}\setminus(k,l)} \frac{a_{n,m}}{2} l_{a\to P}(a_{n,m}, P_{k',l'}) - \frac{1}{2\sigma^2}\left\|P_{k',l'} - a^{[k',l']}\boldsymbol{s}^T\right\|^2\right)}{\sum_{\substack{\boldsymbol{a}^{[k',l']}\in\mathbb{X} \\ a_{k,l}=-1}} \exp\left(\sum_{(n,m)\in\zeta_{k',l'}\setminus(k,l)} \frac{a_{n,m}}{2} l_{a\to P}(a_{n,m}, P_{k',l'}) - \frac{1}{2\sigma^2}\left\|P_{k',l'} - a^{[k',l']}\boldsymbol{s}^T\right\|^2\right)} \tag{17}$$

When LLR is converged or the maximum number of iterations is reached, the final estimated inference of $a_{k,l}$ will be calculated as

$$l_{k,l}(a_{k,l}) = \sum_{(n,m)\in\xi_{k,l}} l_{a\leftarrow P}(a_{k,l}, P_{n,m}) \tag{18}$$

Consequently, the hard decision of $a_{k,l}$ is

$$\hat{a}_{k,l} = \arg\max_{a_{k',l'}\in\mathbb{X}} l_{k',l'}(a_{k,l}) \tag{19}$$

4. Evaluation of IIU

In this section, we evaluate the proposed IIU to show its effectiveness. The Monte Carlo simulations are conducted over a multipath fading channel of type SUI-3 with 3 paths [22]. The FFT size is 64, the block length is 16, the modulation is 4-OQAM and the maximum number of iterations of IIU is 6.

4.1. Performance with Perfect Channel Estimation and Synchronization

Figure 1 shows BER comparisons between conventional FBMC-IOTA and our proposed scheme. Please note that in the simulations, perfect channel estimation and synchronization are assumed. For uncoded systems (without channel coding), the intrinsic interference use technique demonstrates about 1 dB gain over conventional FBMC-IOTA. Moreover, the performance comparison is also provided when a (1024, 512) quasi-cyclic LDPC code (LDPC1 coded in the figure) and a (4096, 2048) irregular repeat accumulate code (LDPC2 coded in the figure) are adopted to the systems. It can be seen that when different LDPC codes are used, the intrinsic interference use still achieves noticeable performance improvement.

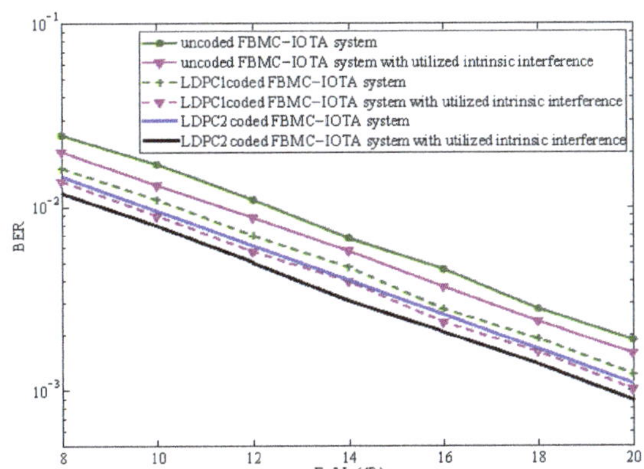

Figure 1. BER performance with perfect channel estimation and synchronization.

4.2. Effect of Carrier Frequency Offset (CFO)

The received signal (after equalizing the channel effect) in the presence of CFO becomes

$$r(t) = s(t)\exp(2\pi j f_0 t) \tag{20}$$

Hence the desired symbol $a_{k,l}$ is demonstrated as

$$
\begin{aligned}
&\hat{a}_{k,l} \\
&= \int r(t) g_{k,l}^*(t) dt \\
&= \int s(t) \exp\left(2\pi j f_0 t\right) g_{k,l}^*(t) dt \\
&= \int \sum_{n,m} a_{n,m} g_{n,m}(t) g_{k,l}^*(t) \exp\left(j 2\pi f_0 t\right) dt \\
&= \sum_{n,m} a_{n,m} \int g(t - n\tau_0) g^*(t - l\tau_0) \exp\left(2\pi j f_0 t\right) \\
&\exp\left(j(m+n-k-l)\pi/2\right) \exp\left(2j\pi(m-k)\nu_0 t\right) dt
\end{aligned} \quad (21)
$$

Denoting $t = x + \frac{(n+l)\tau_0}{2}$, $\hat{a}_{k,l}$ can be calculated by (22). Moreover, considering that $f_0 = \delta\nu_0$, (22) becomes (23) which shows that a frequency offset will change the weights and some of the intrinsic interference will be added to the desired signal. Under this condition, new weights should be calculated. Figure 2 shows how the FBMC-IOTA system with 15 kHz subcarrier spacing are affected by frequency offset of 60 Hz. As indicated by the figure, the gain obtained by IIU reduces due to the frequency offset. This follows from the fact that frequency offset will distort the desired signal and subsequently affects the calculation of parity symbols in the algorithm.

$$
\begin{aligned}
\hat{a}_{k,l} &= \sum_{n,m} a_{n,m} \int g\left(x + \frac{l-n}{2}\tau_0\right) g^*\left(x - \frac{l-n}{2}\tau_0\right) \\
&\exp\left(j(m+n-k-l+(m-k)(n+l))\pi/2\right) \exp\left(2\pi j((m-k)\nu_0 + f_0)x\right) \exp\left(2\pi j f_0 \frac{n+l}{2}\tau_0\right) dx \\
&= \sum_{n,m} a_{n,m} (j)^{m+n-k-l+(m-k)(n+l)} \exp\left(2\pi j f_0 \frac{n+l}{2}\tau_0\right) A_g\left((l-n)\tau_0, (m-k)\nu_0 + f_0\right)
\end{aligned} \quad (22)
$$

$$
\hat{a}_{k,l} = \sum_{n,m} a_{n,m} (j)^{m+n-k-l+(m-k)(n+l)} \exp\left(\pi j \delta \frac{n+l}{2}\right) A_g\left((l-n)\tau_0, (m-k+\delta)\nu_0\right) \quad (23)
$$

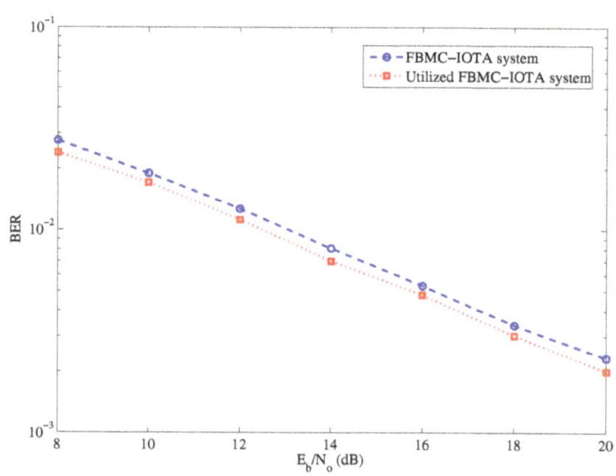

Figure 2. Effect of CFO.

4.3. Effect of Imperfect Channel Estimation

Conventional channel estimation methods used for OFDM cannot be directly applied to FBMC-IOTA due to the intrinsic interference. To address this problem several preamble-based algorithms have been proposed in the literature [15]. We consider that the estimated channel is correlated with the true channel with a correlation factor $\rho_e = E[\hat{H}_{k,l}, H_{k,l}]$, where $H_{k,l}$ represents the channel coefficient at the lth symbol and the kth subcarrier frequency after the IOTA filtering and FFT operation. The estimated channel can be modeled as

$$\hat{H}_{m,n} = \rho_e H_{m,n} + \sqrt{1-\rho_e^2} E_{m,n} \tag{24}$$

where $E_{m,n}$ represents the Gaussian error signal and independent of the real channel. Figure 3 shows that imperfect channel estimation will change parity symbols, consequently the IIU algorithm and its gain are slightly affected. Any discrepancy in the performance is due to the fundamental dependence of the schemes on channel state information (CSI), rather than the particular channel estimation procedures employed. In future we aim to study a scattered pilot-based channel estimation scheme which benefits from the intrinsic interference calculations we carried out in this paper.

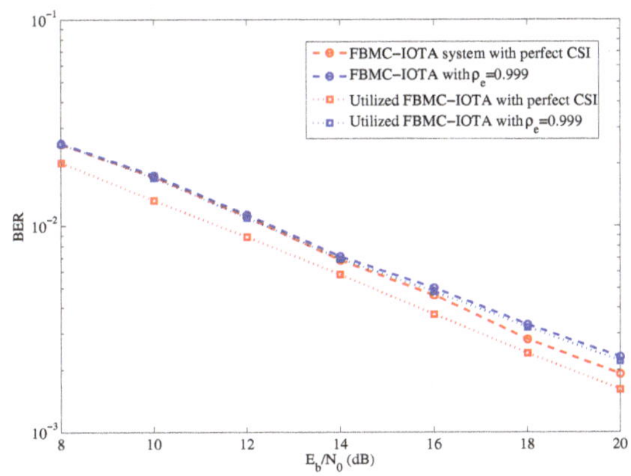

Figure 3. Effect of imperfect channel estimation.

4.4. Complexity of Proposed Algorithm

As explained earlier, we reduced the complexity of the IIU by eliminating some low valued weights. Let us define d_c as the effective number of weights, which corresponds to the effective number of neighboring symbols that interfere with each other on each parity symbol. Denote D as the total number of neighboring symbols (including those with low valued weights) that interfere on each parity symbol. In the proposed algorithm at each received parity symbol, each symbol has only $d_c - 1$ interferers, where $d_c < D$. The complexity order of the proposed IIU algorithm is $\mathcal{O}(|\mathbb{X}|^{d_c})$, which is much less than $\mathcal{O}(|\mathbb{X}|^{D})$ (the optimal case). The complexity of IIU for FBMC-IOTA is shown in (25).

$$\text{Complexity} = |\mathbb{X}|^{d_c} \times (\text{Iteration number}) \times (\text{No. of parity symbols in error}) \tag{25}$$

Please note that the highest possible value for the number of erroneous parity symbols is $N \times L$. Apparently, there is a tradeoff between the accuracy of the algorithm and the complexity. Higher values of d_c will result in more accurate detection but the complexity will increase exponentially with regards to d_c. We set $d_c = 6$, as it is an appropriate value to strike the balance between the accuracy of the algorithm and the complexity.

5. Conclusions

A novel algorithm for FBMC-IOTA systems with IOTA pulse-shaping is proposed to use the intrinsic interference in the demodulation process. The intrinsic interference is treated as parity symbols, which enables us to correct errors using the information offered by these parity symbols.

Based on MPA, IIU is developed to improve the system performance. Effects of CFO and imperfect channel estimation on IIU are investigated.

Author Contributions: Conceptualization, L.W. and R.R.; methodology, J.L.; software, J.L.; validation, R.R. and J.L.; formal analysis, L.W.; investigation, R.R.; resources, J.L.; data curation, R.R.; writing–original draft preparation, J.L.; writing–review and editing, L.W.; visualization, R.R.; supervision, J.L.; project administration, L.W.; funding acquisition, L.W.

Funding: This work was supported in part by the National Defense Technology Foundation under Grant 3101168, in part by the Hunan Natural Science Foundation under Grant 2017JJ2303, and in part by National Natural Science Foundation of China under Grant 61603007 and Grant 61702536.

Conflicts of Interest: The authors declare no conflict of interest.

References

1. Banelli, P.; Buzzi, S.; Colavolpe, G.; Modenini, A.; Rusek, F.; Ugolini, A. Modulation formats and waveforms for 5G networks: Who will be the heir of OFDM: An overview of alternative modulation schemes for improved spectral efficiency. *IEEE Signal Process. Mag.* **2014**, *31*, 80–93. [CrossRef]
2. Wen, L.; Razavi, R.; Imran, M.A.; Xiao, P. Design of joint sparse graph for OFDM system. *IEEE Trans. Wirel. Commun.* **2014**, *14*, 1823–1836. [CrossRef]
3. Nam, H.; Choi, M.; Han, S.; Kim, C.; Choi, S.; Hong, D. A new filter-bank multicarrier system with two prototype filters for QAM symbols transmission and reception. *IEEE Trans. Wirel. Commun.* **2016**, *15*, 5998–6009. [CrossRef]
4. Saeedi-Sourck, H.; Yan, W.; Bergmans, J.W.M. Complexity and performance comparison of filter bank multicarrier and OFDM in uplink of multicarrier multiple access networks. *IEEE Trans. Signal Process.* **2011**, *59*, 1907–1912. [CrossRef]
5. Lopez-Salcedo, J.A.; Gutierrez, E.; Seco-Granados, G.; Swindlehurst, A.L. Unified framework for the synchronization of flexible multicarrier communication signals. *IEEE Trans. Signal Process.* **2013**, *61*, 828–842. [CrossRef]
6. Cui, W.; Qu, D.; Jiang, T.; Farhang-Boroujeny, B. Coded auxiliary pilots for channel estimation in FBMC-OQAM systems. *IEEE Trans. Veh. Technol.* **2016**, *65*, 2936–2946. [CrossRef]
7. Verde, F.; Gelli, G.; Paura, L.; Darsena, D. Widely-linear equalisation and blind channel identification for interference-contaminated multicarrier systems. *IEEE Trans. Signal Process.* **2004**, *51*, 1153–1164.
8. Darsena, D.; Verde, F. Minimum-mean-output-energy blind adaptive channel shortening for multicarrier SIMO transceivers. *IEEE Trans. Signal Process.* **2007**, *19*, 119–134. [CrossRef]
9. Darsena, D.; Gelli, G.; Verde, F. Universal linear precoding for NBI-proof widely-linear equalization in MC systems. *EURASIP J. Wirel. Commun.* **2008**, *19*, 1634–1641. [CrossRef]
10. Darsena, D.; Gelli, G.; Paura, L. A constrained maximum-SINR NBI-resistant receiver for OFDM systems. *IEEE Trans. Signal Process.* **2007**, *43*, 3032–3047. [CrossRef]
11. Bolcskei, H.; Duhamel, P.; Hleiss, R. Design of pulse shaping OFDM/OQAM systems for high data-rate transmission over wireless channels. In Proceedings of the IEEE International Conference on Communications ICC '99, Vancouver, BC, Canada, 6–10 June 1999; Volume 1, pp. 559–564.
12. Vangelista, L.; Laurenti, N. Efficient implementations and alternative architectures for OFDM-OQAM systems. *IEEE Trans. Commun.* **2001**, *49*, 664–675. [CrossRef]
13. Liu, Y.; Chandrasekaran, V.; Anandkumar, A. Feedback message passing for inference in Gaussian graphical models. *IEEE Trans. Signal Process.* **2012**, *60*, 4135–4150. [CrossRef]
14. Farhang, B. OFDM Versus Filter Bank Multicarrier. *IEEE Signal Process. Mag.* **2011**, *28*, 92–112. [CrossRef]
15. Du, J.; Signell, S. Novel Preamble-Based Channel Estimation for OFDM/OQAM Systems. In Proceedings of the 2009 IEEE International Conference on Communications (ICC), Dresden, Germany, 14–18 June 2009.
16. Razavi, R.; Xiao, P.; Tafazolli, R. Information theoretic analysis of OFDM/OQAM with utilized intrinsic interference. *IEEE Signal Process. Lett.* **2016**, *22*, 618–622. [CrossRef]
17. Lee, S.H.; Shamaiah, M.; Vikalo, H.; Vishwanath, S. Message-Passing Algorithms for Coordinated Spectrum Sensing in Cognitive Radio Networks. *IEEE Commun. Lett.* **2013**, *17*, 812–815.
18. Verdu, S. *Multiuser Detection;* Cambridge University Press: Cambridge, UK, 1998.

19. Poon, T.V.; Beaulieu, N.C. Jointly and individually optimum receivers for BPSK signals in cochannel interference plus noise. In Proceedings of the 2003 IEEE Pacific Rim Conference on Communications Computers and Signal Processing (PACRIM 2003) (Cat. No. 03CH37490), Victoria, BC, Canada, 28–30 August 2003; Volume 2, pp. 530–532.
20. Lupas, R.; Verdu, S. Linear multiuser detectors for synchronous code-division multiple-access channels. *IEEE Trans. Inf. Theory* **1989**, *35*, 123–136. [CrossRef]
21. Kschischang, F.; Frey, B. Iterative decoding of compound codes by probability propagation in graphical models. *IEEE J. Sel. Areas Commun.* **1998**, *16*, 219–230. [CrossRef]
22. Erceg, V.; Greenstein, L.; Tjandra, S.; Parkoff, S.; Gupta, A.; Kulic, B.; Julius, A.; Bianchi, R. An empirically based path loss model for wireless channels in suburban environments. *IEEE J. Sel. Areas Commun.* **1999**, *17*, 1205–1211. [CrossRef]

applied sciences

MDPI

Article

Concensus-Based ALADIN Method to Faster the Decentralized Estimation of Laplacian Spectrum

Thi-Minh-Dung Tran [1,*], Luu Ngoc An [1] and Ngoc Chi Nam Doan [2]

1 The Faculty of Electrical Engineering, The University of Danang—University of Science and Technology, 54 Nguyen Luong Bang Street, Danang City 550000, Vietnam; lnan@dut.udn.vn

2 Manufacturing Execution and Control Group, Singapore Institute of Manufacturing Technology, Singapore 138634, Singapore; doanncn@SIMTech.a-star.edu.sg

* Correspondence: ttmdung@dut.udn.vn; Tel.: +84-236-735-112

Received: 13 June 2020; Accepted: 13 July 2020; Published: 13 August 2020

Abstract: With the upcoming fifth Industrial Revolution, humans and collaborative robots will dance together in production. They themselves act as an agent in a connected world, understood as a multi-agent system, in which the Laplacian spectrum plays an important role since it can define the connection of the complex networks as well as depict the robustness. In addition, the Laplacian spectrum can locally check the controllability and observability of a dynamic controlled network, etc. This paper presents a new method, which is based on the Augmented Lagrange based Alternating Direction Inexact Newton (ALADIN) method, to faster the convergence rate of the Laplacian Spectrum Estimation via factorization of the average consensus matrices, that are expressed as Laplacian-based matrices problems. Herein, the non-zero distinct Laplacian eigenvalues are the inverse of the stepsizes $\{\alpha_t, t = 1, 2, \ldots\}$ of those matrices. Therefore, the problem now is to carry out the agreement on the stepsize values for all agents in the given network while ensuring the factorization of average consensus matrices to be accomplished. Furthermore, in order to obtain the entire Laplacian spectrum, it is necessary to estimate the relevant multiplicities of these distinct eigenvalues. Consequently, a non-convex optimization problem is formed and solved using ALADIN method. The effectiveness of the proposed method is evaluated through the simulation results and the comparison with the Lagrange-based method in advance.

Keywords: multi-agent systems; laplacian eigenvalues; augmented Lagrange based Alternating Direction Inexact Newton (ALADIN) method; consensus algorithms; Alternating Direction of Multipliers Method (ADMM)

1. Introduction

Leaders around the world obviously prefer the present era of connectivity as the Fourth Industrial Revolution [2]. Industry 4.0 has been significantly contributing to the transformation of many industries such as transportation, manufacturing, health-care, agriculture, etc. via enabling data transmission and integration between disciplines. Today, we are in the fourth one, a generation of connection between our physical, digital, social, and biological worlds. In particular, data and information from these different areas have been made available and have been connected in complex and dense networks. For better integration and utilization, control aspect of these complex networks, researchers from different communities have brought different contributions varying from topology inference to control strategy, deputizing for interacting systems, which are modeled by graphs, whose vertices represent the components of the system while edges stand for the interactions between these components.

In the last decade there has been dramatic increasing number of publications in the cooperative control of multi-agent systems. In control of multi-agent systems, the performance of the whole system depends on both structure and the connections between individuals of the systems. Here, the total connections can be defined by the graph Laplacian matrix, and its spectrum is involved in some useful properties of the network system [3,4]. For instance, the second smallest graph Laplacian eigenvalue, i.e., the so-called algebraic connectivity of the graph, has the main role in the convergence time of various distributed algorithms as well as the performance and robustness of dynamical systems [5]. Conceptually, agents share their information with each other to achieve common objectives, relative position information, or common control algorithms. This is called consensus problem [6,7], where a group of agents' approaches average consensus in an undirected network under simple linear iteration scheme.

It is well known that in a multi-agent system, consensus is achieved if and only if the network is connected (the algebraic connectivity) being strictly greater than zero [8]. On the other hand, the largest Laplacian eigenvalue is an important factor to decide the stability of the system. For example, minimizing the spectral radius in [9] leads to maximize the robustness of the network to time delays under a linear consensus protocol. Furthermore, to speed up consensus algorithms, the optimal Laplacian-based consensus matrix is obtained with a stepsize which is the inverse of the sum of the smallest and the largest non-zero graph Laplacian eigenvalue [10]. Moreover, the authors in [11,12] have proved that the spectrum of the Laplacian matrix can be used to design consensus matrices to obtain average consensus in finite number of steps.

In order to investigate network efficiency, structural robustness of a network which is related to its performance despite changes in the network topology [13] has been also studied. The concept of natural connectivity as a spectral measure of robustness was introduced in [14]. It is expressed in mathematical form as the average eigenvalue of the adjacency matrix of the graph representing the network topology. The Laplacian spectrum $sp(\mathbf{L}) = \{\lambda_1^{m_1}, \ldots, \lambda_i, \ldots\}$ can also be employed to compute the robustness indices, for instance, the number of spanning trees and the effective graph resistance (Kirchhoff index) [15]. The smaller (or greater) the Kirchhoff index (or the number of spanning trees) is, the more robust the network becomes. In addition, it has been pointed out that adding an edge strictly decreases the Kirchhoff index and hence increases the robustness. In [16], the authors have proposed a method to monitor collaboratively the robustness of the networks partitioned into sub-networks by Kirchhoff index $\mathcal{R}_{\mathbf{L}} = N\sum_{i=2}^{D+1}\frac{m_i}{\lambda_i}$. Here, an Alternating Direction of Multipliers Method (ADMM)-based algorithm was employed to perform the factorization of the averaging matrix and to compute the average degree of the network concurrently. However, the main point in this work was the reformulation into the convex optimization problem, which is convenient to make use of the ADMM method to solve the problem. In addition to that, the impact of the Laplacian spectrum into power systems is expressed through energy management in smart grids [17] and the determination of the grid robustness against low frequency disturbance in [18]. In this work, in the framework of spectral graph theory, the authors reveal that the decomposition of frequency signal along scaled Laplacian spectrum when the damping-inertia ratios are uniform across buses not only makes the system respond faster but also helps lower the system nadir after a disturbance. In dynamic network systems, the spectrum of Laplacian matrix can also be utilized for locally checking the controllability and the observability [19].

From a short literature survey above, it is obvious that the Laplacian spectrum plays an important role in many fields. For instance, Laplacian spectrum can be used to design consensus matrices [11,12], to compute these robustness indices [15–18], or to check the controllability and the observability [19]. Hence, it is desirable to have an efficient method for monitoring the Laplacian spectrum of a dynamics network system.

One thing to remark here is that if the global network topology is known in a-priori, the Laplacian matrix can be easily deduced. However, implementing a centralized structure is an expensive task due to the high computational cost, the heavy communication infrastructure aspect and the problem from large dimensionality. Additionally, if there is a failure problem from one point, it will affect the whole network. Therefore, our study is restricted to the assumption that the network topology (represented by the Laplacian Matrix) is unknown at the first glance. A dominant contribution of this paper is the possibility of implementing this monitoring scheme in a decentralized manner.

In this paper, we present an Augmented Lagrangian based Alternating Direction Inexact Newton (ALADIN) method to estimate the Laplacian spectrum in decentralized scheme for dynamic controlled networks. The key feature of this paper is the direct solution to non-convex optimization for Laplacian spectrum estimation using ALADIN method. To simplify, the scope of this study is restricted to networks performing noise-free as well as the number of the agents N in the network in known in-priori by using random walk algorithm [20]. The network is modeled, then Laplacian eigenvalues and average consensus are retrieved respectively. Since the Laplacian spectrum matrix is not directly computable for undetermined network topology, the decentralized estimation of the Laplacian spectrum has been introduced with three main approaches in the recent literature: Fast Fourier Transform (FFT)-based methods [21,22], local eigenvalue decomposition of given observability-based matrices [23], and distributed factorization of the averaging matrix $\mathbf{J}_N = \frac{1}{N}\mathbf{1}\mathbf{1}^T$ [24]. FFT-based methods require a specific protocol. However, they do not make use of the available measurements coming from the consensus protocol. On the other hand, the method in [23] allows using the transient of the average consensus protocol but for several consecutive initial conditions. The distributed factorization of the averaging matrix in [24] yields the inverses of non-zero Laplacian eigenvalues and can be solved as a constrained consensus problem. The Laplacian eigenvalues can be deduced as the inverse of the stepsizes in each estimating factor, where these factors are constrained to be structured as Laplacian based consensus matrices. In [1], authors have applied a gradient descent algorithm to solve this optimization problem in which only local minima was guarantees accompany with slow convergence rate. In order to solve this annoying issue, in [16,24], the authors have introduced an interesting way by reformulating a non-convex optimization problem in [1] into convex one and solved by applying an ADMM-based method. However, this is an indirect approach obtaining by an adequate re-parameterization. In this paper, we inherit the idea in [1] to form the non-convex optimization for decentralized estimation of Laplacian spectrum and then directly solve it using the ALADIN method that was proposed by [25]. The proposed approach is then evaluated with two network structures for performance evaluation in comparison with gradient descent method [1].

In this paper, we firstly introduce the background of average consensus and state the problem in Section 2, then present the distributed estimation of Laplacian spectrum in Section 3. The structure of this section can be illustrated as in Figure 1. Before concluding the paper, the simulation results are described in Section 4 to evaluate the efficiency of the proposed method.

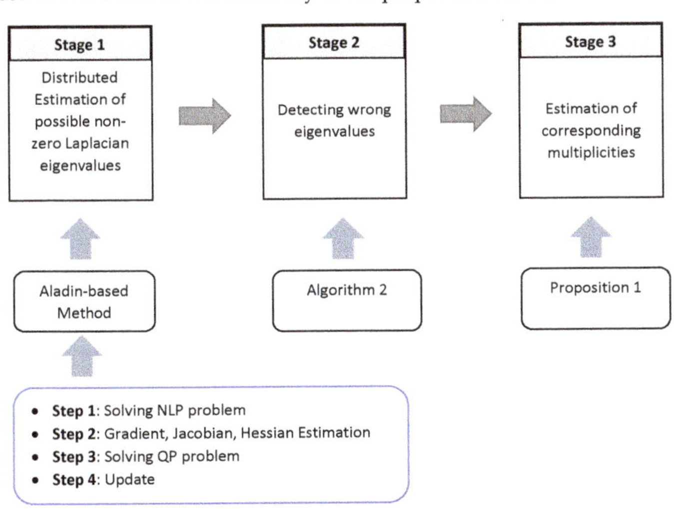

Figure 1. Structure of Section 3.

2. Background and Problem Statement

Consider a dynamic network, in which interconnection is represented by $G(V, E)$, an undirected graph with components' set V and links' set E, consisting of $N = |V|$ nodes, let us denote by $\mathcal{N}_i = \{j \in V : (i, j) \in E\}$ the set of neighbors of node i and $d_i = |\mathcal{N}_i|$ its degree. Interactions between nodes can be captured by the Laplacian matrix $\mathbf{L} \in \Re^N$ with entries $l_{ii} = d_i$, $l_{ij} = -1$ if $j \in \mathcal{N}_i$ and $l_{ij} = 0$ elsewhere. Denote the Laplacian spectrum by $sp(\mathbf{L}) = \{\lambda_1^{m_1}, \lambda_2^{m_2}, \ldots, \lambda_{D+1}^{m_{D+1}}\}$, where the different Laplacian eigenvalues are in increasing order $0 = \lambda_1 < \lambda_2 < \ldots < \lambda_{D+1}$ and superscripts stand for multiplicities $m_i = m(\lambda_i)$, while $\mathbf{S}_2 = \{\lambda_2, \ldots, \lambda_{D+1}\}$ stands for the set of the non-zero distinct Laplacian eigenvalues.

2.1. Average Consensus

For each node $i \in V$, let $x_i(t)$ denotes the value of node i at timestep t. Define $\mathbf{x}(t) = [x_1(t), x_2(t), ..., x_N(t)]^T$, where N is the number of nodes in the network. Average consensus algorithms can be achieved by using the following linear iteration scheme as follows:

$$\mathbf{x}(t) = (\mathbf{I}_N - \alpha\mathbf{L})\mathbf{x}(t-1), \tag{1}$$

where α is an appropriately selecting stepsize [26], by which all nodes converge asymptotically to the same value $\bar{x}$ that is the average of the initial ones $\bar{x}\mathbf{1} = \lim_{t\to\infty} \mathbf{x}(t) = \frac{1}{N}\mathbf{1}\mathbf{1}^T\mathbf{x}(0)$.

On the other hand, it has been shown in [11,12] that the average consensus matrix can be factored as

$$\prod_{t=D}^{1} \mathbf{W}_t = \frac{1}{N}\mathbf{1}\mathbf{1}^T, \tag{2}$$

where $\mathbf{W}_t = \vartheta_t\mathbf{I}_N + \alpha_t\mathbf{L}$, ϑ_t and α_t being parameters to be designed. In [12], the solution was given by $\vartheta_t = 1$ and $\alpha_t = -\frac{1}{\lambda_{t+1}}$, λ_t being a non-zero Laplacian eigenvalue. Owing to the above factorization, average consensus can then be reached in D steps, D being the number of distinct non-zero Laplacian eigenvalues:

$$\bar{\mathbf{x}} = \mathbf{x}(D) = \prod_{t=D}^{1} \mathbf{W}_t\mathbf{x}(0) = \frac{1}{N}\mathbf{1}\mathbf{1}^T\mathbf{x}(0) \text{ for all } \mathbf{x}(0) \in \Re^N. \tag{3}$$

2.2. Problem Statement

It can be noted that by factorizing the average consensus matrix, while constraining the factor matrices to be in the form $\mathbf{I}_N - \alpha_t\mathbf{L}$, the eigenvalues of the Laplacian matrix as the inverse of α_t can be deduced. The uniqueness has been proved in [1].

Lemma 1. *[1] Let $\lambda_2, \cdots, \lambda_{D+1} \neq 0$ be the D distinct non-zero eigenvalues of the graph Laplacian matrix $\mathbf{L}$, then, up to permutation, the sequence $\{\alpha_i\}_{i=1,\cdots,D}$, with $\alpha_i = \frac{1}{\lambda_{i+1}}$, $i = 1, 2, \cdots, D$, is the unique sequence allows getting the minimal factorization of the average consensus matrix as $\frac{1}{N}\mathbf{1}\mathbf{1}^T = \prod_{i=1}^{D}(\mathbf{I}_N - \alpha_i\mathbf{L})$.*

Therefore, in order to implement the proposed method, the knowledge of the network should be known. Meaning that, the number of the components N of the given network should be known by adding a learning mechanism as a configuration step. Practically, in most systems where communications are involved, learning sequences are used for communication channel identification or for synchronization. In [20], the authors have proposed a method using random walks to estimate the global properties of large connected undirected graphs such as number of vertices, edges, etc. However, it is not in the scope of this paper. Indeed, assuming that the number of agents N is known in a-priori, a consensus protocol in [10] is to be uploaded to each agent to compute the average consensus value $\bar{x}$. The main task in our study is to estimate the whole Laplacian spectrum.

3. Distributed Estimation of Laplacian Spectrum

Given an initial input-output pair $\{\mathbf{x}(0), \bar{\mathbf{x}}\}$, with $\bar{\mathbf{x}} = \frac{1}{N}\mathbf{1}\mathbf{1}^T\mathbf{x}(0)$, the matrix factorization problem (3) is equivalent to minimize the cost function $E(\mathbf{W}) = \|\mathbf{x}(D) - \bar{\mathbf{x}}\|^2$ that can also be rewritten as follows:

$$E(\mathbf{W}) = \left\| \prod_{t=D}^{1} \mathbf{W}_t \mathbf{x}(0) - \bar{\mathbf{x}} \right\|^2, \tag{4}$$

where D is the number of steps before reaching average consensus and $\mathbf{W}_t = \mathbf{I}_N - \boldsymbol{\alpha}_t \mathbf{L}$.

Note that there is no need for a central node to set the initial input-output pair. Indeed, such a pair can be obtained after running a standard average consensus algorithm. Each node keeps in memory its own initial value and the consensus value.

Solving this factorization problem consists in finding the sequence of stepsize $\{\boldsymbol{\alpha}_t\}_{t=1,\cdots,D}$. It is obvious that $\boldsymbol{\alpha}_t$ are global parameters. To relax these constraints, define the factor matrices as $\mathbf{W}_t = \mathbf{I}_N - \boldsymbol{\Lambda}_t \mathbf{L}$, where $\boldsymbol{\Lambda}_t = diag(\boldsymbol{\alpha}_t)$, $\boldsymbol{\alpha}_t = [\alpha_{t,1}, \alpha_{t,2}, \ldots, \alpha_{t,N}]$, $t = 1, 2, \ldots, D$. The problem above can be reformulated as a constrained consensus problem, that is to compute the sequence of stepsize $\{\boldsymbol{\alpha}_t\}$ so that $\alpha_{t,1} = \alpha_{t,2} = \ldots = \alpha_{t,N}$. Moreover, in Section 2.1, D is denoted as number of non-zero distinct Laplacian eigenvalues. However, in this work, Laplacian matrix is assumed to be not-known in-a-priori. Therefore, the authors have assigned D as $h = N - 1$ since N can be estimated in the configuration step through the Random Walk Algorithm proposed in [20]. Furthermore, the Laplacian Spectrum estimation procedure is divided in following stages:

- *Stage 1*: Distributed estimation of the set of non-zero Laplacian eigenvalues $\mathcal{S}_1 = \{\lambda_1, \lambda_2, \ldots, \lambda_h\}$, composing of the set of D non-zero distinct Laplacian eigenvalues $\mathcal{S}_2 = \{\lambda_1, \lambda_2, \ldots, \lambda_D\}$.
- *Stage 2*: Eliminating the wrong eigenvalues in the set $\mathcal{S}_1$ to obtain the set $\mathcal{S}_2$
- *Stage 3*: Estimating the multiplicities $\mathbf{m}$ corresponding to each eigenvalues in the set $\mathcal{S}_2$ to achieve the whole Laplacian spectrum $sp(\mathbf{L})$.

3.1. Distributed Estimation of Non-Zero Laplacian Eigenvalues

For distributively carrying out the factorization of the average consensus matrix as factors of Laplacian based consensus matrices, the idea is to minimize the disagreement between neighbors on the value of $\boldsymbol{\alpha}_t$ while ensuring that the factorization of the average consensus matrix is achieved. Such a factorization is assessed by constraining the values of the nodes after h iterations of the consensus algorithm to be equal to the average of the initial values:

$$\begin{array}{ll} \min\limits_{\boldsymbol{\alpha}_t \in R^{N\times 1}, t=1,2,\ldots,h} & \frac{1}{2}\sum\limits_{t=1}^{h}\sum\limits_{i\in V}\sum\limits_{j\in\mathcal{N}_i}(\alpha_{t,j} - \alpha_{t,i})^2 \\ \text{subject to} & \mathbf{x}(h) = \bar{\mathbf{x}} \end{array} \tag{5}$$

or, it can be rewritten in the following form:

$$\begin{array}{ll} \min\limits_{\boldsymbol{\alpha}_t \in R^{N\times 1}} & \frac{1}{2}\sum\limits_{t=1}^{h}\boldsymbol{\alpha}_t^T \mathbf{L} \boldsymbol{\alpha}_t. \\ \text{subject to} & \mathbf{x}(h) = \bar{\mathbf{x}} \end{array} \tag{6}$$

This optimization has been solved by applying Augmented Lagrange Method [1]. However, the disadvantage of this method is the slow convergence rate due to the fact that it is a non-convex optimization problem. To overcome this unexpected issue, the authors have suggested an interesting variant by converting the non-convex function into the convex one. By that, the optimization can be easily and effectively solved by Alternating Direction of Multipliers Methods

(ADMM) [16,27]. In this paper, we proposed a method that can solve a non-convex problem effectively by employing ALADIN method, which is described as following:

(1) **Step 1:** ALADIN solves in parallel a sequence of equality-constrained non-linear problems (NLP) by introducing an augmented variables $\mathbf{y}$ as follows:

$$\begin{aligned} \min_{\boldsymbol{\alpha}_t \in \mathbf{R}^{N\times 1}, t=1,2,\ldots,h} \quad & \frac{1}{2}\boldsymbol{\alpha}_t^T \mathbf{L}\boldsymbol{\alpha}_t + \boldsymbol{\lambda}^T(\boldsymbol{\alpha}_t - \mathbf{y}_t) + \frac{\rho}{2}\|(\boldsymbol{\alpha}_t - \mathbf{y}_t)\|^2_{\Sigma_t} \\ \text{subject to } & \mathbf{x}(h) - \bar{\mathbf{x}} = 0 | \boldsymbol{\beta} \\ & \mathbf{y}_{t,i} = \mathbf{y}_{t,j} \quad i = 1,\ldots,N; j \in \mathcal{N}_i \quad (\mathcal{C}) \end{aligned} \tag{7}$$

where ρ, β are penalty parameter and multiplier of the equality constraint, respectively. $\mathcal{C} = \{\mathbf{y}_t : y_{t,i} = y_{t,j}, i = 1,\ldots,N; j \in \mathcal{N}_i\}$.

One thing to note here is that with the given initial information $x_i(0), i = 1,2,\ldots,N$, running a standard consensus algorithm can determine the average value $\bar{x} = \frac{1}{N}\sum_{i=1}^{N} x_i(0)$. In addition to that, the positive semi-definite scaling matrices Σ_t can be randomly initialized or even be an identity matrix $\mathbf{I}_N$. In this optimization problem (7), augmented variables $\mathbf{y}_t$ are introduced with respect to the constraint $\mathcal{C}$. However, this NLP is to be solved to define the variables $\boldsymbol{\alpha}_t$, hence, the constraint $\mathcal{C}$ is going to be relaxed.

The solution $\boldsymbol{\alpha}_t[k+1], \boldsymbol{\beta}[k+1]$ obtained from (7) is then used to check the stopping criteria the the next steps of the ALADIN-based algorithm procedure with k being an iteration of the optimization process. Herein, if $\frac{1}{2}\sum_{t=1}^{h}\sum_{i\in V}\sum_{j\in\mathcal{N}_i}(\alpha_{t,j} - \alpha_{t,i})^2 < \epsilon$ and $\rho\|\Sigma_t(\boldsymbol{\alpha}_t - \mathbf{y}_t)\|_1 \le \epsilon$, then one can get $\boldsymbol{\alpha}_t^*$ as well as the Algorithm stops.

(2) **Step 2:** The Gradients, Jacobian matrices, and Hessian matrices are estimated for the next quadratic programming (QP) subproblems the as follows:

$$\mathbf{g}_t = \frac{\partial}{\partial \boldsymbol{\alpha}_t}\{\frac{1}{2}\boldsymbol{\alpha}_t^T \mathbf{L}\boldsymbol{\alpha}_t + \boldsymbol{\beta}_t^T(\mathbf{x}(h) - \bar{\mathbf{x}})\} \tag{8}$$

$$\mathbf{C}_t = \frac{\partial}{\partial \boldsymbol{\alpha}_t}(\mathbf{x}(h) - \bar{\mathbf{x}})^T \tag{9}$$

$$\mathbf{B}_t = \frac{\partial^2}{\partial \boldsymbol{\alpha}_t^2}(\frac{1}{2}\boldsymbol{\alpha}_t^T \mathbf{L}\boldsymbol{\alpha}_t). \tag{10}$$

(3) **Step 3:** Analogically to inexact SQP method, the QP problem is solved to find the $\Delta\boldsymbol{\alpha}_t[k]$ and the affine multiplier $\lambda_{QP}[k]$ as follows:

$$\begin{aligned} \min_{\Delta\boldsymbol{\alpha} \in \mathbf{R}^{N\times h}, s \in \mathbf{R}^{N\times 1}, t=1,2,\ldots,h} \quad & \sum_{t=1}^{h}\{\frac{1}{2}\Delta\boldsymbol{\alpha}_t^T \mathbf{B}_t \Delta\boldsymbol{\alpha}_t + \mathbf{g}_t^T \Delta\boldsymbol{\alpha}_t\} + \boldsymbol{\lambda}^T \mathbf{s} + \frac{\mu}{2}\|\mathbf{s}\|^2 \\ \text{subject to } & \sum_{t=1}^{h}(\boldsymbol{\alpha}_t + \Delta\boldsymbol{\alpha}_t - \mathbf{y}_t) = \mathbf{s} | \boldsymbol{\lambda}_{QP} \\ & \mathbf{C}_t \Delta\boldsymbol{\alpha}_t = 0, t = 1,2,\ldots,h | \eta \end{aligned} \tag{11}$$

where $\mathbf{s}$ is the slack variable, introduced into the QP sub-problem to attenuate the numerical reasons when the penalty parameter μ becomes large.

(4) **Step 4:** The final step is to update $\boldsymbol{\lambda}[k+1], \mathbf{y}_t[k+1], \mathbf{B}_t[k+1]$:

$$\boldsymbol{\lambda}[k+1] = \boldsymbol{\lambda}_{QP}[k] \tag{12}$$

$$\hat{\mathbf{y}}_t[k] = \boldsymbol{\alpha}_t[k] + \Delta\boldsymbol{\alpha}_t[k] \tag{13}$$

This update rule is relevant to the full-size step where $a_1 = a_2 = a_3 = 1$ for the steplength computation, which proposed in [25]. Then, projecting $\hat{\mathbf{y}}[k]$ on the constraint $\mathcal{C}$ to derive $\mathbf{y}[k+1]$.

The steps are repeated until the stopping criteria is satisfied.

In order to derive the distributed algorithm, let us take a closer look at each step of the proposed ALADIN-based method.

3.1.1. Implementation of Decoupled Nonlinear Problems

Firstly, in order to solve the decoupled NLP (7) for $t = 1, \ldots, h$, the Augmented Lagrange method is applied here. Hence, we introduce the Augmented Lagrange function with the Lagrange multiplier $\boldsymbol{\beta}$ and penalty parameter $\mathbf{c}$ as below:

$$\mathbf{H}_{1,t} = \frac{1}{2}\boldsymbol{\alpha}_t^T \mathbf{L}\boldsymbol{\alpha}_t + \boldsymbol{\lambda}^T(\boldsymbol{\alpha}_t - \mathbf{y}_t) + \frac{\rho}{2}\|(\boldsymbol{\alpha}_t - \mathbf{y}_t)\|^2_{\Sigma_t} + \beta^T(\mathbf{x}(h) - \bar{\mathbf{x}}) + \frac{c}{2}\|\mathbf{x}(h) - \bar{\mathbf{x}}\|_2^2 \tag{14}$$

The solution of this problem can be obtained by applying a gradient descent method iteratively:

$$\boldsymbol{\alpha}_t[k+1] = \boldsymbol{\alpha}_t[k] - b\frac{\partial \mathbf{H}_{1,t}}{\partial \boldsymbol{\alpha}_t} \tag{15}$$

$$\boldsymbol{\beta}[k+1] = \boldsymbol{\beta}[k] + c(\mathbf{x}(h) - \bar{\mathbf{x}}) \tag{16}$$

where b stands for stepsizes of the gradient descent method, which can be chosen by fix constants or be determined by deploying a line-search algorithms such as Wolfe Condition, Back-stracking, or Armijo ones in [28], while c dedicates for the penalty parameter of the Augmented Lagrange method.

Lemma 2. *The derivatives of this Lagrange function* (14) *is obtained as follows:*

$$\frac{\partial \boldsymbol{H}_{1,t}}{\partial \boldsymbol{\alpha}_t} = \boldsymbol{L}\boldsymbol{\alpha}_t + \boldsymbol{\lambda} + \rho\boldsymbol{\Sigma}_t(\boldsymbol{\alpha}_t - \boldsymbol{y}_t) - diag^{-1}(\boldsymbol{\alpha}_t)diag(\boldsymbol{x}_{t-1} - \boldsymbol{x}_t)\boldsymbol{\delta}_t - diag^{-1}(\boldsymbol{\alpha}_t)diag(\boldsymbol{x}_{t-1} - \boldsymbol{x}_t)\boldsymbol{e}_t \tag{17}$$

where $\boldsymbol{\delta}_h = \beta$, *and* $\boldsymbol{\delta}_t = \boldsymbol{W}_{t+1}\boldsymbol{\delta}_{t+1}$, *while* $\boldsymbol{e}_h = \mathbf{x}(h) - \bar{\mathbf{x}}$ *and* $\boldsymbol{e}_t = \boldsymbol{W}_{t+1}\boldsymbol{e}_{t+1}$.

The proof is showed in the Appendix A.

3.1.2. Implementation of the Coupling Quadratic Programming (QP)

Now, we apply the Karush–Kuhn–Tucker (KKT) conditions for solving the quadratic programming (QP) (11). The Augmented Lagrange Function is described as follows:

$$\begin{aligned}\mathbf{H}_2 = &\sum_{t=1}^{h}\left\{\frac{1}{2}\Delta\boldsymbol{\alpha}_t^T \mathbf{B}_t \Delta\boldsymbol{\alpha}_t + \mathbf{g}_t^T\Delta\boldsymbol{\alpha}_t + \boldsymbol{\lambda}_{QP}^T\Delta\boldsymbol{\alpha}_t\right\} + (\boldsymbol{\lambda}^T - \boldsymbol{\lambda}_{QP}^T)\mathbf{s} + \frac{\mu}{2}\|\mathbf{s}\|^2 \\ &+ \boldsymbol{\lambda}_{QP}^T\sum_{t=1}^{h}(\boldsymbol{\alpha}_t - \mathbf{y}_t) + \sum_{t=1}^{h}\eta_t^T \mathbf{C}_t\Delta\boldsymbol{\alpha}_t\end{aligned} \tag{18}$$

The KKT conditions, which are showed in the Appendix B yields a system of equations as follows:

$$\begin{cases} \mathbf{B}_t\Delta\alpha_t - \boldsymbol{\lambda}_{QP} + \mathbf{C}_t^T\eta_t &= -g_t, t = 1, 2, \ldots, h \\ \sum_{t=1}^{h}\Delta\boldsymbol{\alpha}_t - \frac{1}{\mu}\boldsymbol{\lambda}_{QP} &= -\frac{1}{\mu}\boldsymbol{\lambda} - \sum_{t=1}^{h}(\boldsymbol{\alpha}_t - \mathbf{y}_t) \\ \mathbf{C}_t\Delta\boldsymbol{\alpha}_t &= 0 \end{cases}$$

Solutions of this system of equations are $\Delta\alpha_t^*, \boldsymbol{\lambda}_{QP}^*, \eta_t^*$ respectively in the equivalent matrix form as follows:

$$\begin{pmatrix} \mathbf{B}_1 & \mathbf{0} & \dots & \mathbf{0} & \mathbf{I} & \mathbf{C}_1^T & \mathbf{0} & \dots & \mathbf{0} \\ \mathbf{0} & \mathbf{B}_2 & \dots & \mathbf{0} & \mathbf{I} & \mathbf{0} & \mathbf{C}_2^T & \dots & \mathbf{0} \\ \vdots & \vdots & \ddots & \vdots & \vdots & \vdots & \vdots & \ddots & \vdots \\ \mathbf{0} & \mathbf{0} & \dots & \mathbf{B}_h & \mathbf{I} & \mathbf{0} & \mathbf{0} & \dots & \mathbf{C}_h^T \\ \mathbf{I} & \mathbf{I} & \dots & \mathbf{I} & -\frac{1}{\mu}\mathbf{I} & \mathbf{0} & \mathbf{0} & \dots & \mathbf{0} \\ \mathbf{C}_1 & \mathbf{0} & \dots & \mathbf{0} & \mathbf{0} & \mathbf{0} & \mathbf{0} & \dots & \mathbf{0} \\ \mathbf{0} & \mathbf{C}_2 & \dots & \mathbf{0} & \mathbf{0} & \mathbf{0} & \mathbf{0} & \dots & \mathbf{0} \\ \vdots & \vdots & \ddots & \vdots & \vdots & \vdots & \vdots & \ddots & \vdots \\ \mathbf{0} & \mathbf{0} & \dots & \mathbf{C}_h & \mathbf{0} & \mathbf{0} & \mathbf{0} & \dots & \mathbf{0} \end{pmatrix} \begin{pmatrix} \Delta\alpha_1^* \\ \Delta\alpha_2^* \\ \vdots \\ \Delta\alpha_h^* \\ \boldsymbol{\lambda}_{QP}^* \\ \eta_1^* \\ \eta_2^* \\ \vdots \\ \eta_h^* \end{pmatrix} = \begin{pmatrix} -\mathbf{g}_1 \\ -\mathbf{g}_2 \\ \vdots \\ -\mathbf{g}_h \\ -\frac{\lambda}{\mu} - \sum_{t=1}^{h}(\boldsymbol{\alpha}_t - \mathbf{y}_t) \\ \mathbf{0} \\ \mathbf{0} \\ \vdots \\ \mathbf{0} \end{pmatrix} \tag{19}$$

This linear system can be solved by any linear solver. One thing to note here is the adaptive parameter μ. We can start with a quite small μ and adapt it during the optimization progress to relax the coupling conditions.

3.1.3. Implementation of an ADMM-Based Algorithm

Now, the update steps (12) and (13) are executed. Since the achieved $\hat{y}_{t,i}$ have to agree with the constraint ($\mathcal{C}$), we solve the following optimization problem:

$$\min_{\mathbf{y}_i \in \Re^{h\times 1}} \frac{1}{2}\sum_{i=1}^{N} \|\mathbf{y}_i - \hat{\mathbf{y}}_i\|^2$$
$$\text{subject to} \quad \mathbf{y}_j = \mathbf{y}_i \quad i = 1,\dots,N; j \in \mathcal{N}_i \quad (\mathcal{C})$$

To solve this optimization problem, an Alternating Direction Method of Multipliers (ADMM) in [16,27] can be employed by introducing an augmented parameters $\mathbf{z}_{ij}$. Hence, the optimization can be rewritten as follows:

$$\begin{aligned} &\min_{\mathbf{y}_t} \frac{1}{2}\sum_{i=1}^{N} \|\mathbf{y}_i - \hat{\mathbf{y}}_i\|^2 \\ &\text{subject to} \quad \mathbf{y}_i = \mathbf{z}_{ij} \quad i = 1,\dots,N; j \in \mathcal{N}_i \\ &\qquad\qquad \mathbf{z}_{ji} = \mathbf{z}_{ij} \end{aligned} \tag{20}$$

The Augmented Lagrange Function is defined as follows:

$$\mathbf{H}_3(\mathbf{y},\mathbf{z},\boldsymbol{\tau}) = \frac{1}{2}\sum_{i=1}^{N} \|\mathbf{y}_i - \hat{\mathbf{y}}_i\|^2 + \sum_{j\in\mathcal{N}_i} \boldsymbol{\tau}_{ij}^T(\mathbf{y}_i - \mathbf{z}_{ij}) + \frac{\nu}{2}\sum_{i=1}^{N} \|\mathbf{y}_i - \mathbf{z}_{ij}\|.$$

The ADMM solution acts in three steps repetitively until the tolerance achieves:

- Compute $\mathbf{y}_i$:
$$\mathbf{y}_i[p+1] = (1+\nu d_i)^{-1}\{\hat{y}_i[k] + \nu \sum_{j\in\mathcal{N}_i} \mathbf{z}_{ij}[p] - \sum_{j\in\mathcal{N}_i} \boldsymbol{\tau}_{ij}[p]\}. \tag{21}$$
- Compute $\mathbf{z}_{ij}$:
$$\mathbf{z}_{ij}[p+1] = \frac{\mathbf{y}_i[p+1] + \mathbf{y}_j[p+1]}{2} + \frac{\boldsymbol{\tau}_{ij}[p] + \boldsymbol{\tau}_{ji}[p]}{2\nu}. \tag{22}$$
- Lagrange multiplier update:
$$\boldsymbol{\tau}_{ij}[p+1] = \boldsymbol{\tau}_{ij}[p] + \nu(\mathbf{y}_i[p+1] - \mathbf{z}_{ij}[p+1]) \tag{23}$$

Herein, p is a iteration of the ADMM optimization process, then this output of this process is $\mathbf{y}_i[k+1] = \mathbf{y}_i^*[p+1]$.

The distributed algorithm is illustrated in Algorithm 1.

The convergence analysis of the ALADIN method has been studied clearly for both non-convex and convex optimization problem in [25]. Lemma 3 in [25] has been proven that with the cost function $f(\boldsymbol{\alpha}) = \frac{1}{2}\sum_{t=1}^{h}\sum_{i\in V}\sum_{j\in\mathcal{N}_i}(\alpha_{t,j} - \alpha_{t,i})^2$ being twice continuously differentiable and letting $(\boldsymbol{\alpha}_t^*, \boldsymbol{\lambda}^*)$ for $t = 1,\ldots,h$ of problem (5) be a regular KKT point. On the other hand, the Hessian $\mathbf{B}_t = \frac{\partial^2}{\partial\boldsymbol{\alpha}_t^2}(\frac{1}{2}\boldsymbol{\alpha}_t^T\mathbf{L}\boldsymbol{\alpha}_t) + \rho\Sigma_t \succ 0$ obviously, since $\Sigma_t \succeq 0$. There exists constants χ_1, χ_2 such that for every point $\boldsymbol{\alpha}_t, \lambda$ satisfying the condition convergence of the decoupled minimization problem (7) have unique locally minimizers $\{\mathbf{y}_t, t = 1,\ldots,h\}$ that satisfy $\|\mathbf{y}_t - \boldsymbol{\alpha}_t\| \leq \chi_1\|\boldsymbol{\alpha}_t - \boldsymbol{\alpha}_t^*\| + \chi_2\|\boldsymbol{\lambda} - \boldsymbol{\lambda}^*\|$.

Moreover, if $\frac{1}{\mu} < \mathbf{0}(\|\mathbf{y}_t - \boldsymbol{\alpha}_t\|)$ when solving the QP (11), with the $((\boldsymbol{\alpha}_t^*, \boldsymbol{\lambda}^*))$ is a regular KKT point, then $\chi_1\|\boldsymbol{\alpha}_t - \boldsymbol{\alpha}_t^*\| + \chi_2\|\boldsymbol{\lambda} - \boldsymbol{\lambda}^*\| \leq \frac{(\chi_1+\chi_2)\omega}{2}(\chi_1\|\boldsymbol{\alpha}_t - \boldsymbol{\alpha}_t^*\| + \chi_2\|\boldsymbol{\lambda} - \boldsymbol{\lambda}^*\|)^2$. This is sufficient to prove local quadratic convergence of the algorithm as χ_1, χ_2 are strictly positive constants. As a result, it is effectively applied to our proposed method since it is obviously an equality constrained non-convex optimization problem.

One thing to remark here is that in our study, the penalty parameters ρ, μ can be updated using the following rules:

$$\rho[k+1] = \begin{cases} \iota_\rho\rho[k] & \text{if } \rho[k] < \rho_{max} \\ \rho[k] & \text{elsewhere} \end{cases}$$

$$\mu[k+1] = \begin{cases} \iota_\mu\mu[k] & \text{if } \mu[k] < \mu_{max} \\ \mu[k] & \text{elsewhere} \end{cases}$$

where $\iota_\rho, \iota_\mu > 1$ and $\rho_{max} = 50, \mu_{max} = 30$ obtained by experience to avoid the numerical problem. Moreover, we can use the blockwise and damped Broyden–Fletcher–Goldfarb–Shanno (BFGS) update, which ensures positive definiteness of the $\mathbf{B}_t[k]$ to preserve the convergence properties of ALADIN proposed in [25].

3.2. Retrieving the Non-Zero Laplacian Eigenvalues

As stated before, the set of eigenvalues deriving from the Algorithm 1, denoted $\mathcal{S}_1$, composing of the set of non-zero distinct Laplacian eigenvalues $\mathcal{S}_2$.

Let $\hat{\bar{x}}_i$ be the final consensus value reconstructed by $\hat{\bar{x}}_i = \hat{x}_i(h) = \frac{1}{N}\sum_{i=1}^{N} x_i(0)$ and the iteration scheme of the finite-time average consensus is implemented as in (1). Following the idea of the Proposition 3 in [27], we step-by-step assume to leave one element of the $\mathcal{S}_1$, then, the remaining elements of this set are used to reconstruct $\hat{x}_i(h)$. If $\hat{\bar{x}}_i = \hat{x}_i(h)$ is satisfied, then the left element is not the expected Laplacian eigenvalue. Hence, we can eliminate it out of the set $\mathcal{S}_1$. Otherwise, the left element is one of non-zero distinct eigenvalues. We restore it in the set $\mathcal{S}_1$ and marked as an element in the set $\mathcal{S}_2$. Now, the procedure is continued with another element to the end.

The distributed non-zero Laplacian eigenvalues are described in Algorithm 2.

Algorithm 1 ALADIN-based Laplacian eigenvalues estimation

1. Initialization:
   - Number of nodes N, tolerance ϵ, initial input-output pairs $\{x_i(0), \bar{x}_i, i = 1, 2, \ldots, N\}$, where $\bar{x} = \frac{1}{N}\sum_{i=1}^{N} x_i(0)$ is retrieved from a standard average consensus algorithm.
   - Each node $i, i = 1, \ldots, N$ initializes:
     (a) random stepsizes $\alpha_{t,i}(0)$ for $t = 1, \ldots, h = N-1$.
     (b) random Lagrange multipliers $\beta_{t,i}, \eta_{t,i}$, for $t = 1, \ldots, h$ and $\lambda_i, \lambda_{QP,i}$.
     (c) Semi-positive definite matrices $\Sigma_t \in \Re^N$, learning rates b, penalty parameters ρ, c, μ.
   - Set $k = 0$;
2. Repeat:
   - Set $k := k+1$,
   - Solving decouple NLP problem (7) for $t = 1, \ldots, h = N-1$:
     - Propagate Lagrange multipliers $\beta_{t,i}[k]$ for $t = h, \ldots, 2$ and $i = 1, \ldots, N$:
       (a) Set $\delta_{h,i}[k] = \beta_{t,i}[k]$.
       (b) $\delta_{t-1,i}[k] = \delta_{t,i}[k] + \alpha_{t,i}[k]\sum_{j\in\mathcal{N}_i}(\delta_{t,j}[k] - \delta_{t,i}[k])$.
     - Finite-time average Consensus steps:
$$x_{t,i}[k] = x_{t-1,i}[k] + \alpha_{t,i}[k]\sum_{j\in\mathcal{N}_i}(x_{t-1,j}[k] - x_{t-1,i}[k]).$$
     - Propagate the error $e_{t,i}[k]$ by setting $e_{h,i}[k] = \mathbf{x}_{h,i}[k] - \bar{x}_i[k]$:
$$e_{t-1,i}[k] = e_{t,i}[k] + \alpha_{t,i}\sum_{j\in\mathcal{N}_i}(e_{t,j}[k] - e_{t,i}[k]).$$
     - Update $\alpha_{t,i}$ for $t = 1, \ldots, h$:
$$\begin{aligned}\alpha_{t,i}[k+1] = \alpha_{t,i}[k+1] &- b\sum_{j\in\mathcal{N}_i}(\alpha_{t,j}[k] - \alpha_{t,i}[k]) \\ &- b\lambda_i - b\rho[k]\sum_{j=1}^{N}\Sigma_t(i,j)(\alpha_{t,j}[k] - y_{t,j}[k]) \\ &+ b\sum_{j\in\mathcal{N}_i}(x_{t-1,j}[k] - x_{t-1,i}[k])\delta_{t,i}[k] \\ &+ b\sum_{j\in\mathcal{N}_i}(x_{t-1,j}[k] - x_{t-1,i}[k])e_{t,i}[k]\end{aligned}$$
     - Update NLP Lagrange multipliers $\beta_{t,i}$ for $t = 1, \ldots, h$ by (16).
   - Stopping criteria: if $\|\sum_{t=1}^{h}(\boldsymbol{\alpha}_t - \mathbf{y}_t)\| \leq \epsilon$ and $\|\sum_{t=1}^{h}\boldsymbol{\alpha}_t^T\mathbf{L}\boldsymbol{\alpha}_t\| \leq \epsilon$ are simultaneously satisfied, then stop the optimization procedure.
   - Compute Gradient, Jacobian matrices, and Hessian matrices as in (8)–(10), respectively.
   - Solving the coupling QP problem (11) via solving the linear Equation (19) to define $\Delta\alpha_{t,i}^*[k], \lambda_{QP,i}^*[k]$.
   - Update $\lambda, \hat{y}_{t,i}$:
     (a) $\lambda_i[k+1] = \lambda_{QP,i}^*[k]$
     (b) $\hat{y}_{t,i}[k] = \alpha_{t,i}[k+1] + \Delta\alpha_{t,i}^*[k]$
     (c) Projecting $\hat{y}_{t,i}[k]$ onto the constraint $\mathcal{C}$ by solving an ADMM-based optimization subproblem (20) to derive $y_{t,i}[k+1]$

Algorithm 2 Non-zero eigenvalues Estimation

1. Input: set of stepsizes, obtaining form Algorithm 1 $\mathcal{S}_1$, the input-output pairs $\{\mathbf{x}_i(0), \bar{\mathbf{x}}_i\}$, $i = 1, \ldots, N$, the threshold ϵ.
2. Set $\mathcal{S}_2 = \emptyset, \mathcal{S} = \mathcal{S}_1$.
3. Repeat: **While** $\mathcal{S} \neq \emptyset$, pick an element α_t out of $\mathcal{S}$, hence $\mathcal{S} \setminus \{\alpha_t\}$.
   - Construct average consensus iteration scheme from the remain stepsizes to determine $\hat{x}_i(h), i = 1, \ldots, N$ as follows:
   $$x_{t,i} = x_{t-1,i} + \alpha_{t,i} \sum_{j \in \mathcal{N}_i} (x_{t-1,j} - x_{t-1,i}), t = 1, \ldots, h$$
   - If $\bar{x}_i - \hat{x}_i \leq \epsilon$, then $\mathcal{S} = \mathcal{S} \setminus \alpha_t$ and return to (3)
   - If $\bar{x}_i - \hat{x}_i > \epsilon$, then α_t is included in $\mathcal{S}_2$ and $\mathcal{S} = \mathcal{S} \cup \alpha_t$ and return to (3)
4. Output: If $\mathcal{S}$ is empty, then the non-zero distinct Laplacian eigenvalues can be derived by taking the inverse of the set $\mathcal{S}_2$'s elements.

3.3. Multiplicities Estimation

Now, turning to the last stage, which is the corresponding Laplacian eigenvalues multiplicities estimation. In [16], the authors have proposed a linear integer programming optimization problem to figure out the multiplicities:

Proposition 1 ([16]). *Consider a connected undirected graph of N vertices with degree sequence* $\{d_i\}$ *and Laplacian matrix* $\mathbf{L}$*, having* $D = |\mathcal{S}_2|$ *distinct non-zero Laplacian eigenvalues* $\mathcal{S}_2$*. Let* $\boldsymbol{m} \in \mathbb{Z}^{+D\times 1}$ *be the vector of the corresponding multiplicities and be obtained by solving the integer programming below:*

$$\begin{aligned} \min_{\boldsymbol{m}\in\mathbb{Z}^{+D\times 1}} \quad & \mathcal{S}_2^T\boldsymbol{m} \\ \textit{subject to} \quad & \mathcal{S}_2^T\boldsymbol{m} = \textstyle\sum_{i=1}^{D} d_i \\ & \boldsymbol{1}^T\boldsymbol{m} = N-1 \\ & \boldsymbol{m} \in \mathbb{Z}^{+D\times 1} \end{aligned} \tag{24}$$

The proof was given in [16]. Since all the multiplicities $\mathbf{m}$ are positive integers, then Brand-and-Bound method has been deployed to derive $\mathbf{m}$. Therefore, the problem (24) can be rewritten equivalently in linear integer programming form as follows:

$$\begin{aligned} \min_{\mathbf{m}\in\mathbb{Z}^{+D\times 1}} \quad & \mathcal{S}_2^T\mathbf{m} - \textstyle\sum_{i=1}^{D} d_i \\ \text{subject to} \quad & \mathbf{1}^T\mathbf{m} = N-1 \\ & \mathbf{m} \in \mathbb{Z}^{+D\times 1} \end{aligned} \tag{25}$$

The Algorithm for this problem has been described clearly in [16]. At this step, the whole Laplacian spectrum $sp(\mathbf{L})$ has been obtained.

In fact, the estimation problem can be converted into a convex form and can be solved effectively by using ADMM-based method proposed in [16]. However, the purpose of this study is extremely appropriate for non-convex optimization problem through deploying the promising ALADIN-based method.

4. Simulation Results

In this section, the efficiency of the proposed ALADIN-based method to estimate the Laplacian spectrum is evaluated by considering the two following case studies.

Firstly, it can be said that the Laplacian spectrum can decide the performance of the network since it reveals the connection of the network. For example, the robustness of the network can be estimated before starting the operations to avoid the interruption during these operations and, as a result, enhance the benefits technically and economically.

On the purpose of monitoring the connection of a large network $G^*(V^*, E^*)$, one may face with the numerical issue in step 3 of the ALADIN-based method to solve the linear system (19) due to the huge dimension of the obtained matrix that leads to the common ill-conditioning problem with the inverse matrix calculation.

In [16], the authors have suggested to partition the large network into M disjoint sub-networks $U_\ell, \ell = 1, 2, \cdots, M$, [29]. Let us define $\mathcal{N}_i^* = \{j \in V^* : (i,j) \in E^*\}$ and its cardinality $|\mathcal{N}_i^*|$ as the set of neighbors of node i and its degree in G^*, respectively. Each sub-network is monitored by a super-node i which knows the number N_ℓ of agents in the sub-network and the associated average information state x_ℓ. Node $i \in U_\ell$ is a super-node if it has at least one neighbor in a different subset, i.e., $\exists \ell^* \neq \ell$ s.t. $\mathcal{N}_i \bigcap U_{\ell^*} \neq \emptyset$. Let us consider that two sub-networks are connected if there exist edges linking at least two agents of these sub-networks. If two sub-networks are connected then their super-nodes are linked as showed in Figure 2. Here, the network can be social network, power system network, molecular network, etc.

Let $G = (V, E)$ be the undirected graph representing a network with $N = |V|$ super-nodes, which are black nodes in Figure 2. G captures the interaction between sub-networks of G^*. Therefore, the large network is to be robust if the partitions are strongly linked to each other and if the critical threshold is high enough in [16]. Then, the Laplacian spectrum of network of super-nodes G can monitor the connection of the large network G^* via the robustness index.

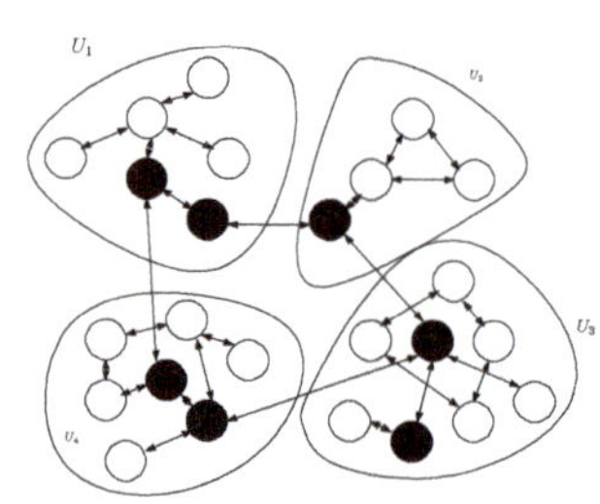

Figure 2. Network partitioned in 4 subsets. Super-nodes are depicted in black.

4.1. Case Study 1

Let us consider a large network partitioning into 4 disjoint sub-networks. Each sub-network has only one super-node. These super-nodes interact with each other by the graph $G = (V, E)$, depicted in Figure 3.

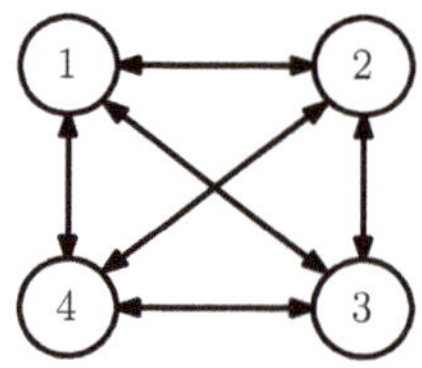

Figure 3. A network constituted by 4 nodes.

This network does have the Laplacian eigenvalues $sp(\mathbf{L}) = \{0, 4, 4, 4\}$. With the parameter of each super-node, denoted as $\mathbf{x}(0) = \{0.7417, 0.7699, 0.3216, 0.5466\}$, after deploying Algorithm 1, the set of $\boldsymbol{\alpha}_t, t = 1, \ldots, 3$ is obtained. The nodes trajectories are described as in Figure 4.

As can be seen, all $\boldsymbol{\alpha}_t$ execute the consensus problem at the beginning of the procedure, and then dig into satisfying the constraint.

Figure 5 illustrate the convergence of the cost function $\frac{1}{2}\sum_{t=1}^{N-1} \boldsymbol{\alpha}_t^T \mathbf{L} \boldsymbol{\alpha}_t$ in according to the constraint $\mathbf{x}(N-1) = \bar{x}\mathbf{1}$.

Furthermore, the Algorithm 2 is applied to eliminate the unexpected eigenvalues. As a result, we receive only one eigenvalue $\lambda = \frac{1}{0.25} = 4$. In order to accomplish the Laplacian spectrum estimation's procedure, we make use the Proposition 1 to pick out $m = 3$. Now, the entire Laplacian spectrum is achieved.

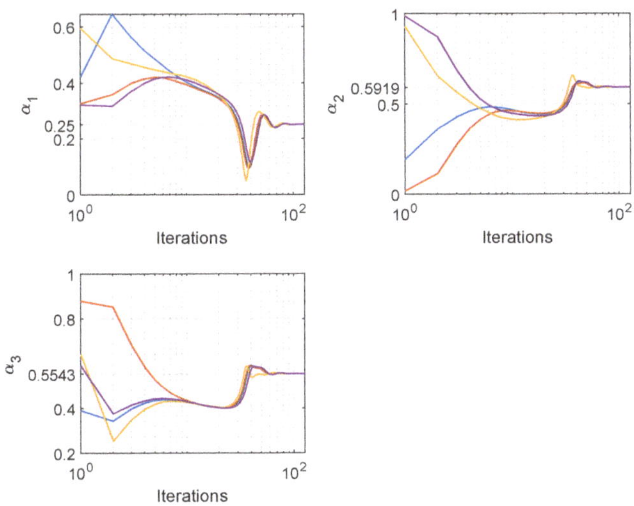

Figure 4. Trajectories convergence of $\boldsymbol{\alpha}_t$.

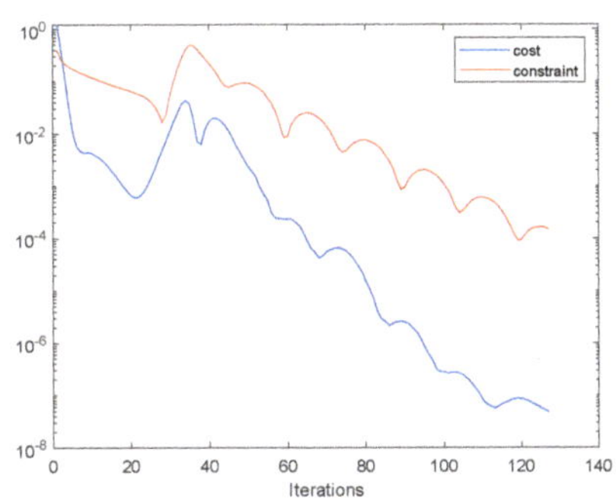

Figure 5. Convergence of the cost function according to its constraint.

At this step, the robustness index such as Kirchhoff index or the number of spanning trees can be calculated.

$$\mathcal{R}_{\mathbf{L}} = N \sum_{i=2}^{D+1} \frac{m_i}{\lambda_i} = 4\frac{3}{4} = 3.$$

Now, let us see how the proposed procedure works via the table below.

Table 1 is obtained after executing the proposed procedure.

As can be seen in Figure 4, at the iteration of around 90 the procedure can be stopped. In order to access the robustness of the whole large network, it is necessary to define the critical threshold, introduced in [16].

Next, we implement a comparison with the Lagrange method described in [1] by considering Case study 2.

Table 1. The achievement of the proposed procedure in the sense of 4-node topology.

S_1	$\{0.25, 0.5919, 0.5543\}$
S_2	$\{0.25\}$
m	3
Iterations	130
$\mathcal{R}_{\mathbf{L}}$	3

4.2. Case Study 2

Let us consider a 6-node network described in Figure 6.

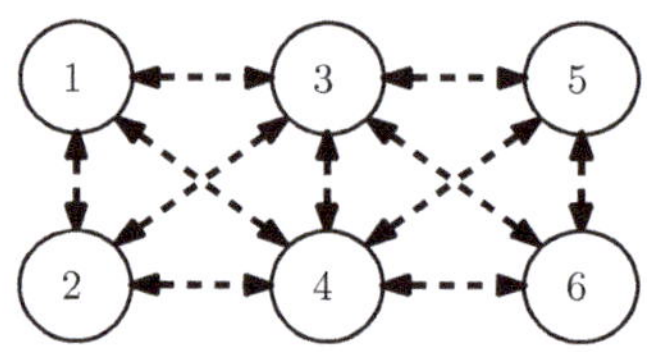

Figure 6. A new network constituted by 6 nodes.

It is known that for this topology, the Laplacian spectrum is $sp(\mathbf{L}) = \{0, 1, 2, 3, 3, 5\}$.

Let us define the same initial information state of each node at time $t = 0$: $\mathbf{x}(0) = \{0.5832, 0.74, 0.2348, 0.7350, 0.9706, 0.8669\}$ for both methods. By using a standard consensus algorithm [26], the consensus value $\bar{x} = 0.6884$ can be easily inferred.

It can be seen in Figure 7 that the Lagrange-based method in [1] takes a long time to achieve the consensus first and then track the constraint to obtain the stepsizes $\alpha_t, t = 1, ..., N-1$.

Now, with the same initial $\boldsymbol{\alpha}_t(0), t = 1, \ldots, 5$, for the iterative procedure of the proposed method, the $\boldsymbol{\alpha}_t$ after using the Algorithm 1 with the convergence trajectories of the $\boldsymbol{\alpha}_t$ are illustrated in Figure 8.

Figures 7 and 8 express the significant pros of our proposed method since the number of iterations is much less than that of Lagrange-based method. The consensus term is executed in advance from the start of the procedure and then try to reach the constraint term to figure out the expected values of the stepsizes $\boldsymbol{\alpha}_t = \{1, 0.5, 0.1027, 0.2, 0.3333\}$. Obviously, since the authors operate the proposed algorithm with the number of $\boldsymbol{\alpha}_t$ being $N - 1 = 5$, it is needed to run the next stage to eliminate the residual values. As can be seen clearly, the executive time for ALADIN-based method is significantly faster than the proposed method as showed in Figure 9.

Figure 9 shows that the ALADIN-based method approaches the destined values earlier than Lagrange-based method. Furthermore, in order to get the non-zero Laplacian eigenvalues of the given network, the Algorithm 2 is carried out to obtain vector of stepsizes $\boldsymbol{\alpha}_t = \{1, 0.5, 0.2, 0.3333\}$.

Finally, by constructing the Brand-and-Bound based method to solve the Problem 1, which has been proposed in [16], we achieve the vector of multiplicities $\mathbf{m} = \{1, 1, 1, 2\}$, hence deduce the Laplacian spectrum $sp(\mathbf{L}) = \{1, 2, 3, 3, 5\}$.

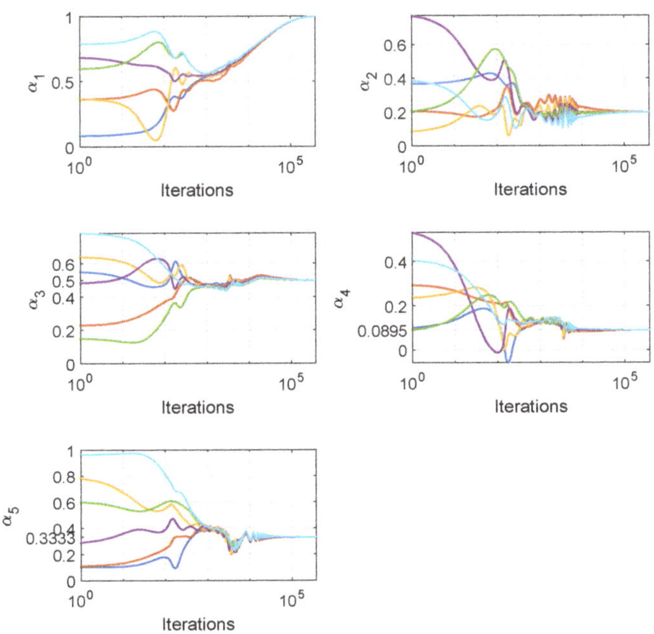

Figure 7. $\boldsymbol{\alpha}_t$ convergence trajectories implemented by Lagrange-based method.

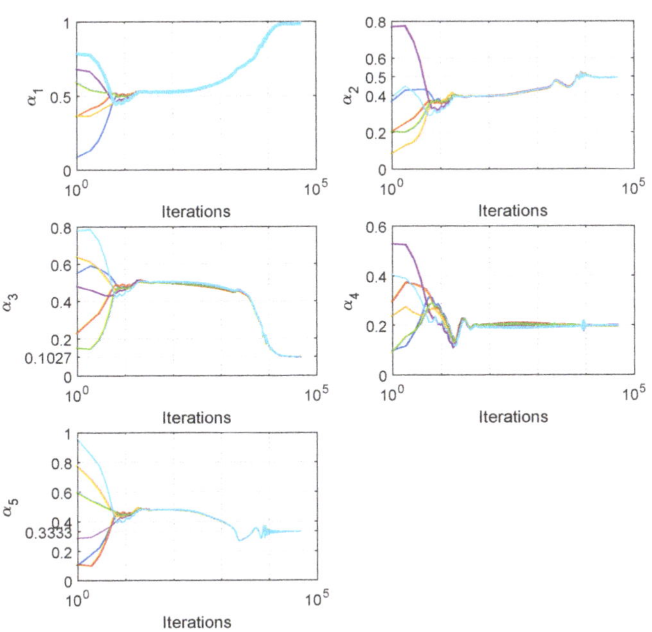

Figure 8. $\boldsymbol{\alpha}_t$ convergence trajectories implemented by the proposed method.

Now, let us see how the proposed procedure works via the table below.
Table 2 is obtained after executing the proposed procedure and the Lagrange-based method.

Table 2. The achievement of two methods in the sense of 6-node topology.

	ALADIN-Based Method	Lagrange-Based Method
$\mathcal{S}_1$	$\{1, 0.5, 0.1027, 0.2, 0.3333\}$	$\{0.9992, 0.2, 0.5001, 0.0895, 0.3333\}$
$\mathcal{S}_2$	$\{1, 0.5, 0.2, 0.3333\}$	$\{0.9992, 0.2, 0.5001, 0.3333\}$
m	$1, 1, 1, 2$	
Iterations	33950	372800
$\mathcal{R}_L$	14.2	

Notice that the proposed method gives results much better than the Lagrange-based method. Let us see that at the iteration of 33,950, the Lagrange-based method gives the set of $\mathcal{S}_1$ that has still not satisfied the constraint.

Recently, besides the Laplacian spectrum estimation basing on the optimization approaches, there are also some works that approximate the Laplacian spectrum via iterative dynamics process (taking random walk for an example). However, from our point of view, the dominant contribution of our proposed method is the possibility of implementation in a decentralized manner. Moreover, another method to faster the convergence rate of the Laplacian spectrum estimation procedure is the ADMM-based method, proposed in [16]. The important step in this work is to re-parameterize adequately the non-convex formulation into convex one. It is hard to compare the efficiency between ADMM-based method and the proposed method. Since, our study focuses on the non-convex formulation.

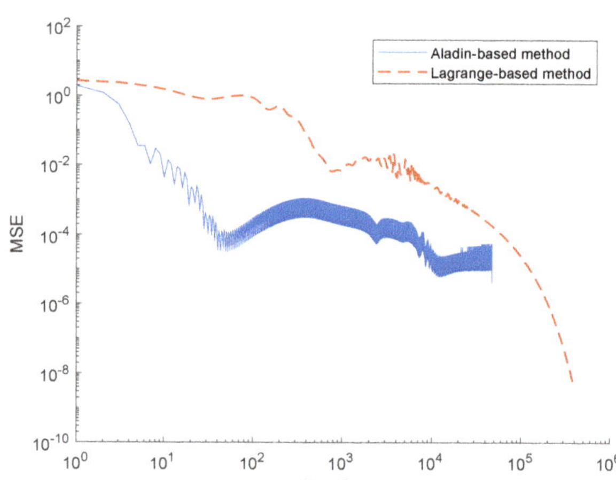

Figure 9. Convergence of the cost function according to its constraint.

5. Conclusions

In this paper, the authors have proposed a promising ALADIN-based method to find out the Laplacian spectrum of a given dynamic network in a distributed way. First and foremost, the study has assumed that the number of agents N in the network can be accumulated by random walk algorithm constructed in the configuration step. Briefly speaking, the proposed procedure is divided into 3 stages. The first stage is to determine the $N-1$ Laplacian eigenvalues. Then, retrieve the non-zero distinct Laplacian eigenvalues in stage 2 before estimating the corresponding multiplicities in stage 3. The ALADIN-based method is appropriate for carrying out the factorization of the average consensus matrix as factors of the Laplacian-based consensus matrices to minimize the disagreement between neighbors on the values of $\alpha_{t,i}$. Then, the Laplacian eigenvalues can be defined by taking the inverse of these $\boldsymbol{\alpha}_t$. Herein, the authors are obviously interested in dealing with the non-convex optimization problems. From the simulation evaluation, it can be concluded that the proposed method converges much faster in comparison with the gradient descent method in [1] for the estimation of Laplacian spectrum.

Author Contributions: Conceptualization, methodology, funding acquisition, T.-M.-D.T.; writing—original draft preparation, T.-M.-D.T. and L.N.A.; resources L.N.A., writing—review and editing, T.-M.-D.T. and N.C.N.D. All authors have read and agreed to the published version of the manuscript.

Funding: This work was supported by The University of Danang - University of Science and Technology, code number of Project: T2019-02-09.

Conflicts of Interest: The authors declare no conflict of interest.

Appendix A. Proof of Lemma 2

From now on, in order to avoid misunderstanding between the iterative step of the consensus algorithm $\mathbf{x}(h)$ and the iterative step of the optimization procedure k, let us denote $\mathbf{x}(h)$ by $\mathbf{x}_h$. From Section 2.1, we have:

$$\begin{aligned}
\mathbf{x}(h) - \bar{\mathbf{x}} &= \prod_{i=h}^{t+1} \mathbf{W}_i \mathbf{W}_t \mathbf{x}_{t-1} - \bar{\mathbf{x}} \\
&= \prod_{i=h}^{t+1} \mathbf{W}_i (\mathbf{I} - diag(\boldsymbol{\alpha}_t)\mathbf{L})\mathbf{x}_{t-1} - \bar{\mathbf{x}} \\
&= \prod_{i=h}^{t+1} \mathbf{W}_i \mathbf{x}_{t-1} - \bar{\mathbf{x}} - \prod_{i=h}^{t+1} \mathbf{W}_i diag(\boldsymbol{\alpha}_t)\mathbf{L}\mathbf{x}_{t-1} \\
&= \prod_{i=h}^{t+1} \mathbf{W}_i \mathbf{x}_{t-1} - \bar{\mathbf{x}} - (\mathbf{x}_{t-1}^T \mathbf{L}^T \odot \prod_{i=h}^{t+1} \mathbf{W}_i)\boldsymbol{\alpha}_t
\end{aligned}$$

with $\odot$ being the Khaitri–Rao product. By employing the property of the Khatri–Rao product (Given matrix $\mathbf{A} \in R^{I\times F}$, and two vectors $\mathbf{b} \in R^{J\times 1}$, $\mathbf{d} \in R^{F\times 1}$, then $\mathbf{A}dieg(\mathbf{d})\mathbf{b} = (\mathbf{b}^T \odot \mathbf{A})\mathbf{d}$) in [1], the derivative of the Lagrange function is described as follows:

$$\frac{\partial \mathbf{H}_{1,t}}{\partial \boldsymbol{\alpha}_t} = \mathbf{L}\boldsymbol{\alpha}_t + \boldsymbol{\lambda} + \rho \boldsymbol{\Sigma}_t(\boldsymbol{\alpha}_t - \mathbf{y}_t) - diag(\mathbf{L}\mathbf{x}_{t-1}) \prod_{i=t+1}^{h} \mathbf{W}_i \boldsymbol{\beta}_t - c.diag(\mathbf{L}\mathbf{x}_{t-1}) \prod_{i=t+1}^{h} \mathbf{W}_i(\mathbf{x}(h) - \bar{\mathbf{x}})$$

Since $\boldsymbol{\delta}_h = \boldsymbol{\beta}_t$ and $\boldsymbol{\delta}_{h-1} = \mathbf{W}_h \boldsymbol{\delta}_h$ then $\prod_{i=t+1}^{h} \mathbf{W}_i \boldsymbol{\beta}_t = \boldsymbol{\delta}_t$.
Analogically, $\mathbf{e}_h = \mathbf{x}(h) - \bar{\mathbf{x}}$ and $\mathbf{e}_t = \mathbf{W}_{t+1}\mathbf{e}_{t+1}$.
On the other hand, $\mathbf{x}_{t-1} - \mathbf{x}_t = diag(\boldsymbol{\alpha}_t)\mathbf{L}\mathbf{x}_{t-1}$ then $\mathbf{L}\mathbf{x}_{t-1} = diag^{-1}(\boldsymbol{\alpha}_t)(\mathbf{x}_{t-1} - \mathbf{x}_t)$
Therefore, $\frac{\partial \mathbf{H}_{1,t}}{\partial \boldsymbol{\alpha}_t} = \mathbf{L}\boldsymbol{\alpha}_t + \boldsymbol{\lambda} + \rho \boldsymbol{\Sigma}_t(\boldsymbol{\alpha}_t - \mathbf{y}_t) - diag^{-1}(\boldsymbol{\alpha}_t) diag(\mathbf{x}_{t-1} - \mathbf{x}_t)\boldsymbol{\delta}_t - diag^{-1}(\boldsymbol{\alpha}_t) diag(\mathbf{x}_{t-1} - \mathbf{x}_t)\mathbf{e}_t$.

Appendix B. KKT Conditions of QP (11)

$$\frac{\delta \mathbf{H}_2}{\Delta \boldsymbol{\alpha}_t} = \mathbf{B}_t \Delta\alpha_t + g_t + \boldsymbol{\lambda}_{QP} + \mathbf{C}_t^T \eta_t = 0$$

$$\frac{\delta \mathbf{H}_2}{\mathbf{s}} = \boldsymbol{\lambda} - \boldsymbol{\lambda}_{QP} + \mu \mathbf{s} = 0$$

$$\frac{\delta \mathbf{H}_2}{\eta_t} = \mathbf{C}_t \Delta\boldsymbol{\alpha}_t = 0$$

$$\frac{\delta \mathbf{H}_2}{\lambda_{QP}} = \sum_{t=1}^{h} \Delta \boldsymbol{\alpha}_t + \sum_{t=1}^{h} (\boldsymbol{\alpha}_t - \mathbf{y}_t) - \mathbf{s} = 0$$

References

1. Tran, T.; Kibangou, A.Y. Consensus-based Distributed Estimation of Laplacian Eigenvalues of Undirected Graphs. In Proceedings of the European Control Conference (ECC), Zurich, Switzerland, 17–19 July 2013; pp. 227–232.
2. Schwab, K. *The Fourth Industrial Revolution*; World Economic Forum: Cologny, Switzerland, 2016.
3. Godsil, C.; Royle, G. *Algebraic Graph Theory*; Springer: Berlin, Germany, 2001.
4. Merris, R. Laplacian matrices of a graph: A survey. *Linear Algebra Appl.* **1994**, *197*, 143–176. [CrossRef]
5. Friedler, M. Algebraic connectivity of graphs. *Czechoslov. Math. J.* **1973**, *23*, 298–305.
6. Olfati-Saber, R.; Fax, J.A.; Murray, R.M. Consensus and cooperation in networked multi-agent systems. *Proc. IEEE* **2007**, *95*, 215–233. [CrossRef]
7. Olfati-Saber, R.; Murray, R. Consensus problems in networks of agents with switching topology and time-delays. *IEEE Trans. Autom. Control* **2004**, *49*, 1520–1533. [CrossRef]
8. Gulzar, M.; Rizvi, S.; Javed, M.Y.; Munir, U.; Asif, H. Multi-Agent Cooperative Control Consensus: A Comparative Review. *Electronics* **2018**, *7*, 22. [CrossRef]
9. Kempton, L.C.; Herrmann, G.; di Bernardo, M. Distributed optimisation and control of graph Laplacian eigenvalues for robust consensus via an adaptive multilayer strategy. *Int. J. Robust Nonlinear Control* **2017**, *27*, 1499–1525. [CrossRef]
10. Xiao, L.; Boyd, S. Fast linear iterations for distributed averaging. *Syst. Control Lett.* **2004**, *53*, 65–78. [CrossRef]
11. Kibangou, A. Finite-time average consensus based protocol for distributed estimation over awgn channels. In Proceedings of the IEEE Conference on Decision and Control (CDC), Orlando, FL, USA, 12–15 December 2011.
12. Kibangou, A. Graph Laplacian based matrix design for finite-time distributed average consensus. In Proceedings of the American Control Conference (ACC), Montreal, QC, Canada, 27–29 June 2012; pp. 1901–1906.
13. Abbas, W.; Egersredt, M. Robust graph Topologies for networked systems. In Proceedings of the 3rd IFAC Workshop on Distributed Estimation and Control in Networked Systems (NeCSYS), Santa Barbara, CA, USA, 13–14 September, 2012; pp. 85–90.
14. Wu, J.; Barahona, M.; Tan, Y.; Deng, H. Spectral Measure of Structural Robustness in Complex Networks. *IEEE Trans. Syst. Man Cybern. Part A Syst. Hum.* **2011**, *41*, 1244–1252. [CrossRef]
15. Klein, D.; Randić, M. Resistance distance. *J. Math. Chem.* **1993**, *12*, 81–95. [CrossRef]
16. Tran, T.; Kibangou, A.Y. Collaborative Network Monitoring by Means of Laplacian Spectrum Estimation and Average Consensus. *Int. J. Autom. Control Syst.* **2019**, *17*, 1826–1837. [CrossRef]
17. Zhao, C.; He, J.; Cheng, P.; Chen, J. Consensus-based energy management in smart grid with transmission losses and directed communication. *IEEE Trans. Smart Grid* **2017**, *8*, 2049–2061. [CrossRef]
18. Guo, L.; Zhao, C.; Low, S.H. Graph Laplacian Spectrum and Primary Frequency Regulation. In Proceedings of the 2018 IEEE Conference on Decision and Control (CDC), Miami Beach, FL, USA, 17–19 December 2018.
19. Franceschelli, M.; Martini, S.; Egerstedt, M.; Bicchi, A.; Giua, A. Observability and controllability verification in multi-agent systems through decentralized Laplacian spectrum estimation. In Proceedings of the IEEE Conference on Decision and Control (CDC), Atlanta, GA, USA, 15–17 December 2010; pp. 5775–5780.
20. Cooper, C.; Radzik, T.; Siantos, Y. Estimating network parameters using random walks. In Proceedings of the 2012 Fourth International Conference on Computational Aspects of Social Networks (CASoN), Sao Carlos, Brazil, 21–23 November 2012; pp. 33–40.
21. Franceschelli, M.; Gasparri, A.; Giua, A.; Seatzu, C. Decentralized Estimation of Laplacian Eigenvalues in Multi-Agent Systems. *Automatica* **2013**, *49*, 1031–1036. [CrossRef]
22. Sahai, T.; Speranzon, A.; Banaszuk, A. Hearing the cluster of a graph: A distributed algorithm. *Automatica* **2012**, *48*, 15–24. [CrossRef]
23. Kibangou, A.Y.; Commault, C. Decentralized Laplacian Eigenvalues Estimation and Collaborative Network Topology Identification. In Proceedings of the 3rd IFAC Workshop on Distributed Estimation and Control in Networked Systems (NecSys'12), Santa Barbara, CA, USA, 14–15 September 2012; pp. 7–12.

24. Tran, T.; Kibangou, A. Distributed estimation of Laplacian eigenvalues via constrained consensus optimization problems. *Syst. Control Lett.* **2015**, *80*, 56–62. [CrossRef]
25. Houska, B.; Frasch, J.; Diehl, M. An Augmented Lagrangian Based Algorithm for Distributed NonConvex Optimization. *SIAM J. Optim.* **2016**, *26*, 1101–1127. [CrossRef]
26. Xiao, L.; Boyd, S.; Kim, S. Distributed Average Consensus with Least-mean-square Deviation. *J. Parallel Distrib. Comput.* **2007**, *67*, 33–46. [CrossRef]
27. Tran, T.; Kibangou, A. Distributed Estimation of Graph Laplacian Eigenvalues by the Alternating Direction of Multipliers Method. In Proceedings of the 19th World Congress of the International Federation of Automatic Control, Cape Town, South Africa, 24–29 August 2014.
28. Chung, F.R.K. *Spectral Graph Theory*; American Mathematical Society: Providence, RI, USA, 1997.
29. Martin, N.; Frasca, P.; Canudas-De-Wit, C. Large-scale network reduction towards scale-free structure. *IEEE Trans. Netw. Sci. Eng.* **2018**, *14*, 1–12. [CrossRef]

MDPI
St. Alban-Anlage 66
4052 Basel
Switzerland
Tel. +41 61 683 77 34
Fax +41 61 302 89 18
www.mdpi.com

Applied Sciences Editorial Office
E-mail: applsci@mdpi.com
www.mdpi.com/journal/applsci

www.ingramcontent.com/pod-product-compliance
Lightning Source LLC
Chambersburg PA
CBHW061822210926
51850CB00006B/778
9783036555751